Multilevel Modeling of Categorical Outcomes Using IBM SPSS

Ronald H. Heck
University of Hawai'i, Mānoa

Scott L. Thomas
Claremont Graduate University

Lynn N. Tabata
University of Hawai'i, Mānoa

Routledge
Taylor & Francis Group
New York   London

Routledge
Taylor & Francis Group
711 Third Avenue
New York, NY 10017

Routledge
Taylor & Francis Group
27 Church Road
Hove, East Sussex BN3 2FA

© 2012 by Taylor & Francis Group, LLC
Routledge is an imprint of Taylor & Francis Group, an Informa business

Printed in the United States of America on acid-free paper
Version Date: 20120309

International Standard Book Number: 978-1-84872-955-1 (Hardback) 978-1-84872-956-8 (Paperback)

Visit the Taylor & Francis Web site at
http://www.taylorandfrancis.com

and the Psychology Press Web site at
http://www.psypress.com

Contents

QUANTITATIVE METHODOLOGY SERIES

George A. Marcoulides, Series Editor

This series presents methodological techniques to investigators and students. The goal is to provide an understanding and working knowledge of each method with a minimum of mathematical derivations. Each volume focuses on a specific method (e.g., Factor Analysis, Multilevel Analysis, Structural Equation Modeling).

Proposals are invited from interested authors. Each proposal should consist of a brief description of the volume's focus and intended market; a table of contents with an outline of each chapter; and a curriculum vita. Materials may be sent to Dr. George A. Marcoulides, University of California – Riverside, george.marcoulides@ucr.edu.

Marcoulides • Modern Methods for Business Research

Marcoulides/Moustaki • Latent Variable and Latent Structure Models

Hox • Multilevel Analysis: Techniques and Applications

Heck • Studying Educational and Social Policy: Theoretical Concepts and Research Methods

Van der Ark/Croon/Sijtsma • New Developments in Categorical Data Analysis for the Social and Behavioral Sciences

Duncan/Duncan/Strycker • An Introduction to Latent Variable Growth Curve Modeling: Concepts, Issues, and Applications, Second Edition

Heck/Thomas • An Introduction to Multilevel Modeling Techniques, Second Edition

Cardinet/Johnson/Pini • Applying Generalizability Theory Using EduG

Creemers/Kyriakides/Sammons • Methodological Advances in Educational Effectiveness Research

Heck/Thomas/Tabata • Multilevel and Longitudinal Modeling with IBM SPSS

Hox • Multilevel Analysis: Techniques and Applications, Second Edition

Heck/Thomas/Tabata • Multilevel Modeling of Categorical Outcomes Using IBM SPSS

Preface

Multilevel modeling has become a mainstream data analysis tool over the past decade, now figuring prominently in a range of social and behavioral science disciplines. Where it originally required specialized software, mainstream statistics packages such as IBM SPSS, SAS, and Stata all have included routines for multilevel modeling in their programs. Although some devotees of these statistical packages have been making good use of the relatively new multilevel modeling functionality, progress has been slower in carefully documenting these routines to facilitate meaningful access to the average user. Two years ago we developed *Multilevel and Longitudinal Modeling with IBM SPSS* to demonstrate how to use these techniques in IBM SPSS Version 18. Our focus was on developing a set of concepts and programming skills within the IBM SPSS environment that could be used to develop, specify, and test a variety of multilevel models with continuous outcomes, since IBM SPSS is a standard analytic tool used in many graduate programs and organizations globally. Our intent was to help readers gain facility in using the IBM SPSS linear mixed-models routine for continuous outcomes. We offered multiple examples of several different types of multilevel models, focusing on how to set up each model and how to interpret the output.

At the time, mixed modeling for categorical outcomes was not available in the IBM SPSS software program. Over the past year or so, however, the generalized linear mixed model (GLMM) has been added to the mixed modeling analytic routine in IBM SPSS starting with Version 19. This addition prompted us to create this companion workbook that would focus on introducing readers to the multilevel approach to modeling with categorical outcomes. Drawing on our efforts to present models with categorical outcomes to students in our graduate programs, we have again opted to adopt a workbook format. We believe this format will prove useful in helping readers set up, estimate, and interpret multilevel models with categorical outcomes and hope it will provide a useful supplement to our first workbook, *Multilevel and Longitudinal Modeling with IBM SPSS*, and our introductory multilevel text, *An Introduction to Multilevel Modeling Techniques*, 2nd *Edition*. Ideal as a supplementary text for graduate level courses on multilevel, longitudinal, latent variable modeling, multivariate statistics, and/or advanced quantitative techniques taught in departments of psychology, business, education, health, and sociology, we believe the workbook's practical approach will also appeal to researchers in these fields. This new workbook, like the first, can also be used with any multilevel and/or longitudinal textbook or as a stand-alone text introducing multilevel modeling with categorical outcomes.

In this workbook, we walk the reader in a step-by-step fashion through data management, model conceptualization, and model specification issues related to single-level and multilevel models with categorical outcomes. We offer multiple examples of several different types of categorical outcomes, carefully showing how to set up each model and how to interpret the output. Numerous annotated screen shots clearly demonstrate the use of these techniques and how to navigate the program. We provide a couple of extended examples in each chapter that illustrate the logic of model development and interpretation of output. These examples show readers the context and rationale of the research questions and the steps around which the analyses are structured. We also provide modeling syntax in the book's appendix for users who prefer this approach for model development. Readers can work with the various examples developed in each chapter by using the corresponding data and syntax files which are available for downloading from the publisher's book-specific website at http://www.psypress.com/9781848729568.

Contents

The workbook begins with a chapter highlighting several relevant conceptual and methodological issues associated with defining and investigating multilevel and longitudinal models with categorical outcomes, which is followed by a chapter on IBM SPSS data management techniques we have found to facilitate working with multilevel and longitudinal data sets. In chapters 3 and 4, we detail the basics of the single-level and multilevel generalized linear model for various types of categorical outcomes. These chapters provide a thorough discussion of underlying concepts to assist with trouble-shooting a range of common programming and modeling problems readers are likely to encounter. We next develop population-average and unit-specific longitudinal models for investigating individual or organizational developmental processes (chapter 5). Chapter 6 focuses on single- and multilevel models using multinomial and ordinal data. Chapter 7 introduces models for count data. Chapter 8 concludes with additional trouble-shooting techniques and thoughts for expanding on the various multilevel and longitudinal modeling techniques introduced in this book. We hope this workbook on categorical models becomes a useful guide to readers' efforts to learn more about the basics of multilevel and longitudinal modeling and the expanded range of research problems that can be addressed through their application.

Acknowledgments

There are several people we would like to thank for their input in putting this second workbook together. First we offer thanks to our reviewers who helped us sharpen our focus on categorical models: Debbie Hahs-Vaughn of the University of Central Florida, Jason T. Newsom of Portland State University, and one anonymous reviewer. We wish to thank Alberto Cabrera, Gamon Savatsomboon, Dwayne Schindler, and Hongwei Yang for helpful comments on the text and our presentation of multilevel models. Thanks also to Suzanne Lassandro, our Production Manager; George Marcoulides, our Series Editor; Debra Riegert, our Senior Editor; and Andrea Zekus, our Editorial Assistant, who have all been very supportive throughout the process. Finally, we owe a huge debt of gratitude to our students who have had a powerful impact on our thinking and understanding of the issues we have laid out in this workbook. Although we remain responsible for any errors remaining in the text, the book is much stronger as a result of support and encouragement of all of these people.

Ronald H. Heck
Scott L. Thomas
Lynn N. Tabata

CHAPTER 1

Introduction to Multilevel Models With Categorical Outcomes

Introduction

Social science research presents an opportunity to study phenomena that are multilevel, or hierarchical, in nature. Examples include college students nested in institutions within states or elementary-aged students nested in classrooms within schools. Attempting to understand individuals' behavior or attitudes in the absence of group contexts known to influence those behaviors or attitudes can severely handicap researchers' ability to explicate the underlying structures or processes of interest. People within particular organizations may share certain properties, including socialization patterns, traditions, attitudes, and goals.

Multilevel modeling (MLM) is an attractive approach for studying the relationships between individuals and their various social groups because it allows the incorporation of substantive theory about individual and group processes into the sampling schemes of many research studies (e.g., multistage stratified samples, repeated measures designs) or into hierarchical data structures found in many existing data sets encountered in social science, management, and health-related research (Heck, Thomas, & Tabata, 2010). MLM is fast becoming the standard analytic approach for examining data and publishing results in many fields due to its adaptability to a broad range of designs (e.g., experiments, quasi-experiments, survey), data structures (e.g., nested data, cross-classified, cross-sectional, and longitudinal data), and outcomes (continuous, categorical). Despite this applicability to many research problems, however, MLM procedures have not yet been fully integrated into research and statistics texts used in typical graduate courses.

Two major obstacles are responsible for this reality. First, no standard language has emerged from this multilevel empirical work in terms of theories, model specification, and procedures of investigation. MLM is referred to by a variety of different names, including random-coefficient, mixed-effect, hierarchical linear, and multilevel regression models. The diversity of names reflects methodological development in several different fields, which has led to differences in the manner in which the methods and analytic software are used in various fields. In general, multilevel models deal with nested data—that is, where observations are clustered within successive levels of a data hierarchy.

Second, until recently, the specification of multilevel models with continuous and categorical outcomes required special software programs such as HLM (Raudenbush, Bryk, Cheong, & Congdon, 2004), LISREL (du Toit & du Toit, 2001); MLwiN (Rasbash, Steele, Browne, & Goldstein, 2009), and Mplus (Muthén & Muthén, 1998–2006). Although the mainstream emergence and acceptance of multilevel methods over the past two decades has been largely due to the development of specialized software by a relatively small group of scholars, other more widely used statistical packages, including IBM SPSS, SAS, and Stata, have in recent years implemented routines that enable the development and specification of a wide variety of multi-level and longitudinal models (see Albright & Marinova, 2010, for an overview of each package).

In IBM SPSS, the multilevel analytic routine is referred to as MIXED, which indicates a class of models that incorporates both fixed and random effects. As such, mixed models imply the existence of data in which individual observations on an outcome are distributed (or vary) across identifiable groups. Repeated observations may also be distributed across individuals and groups. The variance parameter of the random effect indicates its distribution in the population and therefore describes the degree of heterogeneity (Hedeker, 2005). The MIXED routine is a component of the advanced statistics add-on module for the PC and the Mac, which can be used to estimate a wide variety of multilevel models with diverse research designs (e.g., experimental, quasi-experimental, nonexperimental) and data structures. It is differentiated from more familiar linear models (e.g., analysis of variance, multiple regression) through its capability of examining correlated data and unequal variances within groups. Such data are commonly encountered when individuals are nested in social groups or when there are repeated measures (e.g., several test scores) nested within individuals. Because these data structures are hierarchical, people within successive groupings may share similarities that must be considered in the analysis in order to provide correct estimation of the model parameters (e.g., coefficients, standard errors).

If the analysis is conducted only on the number of individuals in the study, the effects of group-level variables (e.g., organizational size, productivity, type of organization) may be over-valued in terms of their contribution to explaining the outcome. This is because there will typically be many more individuals than groups in a study, so the effects of group variables on the outcome may appear much stronger than they really are. If, instead, we aggregate the data from individuals and conduct the analysis between the groups, we will lose all of the variability among individuals within their groups. The optimal solution to these types of problems concerning the unit of analysis is to consider the number of groups and individuals in the analysis. When the research design is multilevel and either balanced or unbalanced (i.e., there are different numbers of individuals within groups), the estimation procedures in MIXED will provide asymptotically efficient estimates of the model's structural parameters and variance components. In short, the MIXED routine provides a nice and effective way to specify models at two or more levels in a data hierarchy.

In our previous IBM SPSS workbook (Heck et al., 2010), our intent was to help readers set up, conduct, and interpret a variety of different types of introductory multilevel and longitudinal models using this modeling procedure. At the time we finished the workbook in April 2010, we noted that the major limitation of the MIXED model routine was that the outcomes had to be continuous. This precluded many situations where researchers might be interested in applying multilevel analytic procedures to various types of categorical (e.g., dichotomous, ordinal, count) outcomes. Although models with categorical repeated measures nested within individuals can be estimated in IBM SPSS using the generalized estimating equation (GEE) approach (Liang & Zeger, 1986), we did not include this approach in our first workbook because it does not support the inclusion of group processes at a level above individuals; that is, the analyst must assume that individuals are randomly sampled and, therefore, *not clustered* in groups.

Today as we evaluate the array of analytic routines available for continuous and categorical outcomes in IBM SPSS, we note that many of these procedures were incorporated over the past 15 years as part of the REGRESSION modeling routine. A few years ago, however, in IBM SPSS various procedures for examining different categorical outcomes were consolidated under the *generalized linear model* (GLM) (Nelder & Wedderburn, 1972), which is referred to as GENLIN. We note that procedures for handling clustered data with categorical outcomes have been slower to develop than for continuous outcomes, due to added challenges of solving a system of nonlinear mathematical equations in estimating the model parameters for categorical outcomes. This is because categorical outcomes result from types of probability distributions other than the normal distribution. Relatively speaking, mathematical equations for linear models with continuous outcomes are much less challenging to solve.

Over the past couple of years, the MIXED modeling routine has been expanded to include several different types of categorical outcomes. The various multilevel categorical models are referred to as generalized linear mixed models (or GENLIN MIXED in IBM SPSS terminology). This capability begins with Version 19, which was introduced in fall 2010, and is refined in Version 20 (introduced in fall 2011). The inclusion of this new categorical multilevel modeling capability prompted us to develop this second workbook. We wanted to provide a thorough applied treatment of models for categorical outcomes in order to finish our original intent of introducing multilevel and longitudinal analysis using IBM SPSS. Our target audience has been and remains graduate students and applied researchers in a variety of different social science fields. We hope this presentation will be a useful addition to our readers' repertoire of quantitative tools for examining a broad range of research problems.

Our Intent

In this second workbook, our intent is to introduce readers to a range of single-level and multilevel models for cross-sectional and longitudinal data with categorical outcomes. One of our motivations for this book was our observation that introductory and intermediate statistics courses typically devote an inordinate amount of time to models for continuous outcomes and, as a result, graduate students in the social sciences have relatively little experience with various types of quantitative modeling techniques for categorical outcomes. There are many good reasons for an emphasis on models for continuous outcomes, but we believe this has left students and, ultimately, their fields ill prepared to deal with the wide range of important questions that do not accommodate continuously measured outcomes.

There are a number of important conceptual and mathematical differences between models for continuous and categorical outcomes. Categorical responses result from probability distributions other than the normal distribution and therefore require different types of underlying mathematical models and estimation methods. Because of these differences, they are often more challenging to investigate. First, they can be harder to report about because they are in different metrics (e.g., log odds, probit coefficients, event rates) from the unstandardized and standardized multiple regression coefficients with which most readers may be familiar. In other fields, such as health sciences, however, beginning researchers are more apt to encounter categorical outcomes more routinely—one example being investigating the presence or absence of a disease.

Second, with respect to multilevel modeling, models with categorical outcomes require somewhat different estimation procedures, which can take longer to converge on a solution and, as a result, may require making more compromises during investigation than typical continuous-outcome models. Despite these added challenges, researchers in the social sciences often encounter variables that are not continuous because outcomes are often perceptual (e.g., defined on an ordinal scale) or dichotomous (e.g., deciding whether or not to vote, dropping out or persisting) or refer to membership in different groups (e.g., religious affiliation, race/ethnicity). Of course, depending on the goals of the study, such variables may be either independent (predictor) or dependent (outcome) variables. Therefore, building skills in defining and analyzing single-level and multilevel models should provide opportunities for researchers to investigate different types of categorical dependent variables.

In developing this workbook, we, of course, had to make choices about what content to include and when we could refer readers to other authors for more extended treatments of issues we raise. There are many different types of quantitative models available in IBM SPSS for working with categorical variables, beginning with basic contingency tables and related measures of association, loglinear models, discriminant analysis, logistic and ordinal regression, probit regression, and survival models, as well as multilevel formulations of many of these basic single-level models. We simply cannot cover all of these various types of analytic approaches for categorical outcomes in detail. Instead, we chose to highlight some types of categorical outcomes researchers are likely to encounter regularly in investigating multilevel models with various types of cross-sectional and repeated measures designs. We encourage readers also to consult other discussions of the various analytic procedures available for categorical outcomes to widen their understanding of the assumptions and uses of these types of models.

As in our first workbook, we spend considerable time introducing and developing a general strategy for setting up, running, and interpreting multilevel models with categorical outcomes. We also devote considerable space to the various types of categorical outcomes that are frequently encountered, some of the differences involved in estimating single-level and multilevel models with categorical versus continuous outcomes, and what the meaning of the output is for various categorical outcomes. We made this decision because we believe that students are generally less familiar with models having categorical outcomes and the various procedures available in IBM SPSS that can be used to examine them. Our first observation is that the nature of categorical outcomes themselves (e.g., their measurement properties, sampling distributions, and methods to estimate model parameters) requires some modification in our typical ways of thinking about single-level and multilevel investigation.

As we get into the meat of the material in this workbook, you will see that we assume that people using it are familiar with our previous one; that is, we attempt to build on the information included there for continuous outcomes. We emphasize that it is not required reading; however, we do revisit and extend several important issues developed in our first workbook. Readers will recognize that Chapters 1 and 2 in this workbook, which focus on general issues in estimating multilevel models and preparing data for analysis in IBM SPSS, are quite similar to the same chapters in the first workbook. Our discussions of longitudinal models, multivariate outcomes, and cross-classified data structures in the first workbook would be useful in thinking about possible extensions of the basic models with categorical outcomes that we introduce in this workbook.

Readers familiar with our first workbook may also note that we spend a fair amount of energy tying up a few loose ends from that other workbook—one being the use of the add-on multiple imputation routine available in IBM SPSS to deal with missing data in a more satisfactory manner. We provide a simple illustration of how multiple imputation can be used to replace missing data in the present workbook. Another is the issue of sample weights for single-level and multilevel data. IBM SPSS currently does not support techniques for the adjustment of design effects, although an add-on module can be purchased for conducting single-level analyses of continuous and (some) categorical outcomes from complex survey designs. In this workbook, we update our earlier coverage of sample weights, offering considerations of different weighting schemes that help guide the use of sample weights at different levels in the analysis. This is an area in which we feel that more attention will be paid in subsequent versions of IBM SPSS and other programs designed to analyze multilevel data. We attempt to offer some clear guidance in the interim.

Veteran readers will note that we continue with the model building logic we developed and promoted in that first workbook. Readers of our earlier work know that we think it is important to build the two-level (and three-level) models from the variety of *single-level* analytic routines available in IBM SPSS. We take this path so that readers have a working knowledge of the extensive analytic resources available for examining categorical variables in the program. This approach seemed logical to us because, for a number of years, as we noted, several types of categorical analytic routines have been added to the program that can be used to examine categorical outcomes for cross-sectional and longitudinal data. Single-level, cross-sectional analyses with dichotomous or ordinal outcomes can be conducted using the regression routine (ANALYZE: Regression), including binary logistic regression, ordinal regression, and probit regression.

A more extensive array of single-level categorical analytic techniques has been consolidated under the generalized linear modeling (GLM) framework, which includes continuous, dichotomous, nominal, ordinal, and count variables. Also subsumed under the GLM framework within GENLIN is the longitudinal (repeated measures) approach for categorical outcomes, which can be conducted using the GEE procedure. As readers may realize, these various analytic routines provide a considerable number of options for single-level analyses of categorical outcomes. Single-level analyses refer to analyses where either individuals or groups are the unit of analysis. In our presentation, we emphasize the program's GENLIN analytic routines for cross-sectional and repeated measures data because they form the foundation for the multilevel generalized linear mixed modeling approach that is now operational (GENLIN MIXED). We will take up the similarities, differences, and evolution of these various procedures as we make our way through this introductory chapter.

Although we introduce a variety of two- and three-level categorical models, we note at the beginning that in a practical sense, running complex models with categorical outcomes and more than two levels can be quite demanding on current model estimation procedures. Readers should keep in mind that multilevel modeling routines for categorical outcomes are still relatively new in most existing software programs. Because multilevel models with categorical outcomes require quasilikelihood estimation (i.e., a type of *approximate* maximum likelihood) or numerical integration (which becomes more complex with more random effects) to solve complex nonlinear equations, model convergence is more challenging than for continuous outcomes.

These problems can increase with particular types of data sets (i.e., large within-group sample sizes, repeated measures with several hierarchical levels, more complex model formulations with several random slopes). Estimating these models is very computationally intensive and can require a fair amount of processing time. We identify a number of tips for reducing the computational demand in estimation. For example, one technique that can reduce the time it takes to estimate the model is to recode individual identifiers within their units (1,…, *n*). We show how to do this in Chapter 2. Lengthy estimation times for multilevel categorical models are not a problem that is unique to IBM SPSS; we have previously encountered other complex multilevel categorical models that can take hours to estimate, so remember that patience is a virtue!

We run the categorical models in this workbook using the 32-bit versions of IBM SPSS Version 20 (statistics base with the advanced statistics module add-on) and Windows 7 Professional. Users running the model with other operating systems or older versions of the program may notice slight differences between their screen displays and our screenshots, as well as slight differences in output appearance (and perhaps even estimates).

What follows is an introduction to some of the key conceptual and technical issues in categorical data analysis, generally, and multilevel categorical modeling, specifically. Our coverage here is wide ranging and allows us to set the stage for more focused treatment in subsequent chapters. Key to our success here, however, is defining the central conceptual elements in the linear model for categorical data. In the end, we think that the reader will see clearly that these models simply transform categorical (nonlinear) outcomes into linear functions that can be modeled with generalized forms of the general linear model. Let us now provide some important background and offer a few critical distinctions that will help make the preceding point understandable. We start at the beginning.

Analysis of Multilevel Data Structures

We begin with the observation that quantitative analysis deals with the translation of abstract theories into concrete models that can be empirically investigated. Our statistical models represent a set of proposed theoretical relations that are thought to exist in a population—a set of theoretical relations that are proposed to represent relationships actually observed in the sample data from that population (Singer & Willett, 2003). Decisions about research questions, designs and data structures, and methods of analysis are therefore critical to the credibility of one's results and the study's overall contribution to the relevant knowledge base. Multilevel models open up opportunities to examine relationships at multiple levels of a data hierarchy and to incorporate a time dimension into our analyses. But the ability to model more complex relationships comes at a computational cost. The more complex models can easily bog down, fail to converge on a solution, or yield questionable results.

Multilevel models are more data demanding in that adequate sample sizes at several levels may be required to ensure sufficient power to detect effects; as a result, the models can become quite complicated, difficult to estimate, and even more difficult to interpret (Heck et al., 2010). Even in simple two-level analyses, one might allow the intercept and multiple slopes to vary randomly across groups while employing several group-level variables to model the variability in the random intercept and each random slope. These types of "exploratory" models are usually even more difficult to estimate with categorical outcomes than with continuous outcomes. As Goldstein (1995) cautions, correct model specification in a single-level framework is one thing; correct specification within a multilevel context is quite another. For this reason, we emphasize the importance of a sound conceptual framework to guide multilevel model development and testing, even if it is largely exploratory in nature.

One's choice of analytic strategy and model specification is therefore critical to whether the research questions can be appropriately answered. More complete modeling formulations may suggest inferences based on relationships in the sample data that are not revealed in more simplistic models. At the same time, such modeling formulations may lead to fewer findings of substance than have often been claimed in studies that employ more simplistic analytic methods (Pedhazur & Schmelkin, 1991). In addition to choices about model specification, we also note that in practice our results are affected by potential biases in our sample (e.g., selection process, size, missing cases). Aside from characteristics of the sample itself, we emphasize that when making decisions about how to analyze the data, the responsible researcher should also consider the approach that is best able to take advantage of the features of the particular data structures with respect to the goals of the research.

Multilevel data sets are distinguished from single-level data sets by the nesting of individual observations within higher level groups, or within individuals if the data consist of repeated measures. In single-level data sets, participants are typically selected through simple random sampling. Each individual is assumed to have an equal chance of inclusion and, at least in theory, the participants do not belong to any higher order social groups that might influence their responses. For example, individuals may be differentiated by variables such as gender, religious affiliation, or membership in a treatment or control group; however, in practice, individual variation within and between subgroups in single-level analyses cannot be considered across a large number of groups simultaneously. The number of subgroups would quickly overwhelm the capacity of the analytic technique.

In multilevel data analyses, the grouping of participants, which results from either the sampling scheme (e.g., neighborhoods selected first and then individuals selected within neighborhoods) or the social groupings of participants (e.g., being in a common classroom, department, organization, or political district), is the focus of the theory and conceptual model proposed in the study (Kreft & de Leeuw, 1998). These types of data structures exist in many different fields and as a result of various types of research designs. For nonexperimental designs, such as survey research, incorporating the hierarchical structure of the study's sampling design into the analysis opens up a number of different questions that can be asked about the relationships between variables at a particular level (e.g., individual level, department level, organizational level), as well as how activity at a higher organizational level may impact relationships at a lower level.

We refer to the lowest level of the data hierarchy (Level 1) as the *micro* level, with all successive levels referred to as *macro* levels (Hox, 2002). The relationships among variables observed for the microlevel often refer to individuals within a number of macrolevel groups or contexts (Kreft & deLeeuw, 1998). Repeated-measures experimental or quasi-experimental designs with treatment and control groups are generally conceptualized as single-level multivariate models (i.e., using repeated measures ANOVA or MANOVA); however, they can also be conceptualized as two-level, random-coefficient models where time periods are nested within subjects. One of the primary advantages of the multilevel formulation is that it opens up possibilities for more flexible treatment of unequal spacing of the repeated measurements, more options for incorporation of missing data, and the possibility of examining randomly varying intercepts and regression slopes across individuals and higher level groups.

Figure 1.1 presents a hierarchical data structure that might result from a survey conducted to determine how organizational structures and processes affect organizational outcomes such as productivity. The proposed conceptual model implies that organizational outcomes may be influenced by combinations of variables related to the backgrounds and attitudes of individuals (e.g., demographics, experience, education, work-related skills, attitudes), the processes of organizational work (e.g., leadership, decision making, professional development, organizational values, resource allocation), the context of the organization (demands, changing economic and political conditions), or the structure of the organization (e.g., size, managerial arrangements within its clustered groupings, etc.).

The multilevel framework also implies that we may think about productivity somewhat differently at each level. For example, at the micro (or individual) level, productivity might be conceived in our study as the probability that an individual employee meets specific productivity goals. We might define the variable as dichotomous (met = 1, not met = 0) or ordinal (exceeded = 2, met = 1, not met = 0). At Level 2 (the subunit level within each organization), the outcome can focus on variability in productivity between subunits (e.g., departments, work groups) in terms of having a greater or lesser proportion of employees who meet their productivity goals. At Level 3, the focus might be on differences between various organizations in terms of the collective employee productivity.

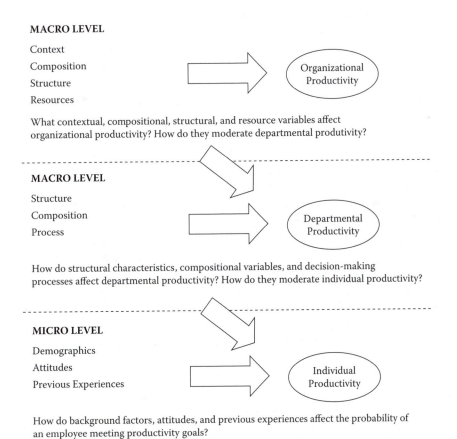

FIGURE 1.1   Defining variables in a multilevel categorical model.

In the past, analytic strategies for dealing with the complexity of hierarchical data structures were somewhat limited. Researchers did not always consider the implications of the assumptions that they made about moving variables from one level to another. As we noted earlier, one previous strategy was to aggregate data about individuals to the group level and conduct the analysis based on the number of groups. This strategy was flawed, however, because it removed the variability of individuals within their groups from the analysis.

A contrasting strategy was to disaggregate variables conceptualized at a higher level (such as the size of the organization) and include them in an analysis conducted at the microlevel. This strategy was also problematic because it treated properties of organizations as if they were characteristics of individuals in the study. The implication is that analyses conducted separately at the micro- or macrolevel generally produce different results. Failure to account for the successive nesting of individuals within groups can lead to underestimation of model parameters, which can result in erroneous conclusions (Raudenbush & Bryk, 1992). Most important, simultaneous estimation of the micro- and macrolevels in one model avoids problems associated with choosing a specific unit of analysis.

Aside from the technical advantages that can be gained from MLM, these models also facilitate the investigation of variability in both Level-1 intercepts (group means) and regression coefficients (group slopes) across higher order units in the study. When variation in the size of these lower level relationships across groups becomes the focus of modeling building, we have a specific type of multilevel model, which is referred to as a *slopes-as-outcome* model (Bryk & Raudenbush, 2002). This type of model concerns explaining variation in the random Level-1 slope across groups. Relationships between variables defined at different organizational levels are referred to as *cross-level interactions*. Cross-level interactions, as the term suggests, extend from a macrolevel in a data hierarchy toward the microlevel; that is, they represent vertical relationships between a variable at a higher level that *moderates* (i.e., increases or diminishes) a relationship of theoretical interest at a lower level.

In Figure 1.1, such relationships are represented by vertical arrows extending from a higher to lower level. An example might be where greater input and participation in departmental decision making (at the macrolevel) strengthens the microlevel relationship between employee motivation and probability of meeting individual productivity goals. In such models, cross-level interactions are proposed to explain variation in a random slope. We might also hypothesize that organizational-level interventions (e.g., resource allocation, focus on improving employee professional development) might enhance the productivity of work groups within the organization. Such relationships between variables at different organizational levels may also be specified as multilevel mediation models (MacKinnon, 2008).

Specifying these more complex model formulations represents another advantage associated with MLM techniques. As readers may surmise, however, examining variables on several levels of a data hierarchy simultaneously requires some adjustments to traditional linear modeling techniques in order to accommodate these more complex data structures. One is because individuals in a group tend to be more similar on many important variables (e.g., attitudes, socialization processes, perceptions about their workplace). For multilevel models with continuous outcomes, a more complex error structure must be added to the model at the microlevel (Level 1) in order to account for the correlations between observations collected from members of the same social group. Simply put, individuals within the same organization may experience particular socialization processes, hold similar values, and have similar work expectations for performance that must be accounted for in examining differences in outcomes between groups. For continuous outcomes, it is assumed that the Level-1 random effect has a mean of zero and homogeneous variance (Randenbush & Bryk, 2002). These problems associated with multilevel data structures are well discussed in a number of other multilevel texts (e.g., Bryk & Raudenbush, 2002; Heck & Thomas, 2009; Hox, 2010; Kreft & de Leeuw, 1998).

For multilevel models with categorical outcomes, however, we generally cannot add a separate residual (error term) to the Level-1 model because the Level-1 outcome is assumed to follow a sampling distribution that is different from a normal distribution. Because of this difference in sampling distributions, the Level-1 residual typically can only take on a finite set of values and therefore is not normally distributed. For example, with a dichotomous outcome, the residuals can only take on two values; that is, either an individual is incorrectly predicted to pass a course when she or he failed or vice versa. Moreover, the residual variance is not homogenous within groups; instead, it depends on the predicted value of the outcome (Raudenbush et al., 2004).

The lack of a separately estimated Level-1 random effect is one primary difference between multilevel models with continuous outcomes and those with discrete outcomes. A second difference, as we have mentioned on more than one occasion already, is that model estimation for categorical outcomes requires more complex procedures, which can take considerable computational time, with more variables, random effects, and increased sample size. Atheoretical exploratory analyses (or fishing expeditions) with multilevel models can quickly prove to be problematic. Given the preceding points, such analyses with categorical outcomes are more perilous.

Theory needs to guide the development of the models, and we suggest that when they build models with categorical outcomes, researchers start first with *fixed* slope effects (i.e., they do not vary across groups) at the lower level because they are less demanding on model estimation. We suggest adding random slopes sparingly at the latter stage of analysis—and only when there is a specific theoretical interest in such a relationship associated with the purposes of the research.

Scales of Measurement

Empirical investigation requires the delineation of theories about how phenomena of interest may operate, a process that begins with translating these abstractions into actual classifications and measures that allow us to describe proposed differences between individuals' perceptions and behavior or those of characteristics of objects and events. Translating such conceptual dimensions into operationalized variables constitutes the process of measurement, and the results of this process constitute quantitative data (Mueller, Schuessler, & Costner, 1977). If we cannot operationalize our abstractions adequately, the empirical investigation that follows will be suspect. Conceptual frameworks "may be understood as mechanisms for comprehending empirical situations with simplification" (Shapiro & McPherson, 1987, p. 67). A conceptual framework (such as Figure 1.1) identifies a set of variables and the relationships among them that are believed to account for a set of phenomena (Sabatier, 1999). Theories encourage researchers to specify which parts of a conceptual framework are most relevant to certain types of questions. For multilevel investigations, there is the added challenge of theorizing about the relationships between group and individual processes at multiple levels of a conceptual framework and perhaps over time.

One definition of measurement is "the assignment of numbers to aspects of objects or events according to one or another rule or convention" (Stevens, 1968, p. 850). Stevens (1951) proposed four broad levels of measurement "scales" (i.e., nominal, ordinal, interval, and ratio), which, although widely adopted, have also generated considerable debate over the years as to their meaning and definition. Nominal (varying by distinctive quality only and not by quantity) and ordinal (varying by their rank ordering, such as least to most) variables are often referred to as *discrete* variables, and interval (having equal distance between numbers) and ratio (expressing values as a ratio and having a fixed zero point) variables constitute *continuous* variables. Continuous variables can take on any value between two specified values (e.g., "The temperature is typically between 50° and 70° during March"), but discrete variables cannot (e.g., "How often do you exercise? [Never, Sometimes, Frequently, Always]"). Discrete variables can only take on a finite number of values. For example, in the exercise case, we might assign the numbers 0 (never), 1 (sometimes), 2 (frequently), and 3 (always) to describe the ordered categories, but 2.3 would not be an acceptable value to describe an individual's frequency of exercise.

This previous discussion suggests that a continuous variable results from a sampling distribution where it can take on a continuous range of values; hence, its probability of taking on any *particular* value is 0 because there are an infinite number of other values it can take on within a given interval. As we have noted, one frequently encountered probability distribution for continuous variables is the normal distribution. For an outcome Y that is normally distributed, we generally find that a linear model will adequately describe how a change in X is associated with a change in Y. One simple illustration is that as the length of the day with sunlight (X) increases by some unit over the first several months of the year in the Northern Hemisphere, the average temperature (Y) will increase by some amount. We can see that Y can take on a continuous range of values generally between 0° and 100° between January and July (e.g., 0°–73° in Fairbanks, Alaska; 37°–69° in Denmark; 68°–84° in Los Angeles; 67°–107° in Phoenix).

Because a discrete variable (e.g., failing or passing a course) cannot take on any value, predicting the probability of the occurrence of each specified outcome (among a *finite* set) represents an alternative. The probability of passing or failing a course, however, falls within a restricted numerical range; that is, the probability of success cannot be less than 0 or more than 1. An event such as passing a course may never occur, may occur some proportion of the time, or may always occur, but it cannot exceed those boundaries. Moreover, the resultant shift in the probability of an event Y occurring will not be the same for different levels of X; that is, the relationship between Y and X is not linear.

To illustrate, the effect of increasing a unit of X, such as hours studied, on Y will be greatest at the point where passing or failing the course is equally likely to occur. Increasing one's hours of study will have very little effect on passing or failing when the issue is no longer in doubt! Although the normal distribution often governs continuous level outcomes, common discrete probability distributions that we will encounter when modeling categorical outcomes include the Bernoulli distribution, binomial distribution, multinomial distribution, Poisson distribution, and negative binomial distribution.

Methods of Categorical Data Analysis

As this fundamental difference between discrete and continuous variables suggests, the relationship between the measurement of outcome variables and "permissible" statistics depending on their measurement characteristics has generated much debate in the social sciences over the past century (Agresti, 1996; Mueller et al., 1977; Nunnally, 1978; Pearson & Heron, 1913; Stevens, 1951; Yule, 1912). Categorical data analysis concerns response variables that are almost always discrete—that is, measured on a nominal or ordinal scale. The development of methods to examine categorical data stems from early work by Pearson and Yule (Agresti, 1996). Pearson argued that categorical variables were simply proxies for continuous variables, and Yule contended that they were inherently discrete. As we can see, both views were partly correct. Certainly, ordinal variables represent difference in quantity, such as how important voting may be to an individual; assigning numbers to variables such as religious affiliation, gender, or race/ethnicity merely represents a convenient way of classifying individuals into similar groups according to distinctive characteristics (Azen & Walker, 2011).

Until perhaps the last decade or so, methods of categorical data analysis generally had marginal status in statistics texts, often relegated to the back of each as the obligatory chapter on "nonparametric" statistics. Despite this limited general coverage, categorical methods have always garnered some interest in social science research. For example, in psychology, rank-order associations such as gamma and Somer's D provide a means of examining associations between perceptual variables; in sociology, interest in the association between categorical variables (e.g., socioeconomic status, gender, and likelihood to commit a crime) often focused on cross-classifying individuals by two or more variables in contingency tables.

A contingency table summarizes the joint frequencies observed in each category of the variables under consideration. An example might be the relationship between gender and voting

TABLE 1.1   Perception of Undergraduate Peer Alcohol Use by Residence Location

| | | | Perception of Peer Alcohol Use | | |
			Low Use	Moderate or High Use	Total
Currently live	Off campus	Count	131.0	110.0	241.0
		Expected count	143.9	97.1	241.0
	On campus	Count	152.0	81.0	233.0
		Expected count	139.1	93.9	233.0
Total		Count	283.0	191.0	474.0
		Expected count	283.0	191.0	474.0

Notes: Pearson chi-square = 5.828, *p* = .016; contingency coefficient = 0.11.

behavior, perhaps controlling for a third variable such as political party affiliation. The term *contingency table,* in fact, dates from Pearson's early twentieth century work on the association between two categorical variables (i.e., the Pearson chi-square statistic $\sum_{i=1}^{c} \frac{(O_i - E_i)^2}{E_i}$ and the corresponding contingency coefficient $\sqrt{\frac{\chi^2}{\chi^2 + n}}$ for testing the strength of the association). The goal of the analysis was to determine whether the two categorical variables under consideration were statistically independent (i.e., the null hypothesis) by using a test statistic such as the chi-square statistic, which summarizes the discrepancy between the observed (*O*) frequencies we see in each cell of a cross-classified table against the frequencies we would expect (*E*) by chance alone (i.e., what the independence model would predict), given the distribution of the data.

To illustrate this basic approach for categorical analysis, in Table 1.1 we examine the proposed relationship between undergraduates' place of residence (off campus or on campus) and their perceptions of their peers' use of alcohol during a typical month. We might propose that students who live on campus will perceive that their peers use alcohol less frequently during a typical month than students who live off campus perceive this use. Although only two categories of perceived alcohol use are used to illustrate the approach, there were originally four possible responses (none, some, moderate, high).

We see, for example, that students living off campus do perceive that their peers engage in moderate to heavy use of alcohol more frequently than we would expect by chance alone if the variables were statistically independent (i.e., 110 perceive the use to be moderate or high compared against the expected count of 97.1). In contrast, students living on campus are more likely to perceive alcohol use by their peers as low compared to what we would expect by chance alone (152 observed to 139.1 expected). The significant chi-square coefficient suggests that we reject the null hypothesis of statistical independence (χ^2 = 5.828, *p* = .016); hence, we conclude that there is a statistical association between the two variables. The resulting contingency coefficient suggests a weak association between them (CC = 0.11).

We might next investigate whether gender affects this primary relationship of interest. Adding the third variable to the model is the essence of contingency-table testing of proposed models. More specifically, gender might "wash out" (i.e., make *spurious*) the relationship between residence setting and perceptions of peer alcohol use; it might *specify* the relationship between residence and perceptions of alcohol use (i.e., increase or decrease the strength of relationship on a particular subset of residents) or it might not affect the primary relationship at all. When we actually test for the effect of gender on the original relationship of interest (see Table 1.2), we find that, for females, the relationship between residence and perceptions of peer alcohol use is significant (χ^2 = 4.27, *p* = .039), but for men it is not (χ^2 = 2.45, *p* = .118).

Hence, based on the cross classification, we must alter our original proposition because the relationship between place of residence and perceptions of peers' use of alcohol is *specified* by the

TABLE 1.2   Perception of Peer Alcohol Use by Residence and Gender

Female				Low Use	Moderate or High Use	Total
				Perception of Peer Alcohol Use		
Male	Currently live	Off campus	Count	78.0	67.0	145.0
			Expected count	84.7	60.0	145.0
		On campus	Count	99.0	59.0	158.0
			Expected count	92.3	65.0	158.0
	Total		Count	177.0	126.0	303.0
			Expected count	177.0	126.0	303.0
Female	Currently live	Off campus	Count	53.0	43.0	96.0
			Expected count	59.5	36.0	96.0
		On campus	Count	53.0	22.0	75.0
			Expected count	46.5	28.5	75.0
	Total		Count	106.0	65.0	171.0
			Expected count	106.0	65.0	171.0
Total	Currently live	Off campus	Count	131.0	110.0	241.0
			Expected count	143.9	97.0	241.0
		On campus	Count	152.0	81.0	233.0
			Expected count	139.1	93.0	233.0
	Total		Count	283.0	191.0	474.0
			Expected Count	283.0	191.0	474.0

respondent's gender. From a technical standpoint, because adding the third variable requires additional chi-square tests for each condition of the added variable, we note that the number of individuals in each subtest conducted will be considerably smaller with each variable added to the analysis. We can see that adding more variables to this type of analysis will soon make it very complex and increasingly suspect due to small cell sizes. In fact, as Lancaster (1951) observed, "Doubtless little use will ever be made of more than a three-dimensional classification" (cited in Agresti, 2002, p. 626).

Although this type of analysis of cross-tabulation tables was limited in terms of the complexity of relationships that could be examined in one analysis, it was the typical analytic approach for examining relationships between categorical variables until about 1970. During the 1970s, cross-classified data analysis was extended considerably with the introduction of loglinear models (Goodman, 1970, 1972; Knoke & Burke, 1980), which facilitated the analysis of the association between categorical variables by taking the natural logarithms of the cell frequencies within the cross classification. In recent decades, interest in investigating relationships between categorical outcomes has grown considerably in many fields. Advancements in formulating models to examine dependent variables that arise from sampling distributions other than the normal distribution and computer software programs that can analyze more complex types of quantitative models with categorical outcomes and larger sets of predictors have greatly expanded possibilities for investigating categorical outcomes.

To summarize some of these advancements briefly, dating from the mid-1930s, probit and logit models facilitated the use of linear modeling techniques such as multiple regression and analysis of variance with dichotomous outcomes through applying a mathematical model that produced an underlying continuous predictor of the outcome. This then could be reformulated as a predicted probability of the outcome event occurring. These dichotomous-outcome models were extended to ordinal outcomes in the 1970s (Agresti, 1996). Nelder and Wedderburn (1972) unified several of these models for examining categorical outcomes (e.g., logistic, loglinear, probit models) as special cases of GLMs.

The GLM facilitates the examination of different probability distributions for the dependent variable Y using a mathematical "link" function—that is, a mathematical model that specifies the relationship between the linear predictor and the mean of the distribution function. The link function transforms the categorical outcome into an underlying continuous predictor of Y. The generalized linear model assumes that the observations are uncorrelated. Extensions of the GLM have been developed to allow for correlation between observations as occurs, for example, in repeated measures and clustered designs. More specifically, categorical models for clustered data, referred to as the GEE approach for longitudinal data (Liang & Zeger, 1986), and mixed models for logistic (e.g., Pierce & Sands, 1975) and probit (Ashford & Sowden, 1970) scales enable applications of GLM to the longitudinal and multilevel data structures typically encountered with continuous and noncontinuous outcomes.

Sampling Distributions

A sampling distribution provides the probability of selecting a particular value of the variable from a population (Azen & Walker, 2011). Their properties are related to the measurement characteristics of the *random* outcome variable. In this instance, by "random" we mean the level of the outcome that happens by chance when drawing a sample, as related to characteristics of its underlying probability distribution. More specifically, a probability distribution is a mathematical function that describes the probability of a random variable taking on particular values. Each type of outcome has a set of possible random values it can take on. For example, a continuous outcome can take on any value between a chosen interval such as 0 and 1, but a dichotomous outcome can take on only the value of 0 or 1. In this latter case, the expected value, or mean, of the outcome will simply be the proportion of times the event of interest occurs.

We assume that readers are generally familiar with multiple regression and analysis of variance (ANOVA) models for explaining variability in continuous outcomes. These techniques assume that values of the dependent variable are normally distributed about its mean with variability similar to a bell curve; that is, they result from a *normal* sampling distribution. There are no restrictions on the predicted values of the Level-1 outcome; that is, they can take on any real value (Raudenbush et al., 2004). As we noted previously, the residuals, or errors associated with predicting Y from X at Level 1, are approximately normally distributed with a mean of 0 and some homogenous variance. In the multilevel setting, we can often transform the continuous outcome that is considerably skewed such that the residuals at Level 1 will be approximately normal.

When outcome variables are categorical, however, we have noted that different types of sampling distributions need to be considered at Level 1. One of the key issues is that the assumption of a normal distribution is likely to be violated. Consider the probability of flipping a coin and encountering a heads or tails. We can assume that the likelihood of encountering each possibility is 1:1 (discounting any potential flaw in the coin). With a dichotomous variable, where responses are usually coded 0 and 1, there can be no normal distribution of responses around the mean. The mean is simply the proportion of "successes," or favored events (i.e., usually coded 1), against the other event (coded 0). Over repeated trials, in the coin example, the mean will tend to be 0.5, suggesting equal heads and tails responses. Of course, in the short run, we might observe five or six heads or tails in a row. The probability of either outcome occurring is defined as the number of ways it can occur out of the total number of possible outcomes (Azen & Walker, 2011). It follows that such a random variable can assume only a finite (or countably infinite) number of values.

If we think about the probability of obtaining any particular number 1 to 6 when we roll a die, the probability for each event (or number of ways it can occur out of the total possible outcomes) is 1/6. This is an example of a discrete uniform distribution—that is, when each possible

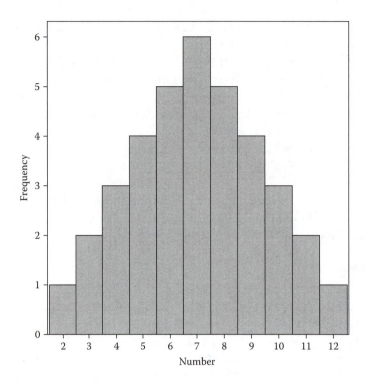

FIGURE 1.2 Distribution of possible combinations of rolling two dice.

outcome is equally likely to occur. The predicted probability of obtaining any number, however, cannot be less than zero or more than unity. For rolling two dice, however, when the possible combinations range from 2 to 12, we are more likely to roll a value near the mean of 7 of the distribution (e.g., perhaps 6, 7, or 8) than either a 2 or 12. Although we certainly could roll a 2 or 12 on our first trial, often we will have to roll 40 or more times before we obtain a 2 or a 12. In this latter case, the probability of either event occurring is 1/36 (1/6*1/6 = 1/36).

In Figure 1.2, we can quickly see that the total set of probabilities is not evenly distributed but, rather, is distributed in a triangular curve. For example, there are six possible combinations (6/36 = 1/6) leading to a 7 (i.e., 1,6; 2,5; 3,4; 6,1; 5,2; 4,3), but only one possible combination leading to a 2 or a 12.

With each successive die we might add, the curve of probabilities becomes more bell shaped, as suggested by the central tendency theorem. The curve would become gradually smoother than in the two-dice case, and it would appear to be closer to a bell curve, or Gaussian function. The normal distribution, therefore, implies that values near the mean will likely be encountered more frequently in the population. Keep in mind, however, that a continuous distribution assigns probability 0 to every individual outcome because it can take on any real number value.

A probability distribution, therefore, is a mathematical model that links the actual outcome obtained from empirical investigation to the probability of its occurrence (Azen & Walker, 2011). If we think about a dichotomous outcome again, because the predicted value of the outcome Y can only take on two values, the random effect at Level 1 also can only be one of two values; therefore, the residuals cannot be normally distributed and no empirical data transformation can make them that way (Hox, 2010). For a dichotomous variable, this type of sampling distribution is referred to as binomial (representing the number of successes in a set of independent trials). A binomial distribution can be thought of as an experiment with a fixed number of independent trials, each of which can have only two possible outcomes, where, because each trial is independent, the probabilities remain constant. Our previous example of tossing a coin a number of times to determine how many times a heads occurs is an example of a binomial experiment. One commonly encountered type of binomial

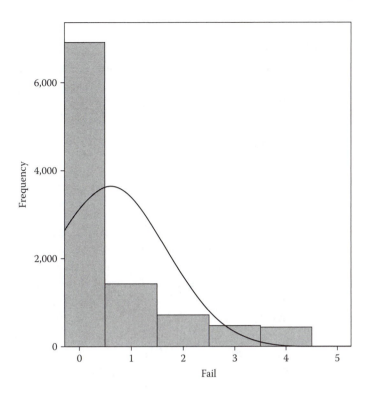

FIGURE 1.3   Distribution of core course failures among ninth grade students.

distribution is the Bernoulli distribution, where *Y* can take on values of 0 or 1. Examples of dichotomous variables include having a particular disease or not and being proficient or not in math. In this simplified binomial distribution, the focus is on the probability of success in one trial; that is, the proportion of cases coded 0 and 1 must add up to 1.0.

Count variables are another type of sampling distribution because there may be so many zeros in the data (perhaps 70% of the data). They can often be represented as a Poisson distribution, which is useful predicting relatively rare events over a fixed period of time. The histogram in Figure 1.3 provides an illustration of this latter type of distribution. The figure summarizes a distribution of students who failed at least one core course (English, math, science, or social studies) during their first year of high school. We have placed a "normal curve" over the distribution to illustrate that the probability of course failure does not follow a typical normal distribution. Although we can calculate a mean of 0.60 and a standard deviation, we can see that they do not adequately represent the fact that most students did not fail any course. In this case, the standard deviation is not very useful because the distribution of values cannot extend below 0. As the average number of times the outcome event of interest would increase, however, the Poisson distribution tends to resemble a normal distribution more closely and, at some sufficient mean level, the normal distribution would be a good approximation to the Poisson distribution (Azen & Walker, 2011; Cheng, 1949).

Similarly, nominal or ordered categorical data result from a multinomial probability distribution. The multinomial distribution represents a multivariate extension of the binomial distribution when there are *c* possible outcome categories rather than only two. This probability distribution indicates the probability of obtaining a specific outcome pattern across *c* categories in *n* trials. The sum of the probabilities across the five categories will be 1.0. An example would be an item asking employees the likelihood that they might leave the organization this year on a five-point ordered scale (i.e., very unlikely, unlikely, neutral, likely, very likely).

We develop each probability distribution in more detail in subsequent chapters. We then provide single-level and multilevel models illustrating each type.

Link Functions

The probability distribution for the random component (outcome) is linked to the explanatory model for the categorical outcome through a specific link function, which is a mathematical function that is used to transform the dependent outcome Y so that it can be modeled as a linear function of a set of predictors (Azen & Walker, 2011; Hox, 2010). Generalized linear models all share the following form:

$$g(E(Y)) = \beta_0 + \beta_1 X_1 + \beta_2 X_2 + \cdots + \beta_q X_q. \tag{1.1}$$

The link function (g) links the expected value of the random component of Y ($E(Y)$) to the deterministic component (i.e., the linear model). As Hox (2010) notes, the GLM approach to outcomes that result from sampling distributions other than the normal distribution is to incorporate the necessary transformation of the dependent variable and choice of appropriate sampling distribution *directly* into the statistical model. In other words, if we imposed a linear model on a binary outcome (which is assumed to be sampled from a binomial distribution) without transformation, the set of predictors would be unlikely to provide good predicted values of Y.

Equation 1.1 implies that the link function transforms the outcome Y in some appropriate manner, depending on its sampling distribution, so that its expected value can be predicted as a *linear* function of a set X predictors (Azen & Walker, 2011). For a continuous outcome, there is no need to transform Y because it is assumed to be sampled from a normal distribution, with a corresponding identity link function that results in the same expected value for Y. Thus, through the use of link functions and alternative forms of sampling distributions, we enable the use of the GLM to examine noncontinuous, categorical outcomes.

Developing a General Multilevel Modeling Strategy

In this workbook we apply a general strategy for examining multilevel models (e.g., Bryk & Raudenbush, 1992; Heck & Thomas, 2009; Hox, 2002). We have found that in many instances multilevel investigations unfold as a series of analytic steps. Of course, there may be times when the analyst might change the specific steps, but, in general, this overall model development strategy works pretty well. Multilevel models are useful and necessary only to the extent that the data being analyzed provide sufficient variation at each level. "Sufficiency" of variation is relative and depends as much on theoretical concerns as it does on the structure and quality of data. Multilevel modeling can be used to specify a hierarchical system of regression equations that take advantage of the clustered data structure (Heck & Thomas, 2009).

The mathematical equations that specify multilevel models are typically presented in one of two ways: (1) by presenting separate equations for each of the levels in a data hierarchy (e.g., employees, workgroups, departments, divisions, corporations, etc.), or (2) by laying out the separate equations and then combining all equations through substitution into a single-model equation (Hox, 2002). For readers already familiar with HLM (Raudenbush et al., 2004), the software uses separate equations specified at each level to build the multilevel model. This approach first requires constructing separate data sets for each level (e.g., individuals, classrooms, schools, etc.), which are then "combined" within the software program to make the final data file (called a multivariate data matrix, or .mdm file). However, the user can neither see nor edit the case-specific contents of this final data set.

The separate-equations approach has the advantage of showing how the model is built at each level and is likely more intuitive to those analysts coming from a regression-based framework. But, because the final data set remains hidden, the disadvantage is that this approach obscures how examining variability in regression slopes results from adding cross-level interaction terms to the model (Hox, 2002).

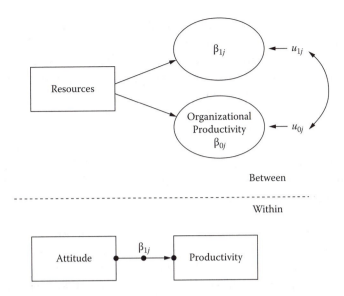

FIGURE 1.4   Proposed two-level model with categorical outcome examining a random intercept and slope.

The other approach is to use algebraic substitution to combine the separate equations at each level into a single equation. Most software packages, such as IBM SPSS MIXED, use single-equation representation of mixed (multilevel) models, so all analyses can be conducted from within a single data set that combines observations from multiple levels of analysis. As we will show in Chapter 2, however, we sometimes need to reorganize the single-level data set for particular types of analyses (e.g., longitudinal and multivariate analyses). The single-level approach emphasizes how interactions are added, but it tends to conceal the addition of error components that are created when modeling randomly varying slopes (Hox, 2002). We will walk readers through this process of building equations and combining them into the single-equation model in the following section. In successive chapters, however, we primarily adopt the approach of laying out separate equations for each level for clarity in building models. We provide the relevant equations in our model-building screenshots. However, we also provide for users the single-model equations in which substitution is applied. We believe that this process of specifying the models as they are being built facilitates understanding the parameters estimated at each step and the output reported in the fixed-effect and random-effect tables.

In Figure 1.4, we provide an illustration of how a simple two-level model to explore a random intercept describing employees' productivity and a random slope describing the effect of employee attitudes on their productivity might look. In this case, we will define productivity as some type of categorical outcome, such as meeting targeted employee productivity goals or not. For example, the outcome could be coded 0 = did not meet goals and 1 = met goals. Alternately, productivity might be ordinal and coded 0 = below productivity standard, 1 = met standard, and 2 = exceeded standard. If we were to use a single-level analysis, we would focus on the relationship between employee attitudes and the probability that they meet productivity goals without reference to their Level-2 units. For example, in a logistic regression model, we would estimate a *fixed* slope expressing the relationship between individuals' attitudes and their probability of meeting productivity goals and a fixed intercept representing the adjusted value of their probability of meeting productivity goals that controls for the level of their attitudes. By *fixed,* we mean that the estimates of the intercept and slope are estimated as the *average* for the sample. This would be fine if our interest were limited to describing an overall relationship between the two variables.

If our interest were also in understanding how employee groupings (e.g., workgroups or departments) may introduce variability in the levels of individuals' productivity or how a particular group variable (e.g., participation in decision making) might moderate the slope describing the effect of individuals' attitudes on their productivity, however, we need to incorporate information about the organizational grouping of employees in our model. Once we shift into thinking about such multilevel relationships, where employees are now nested within groups of some kind, a range of new analytical possibilities emerges. Within groups we might define a randomly varying intercept (β_{0j}) describing the individual's likelihood to achieve productivity targets and a randomly varying slope describing the effect of employee attitudes on the likelihood of meeting these productivity targets. The random slope (β_{1j}) is shown in the figure with a filled dot on the line representing the relationship between attitudes and productivity. The subscript j (without i) is used to describe a randomly varying parameter at Level 2. We point out again that there is no Level-1 residual term shown in the figure. Because the probability distribution is binomial, the variance is a function of the population proportion and, therefore, cannot be estimated separately (Hox, 2010).

At the organizational level, we can formulate a model to explain variability in the random intercept and slope across organizations. First, we propose that differences in resource allocation for employee support affect the probability of meeting productivity goals (β_{0j}) between organizations. Second, we propose that differences in organizational-level resource allocation moderate the size of the effect of employee attitudes on the probability that they meet productivity goals (β_{1j}). More specifically, we might hypothesize that organizational resource allocations considerably above the mean of such allocations in the sample of organizations enhance the effect of employee attitudes on their probability to attain productivity targets.

This latter type of cross-level interaction effect implies that the magnitude of a relationship observed within groups is dependent on contextual, structural, or process features of the groups. We therefore can formulate a Level-2 model to explain variability in intercepts and slopes (shown as ovals representing unknowns in Figure 1.2). We note that each random effect at Level 2 has its own residual variance term, which may covary (shown by the curved arrow) or not, depending on our model specification. As we have noted, this ability to examine how group processes moderate individual-level slope relationships is one of the key conceptual features of multilevel models.

In the sections that follow we offer an example of the logic we use in developing the multilevel model for categorical outcomes. The core logic is consistent with the approach that we adopted in our first workbook. We elaborate that approach to accommodate the demands of the categorical models that are our focus here. We also introduce important notation conventions and several key features that are unique to the development of models with categorical outcomes. This is merely an introduction, as our purpose here is to reveal the broad contours in the construction and evolution of these models. This begins to set our foundation, and we pick up specific pieces of this presentation for more detailed treatment in later chapters.

Determining the Probability Distribution and Link Function

A good first step is to determine the particular sampling distribution and link function to be used in building a set of models. For generalized linear models, the choice of link function must be made initially in order to develop a model that will predict the likelihood of the outcome event of interest occurring. For some types of categorical outcomes, several choices regarding link functions can be made. Selection often hinges on several factors, such as the measurement characteristics of the outcome and its distribution in the data sample that one has available, the characteristics of the model that one is building, and the metric that one wishes to use to report the results.

As part of this initial process, it is also important that the outcome variable be defined appropriately in terms of its scale of measurement (e.g., nominal, ordinal). For example, a correct identification of these measurement issues will generally determine the sampling distribution and link function that are specified for a particular model. Mistakenly identifying the level of measurement of dependent variables may prevent the program from processing the proposed

TABLE 1.3 Model Estimation Criteria and Fit Indices

Target	Productivity
Measurement level	Nominal
Probability distribution	Binomial
Link function	Logit
Information criterion:	
Akaike corrected	29,610.042
Bayesian	29,616.831

Note: Information criteria are based on the –2 log pseudo-likelihood (29,608.041) and are used to compare models. Models with smaller information criterion values fit better. When comparing models using pseudolikelihood values, caution should be used because different data transformations may be used across the models.

model or cause the user to misinterpret the output. In this case, we will assume that productivity is measured as a dichotomous (nominal) variable. It will therefore have a binomial sampling distribution, and we will use a logit link function for this example. We can confirm our choices from the output in Table 1.3, which summarizes the model estimation criteria. This will be constant throughout the model development. We note that the output provides a caution regarding subsequent model comparison using the indices provided (because of the rescaling that takes place each time variables are added).

Developing a Null (or No Predictors) Model

Often in models with continuous outcomes, we are first interested in partitioning the variance in the outcome. We can examine the extent to which variation in a Level-1 outcome exists *within* Level-2 units relative to its variation *between* Level-2 units. Because there is no separate variance term at Level 1 for categorical variables (i.e., it is fixed to a scale factor of 1.0), it may or may not be of interest to the analyst whether to examine a model with no predictors. Hence, our primary interest may be whether there is sufficient variability in intercepts present at Level 2 (across groups) in the sample. In our example, we want to know if significant variance exists in average likelihood of individuals to meet productivity targets across groups—something that would be invisible to us in a single-level model. Little variability between the Level-2 units would suggest little need to conduct a multilevel analysis.

If we do develop such a model with no predictors, we can determine what the average probability of meeting productivity goals is at the organizational level. Notice that, in Equation 1.2, we add subscripts for individuals (i) and for organizations (j). The probability of an event occurring is generally referred to using the Greek letter pi, π_{ij}. At Level 1, the null model for individual i in organization j can be represented as follows:

$$\eta_{ij} = \log(\pi_{ij}/(1 - \pi_{ij})) = \beta_{0j}, \tag{1.2}$$

where β_{0j} is the intercept for the jth group, and $\log(\pi_{ij}/(1 - \pi_{ij}))$ is the mathematical function (g) that links the expected value of the dichotomous outcome Y_{ij} to the predicted values for variate η_{ij} (McCullagh & Nelder, 1989).

In this case, as noted in Table 1.3, we chose the *logit* link function, which is the natural logarithm (i.e., abbreviated as log, $\log_e$, or ln) of the odds that $Y = 1$ (π_{ij}) versus $Y = 0$ ($1 - \pi_{ij}$). We could make other choices, but the logit link is most commonly used with dichotomous outcomes. We describe sampling distributions and link functions in more detail in Chapter 3. Here we emphasize that the Greek letter eta (η_{ij}) is generally used to represent the predicted values of the transformed continuous outcome (i.e., in this case, the natural logarithm of the odds $Y = 1$) resulting from the multiple regression equation, which in the first model only consists of the intercept. Usually, the link function used for transforming the outcome extends over the entirety of real numbers (Hox, 2010).

In a two-level model, the fixed effects are typically expressed as unstandardized β coefficients. In this case, *unstandardized* means that the predictor coefficients at each level are in their original metrics and indicate the change in the log odds of the underlying variate η_{ij} (rather than the categorical outcome *productivity*) per unit change in the predictor (while holding any other predictors in the model constant). The subscript j indicates that the intercept varies across groups. Again, there is no residual variance at Level 1 because the variance of a dichotomous outcome is tied to its mean value and, therefore, cannot be modeled separately at Level 1 (Raudenbush et al., 2004).

Between groups, variation in random intercepts (β_{0j}) can be represented as

$$\beta_{0j} = \gamma_{00} + u_{0j}. \tag{1.3}$$

Level-2 fixed-effect coefficients are generally expressed as the Greek lowercase letter gamma (γ). Variability in organizational intercepts is represented as u_{0j}.

Through substitution, we can arrive at the corresponding single-equation model. In this case, by substituting Equation 1.3 into Equation 1.2, we arrive at the combined two-level intercept model, which can be written as

$$\eta_{ij} = \gamma_{00} + u_{0j}. \tag{1.4}$$

Readers should keep in mind that, in Equation 1.4 and subsequent equations in this section, η_{ij} is the predicted log odds resulting from the regression equation linked by the logistic transformation $\log(\pi / (1 - \pi))$ of the odds of the outcome event occurring versus not occurring. The null model therefore provides an estimated intercept (mean) log odds of meeting employee productivity targets for all organizations. Equation 1.4 indicates there are two parameters to estimate: the intercept and the between-organization variability, or deviation, from the intercept (u_{0j}). The estimated intercept is considered a fixed component, and the between-group variation in intercepts is considered the random effect. Information about the model's initial parameters can be useful in examining the baseline (no predictors) model with two estimated parameters against subsequent models with more estimated parameters.

Selecting the Covariance Structure

One of the advantages of IBM SPSS MIXED is the ease in which users can specify alternative covariance structures for Level 1 and Level 2. In this case, as we noted, there is no residual component for Level 1. It is scaled to be 1.0 as the default, which provides a metric for the underlying continuous predictor η_{ij} (Hox, 2010). The default specification is "variance components." With only one random effect at Level 2, this is the same as specifying an identity covariance matrix or a diagonal covariance matrix. It is always useful to check the specification of the proposed model (Model 1) against the model dimension output, as summarized in Table 1.4, to make sure the model proposed is the one *actually* tested. The table confirms there is one fixed effect (i.e., the organization-level intercept) and one random effect (i.e., the variance in intercepts across organizations at Level 2) to be estimated. "Common subjects" refers to the number of Level-2 units in the analysis ($N = 122$). Table 1.4 also confirms there will be one variance component (i.e., the Level-2 variance in intercepts) and no Level-1 residual variance component.

TABLE 1.4 Model 1: Covariance Parameters

Covariance Parameters	
Residual effect	0
Random effects	1
Design Matrix Columns	
Fixed effects	1
Random effects	1[a]
Common subjects	122

Note: Common subjects are based on the subject specifications for the residual and random effects and are used to chunk the data for better performance.

[a] This is the number of columns per common subject.

Readers will note that, in the typical null model for continuous outcomes, in the single equation there are three parameters to estimate, as in the following:

$$Y_{ij} = \gamma_{00} + u_{0j} + \varepsilon_{ij}. \tag{1.5}$$

The two variance components from Level 1 (ε_{0j}) and Level 2 (u_{0j}) are used to partition the variance in an outcome into its within- and between-group components (referred to as the intraclass correlation). Because there is no separate Level-1 residual variance component (ε_{0j}) in the multilevel GLM, the intraclass correlation (ICC), which describes the portion of variability in outcomes that lies between groups compared to the total variability, cannot be directly calculated in a manner similar to multilevel models with continuous outcomes. We take this issue up further in Chapter 4.

Analyzing a Level-1 Model With Fixed Predictors

Assuming that a sufficient amount of variance in *Y* exists at Level 2, we can investigate a model with fixed predictors at Level 1. Level-1 predictors are often referred to as *X*. In this case, we have one individual-level predictor (attitudes). For each individual *i* in organization *j*, a proposed model similar to Equation 1.1—in this case, summarizing the effect of employee attitudes on productivity—can be expressed as

$$\eta_{ij} = \beta_{0j} + \beta_1 X_{ij}, \tag{1.6}$$

where η_{ij} is again the predicted log odds from the logistic regression equation.

Equation 1.6 suggests that, within groups, X_{ij} (employee attitude) is related to probability of an individual to meet productivity targets. Typically, the unstandardized within-group predictors ($\beta_1 X_{ij}$) are either grand-mean or group-mean centered to facilitate interpretation of the coefficients. This is needed because, in multilevel models, the intercept is interpreted as the value on the outcome when all the predictors are equal to 0. Often, however, the value of 0 in its raw metric on a predictor may not be very useful in interpreting the estimates. For example, if we are examining the effect of an employee attitude (perhaps coded from 0 to 10) on productivity, there may be no individual in the sample with an attitudinal score of 0.

Grand-mean centering recenters the individual's standing on the predictor against the mean for the predictor in the sample. For example, if the sample mean for employee attitude were 5.4, an individual on the grand mean would have her or his score rescaled to 0. An individual with a value below the mean would be rescaled to a negative number; an individual above the mean would be rescaled to a positive number. We grand-mean centered attitude and saved it in the data set (gmattitude). In contrast, group-mean centering implies that the individual's attitude score is rescaled against the mean for her or his group, with the group mean now equal to 0. In Chapter 2, we describe how grand-mean and group-mean centered variables can be developed in IBM SPSS.

At Level 2, Equation 1.7 implies that variation in intercepts can be described by an organization-level intercept (γ_{00}), or grand mean, and a random parameter capturing variation in individual organization means (u_{0j}) from the grand mean:

$$\beta_{0j} = \gamma_{00} + u_{0j}. \tag{1.7}$$

In the case where we wish to treat the within-group slope describing employee attitudes as *fixed* (i.e., it does not vary across schools), Equation 1.8 would be written as

$$\beta_{1j} = \gamma_{10}. \tag{1.8}$$

This suggests that the variance component of the slope is fixed at zero. As the equation indicates, there is no corresponding random component (u_{1j}), so the slope coefficient is fixed to one value for the sample. Through substitution of β_{0j} and β_{1j} into Equation 1.6, the single-equation model can be summarized as

$$\eta_{ij} = \gamma_{00} + u_{0j} + \gamma_{10}X_{ij} \tag{1.9}$$

and then reorganized with fixed parameters (γs) and random parameters as

$$\eta_{ij} = \gamma_{00} + \gamma_{10}gmattitude_{ij} + u_{0j}, \tag{1.10}$$

where we have replaced X with the individual-level variable name.

We find that it is sometimes useful to write in the variable names as we are building the models to provide easy recognition with the predictors in the output. Equation 1.10 suggests that there are three parameters to estimate. The fixed effects are the intercept and Level-1 predictor gmattitude. The random effect is the variation in intercepts at the group level. We can confirm this in the model dimension in Table 1.5.

We note that if the Level-1 predictor employee attitudes were instead defined as a dichotomous factor (e.g., 0 = unsupportive, 1 = supportive) or as having several categories, it will change the calculation of the fixed effects in the design matrix printed in the covariance parameter output. For example, with attitudes defined as dichotomous, there will be two fixed effects assigned for *attitude* (one for each category of the predictor). This makes three total fixed effects (i.e., the

TABLE 1.5 Model 2: Covariance Parameters

Covariance parameters	
Residual effect	0
Random effects	1
Design matrix columns	
Fixed effects	2
Random effects	1
Common subjects	122

TABLE 1.6   Model 2 (With Attitude as Dichotomous Factor): Covariance Parameters

Covariance parameters	
Residual effect	0
Random effects	1
Design matrix columns	
Fixed effects	3
Random effects	1
Common subjects	122

intercept and two categories of attitude). Of course, one fixed effect for attitude is redundant, so there will still be only two fixed-effect parameters in the output. We illustrate this subtle difference in the model specification in Table 1.6.

Adding the Level-2 Explanatory Variables

Next, it is often useful to add the between-group predictors of variability in intercepts. Group variables are often referred to as W (or Z). From Figure 1.2, the Level-2 model with resources added will look like the following:

$$\beta_{0j} = \gamma_{00} + \gamma_{01}W_j + u_{0j}, \tag{1.11}$$

where W_j refers to the level of resources (grand-mean centered) in the organization. Substituting Equation 1.11 and 1.8 into Equation 1.6, the combined single-equation model is now the following:

$$\eta_{ij} = \gamma_{00} + \gamma_{01}gmresources_j + \gamma_{10}gmattitude_{ij} + u_{0j}, \tag{1.12}$$

where we have again included the names of the individual- and group-level predictors for W_j and X_{ij}, respectively.

Readers will notice that in a Level-2 model, after substitution, all estimates of group-level and individual-level predictors are expressed as γ coefficients. There are now four parameters to estimate (three fixed effects and one random effect). Table 1.7 confirms this model specification.

Examining Whether a Particular Slope Coefficient Varies Between Groups

We may next assess whether key slopes of interest have a significant variance component between the groups. Our theoretical model (Figure 1.2) proposes that the relationship between employee attitudes and likelihood to meet productivity targets may vary across organizations. Testing random slopes is best accomplished systematically, one variable at a time, because if we were to test several slopes simultaneously, we would be unlikely to achieve a solution that converges (Hox,

TABLE 1.7   Model 3: Covariance Parameters

Covariance parameters	
Residual effect	0
Random effects	1
Design matrix columns	
Fixed effects	3
Random effects	1
Common subjects	122

2002). As we suggested, if the within-unit slope (e.g., gmattitude-productivity) is defined to be randomly varying across units, the Level-2 slope model can be written as

$$\beta_{1j} = \gamma_{10} + u_{1j}. \tag{1.13}$$

Equation 1.13 suggests that variability in slopes can be described by a group-level average intercept coefficient (γ_{10}), or grand mean, and a random parameter capturing variation in individual school coefficients (u_{1j}) from the grand mean.

Through substitution, the combined single-equation model will be the following:

$$\eta_{ij} = \gamma_{00} + \gamma_{01} gmresources_j + \gamma_{10} gmattitude_{ij} + u_{1j} gmattitude_{ij} + u_{0j}. \tag{1.14}$$

As we suggested previously, notice that the substitution of β_{1j} in the within-group (Level 1) model (Equation 1.6) results in the addition of the interaction $u_{1j}X_{1ij}$ (i.e., when X_{1ij} is employee attitude) to the single-equation model. This interaction is considered as a random effect, which is defined as the deviation in slope for cases in group j multiplied by the Level-1 predictor score (X_1) for the ith case in group j (Tabachnick, 2008). We note that when we just specify the equation at Level 2 (as in Equation 1.13), the addition of the random effect as an interaction is hidden (Hox, 2010).

Covariance Structures

With a random slope and intercept, it is often easiest first to specify a diagonal covariance structure at Level 2 (using either diagonal or variance components). A diagonal covariance matrix provides an estimate of the variance for each random effect, but the covariance between the two random effects is restricted to be 0:

$$\begin{bmatrix} \sigma_I^2 & 0 \\ 0 & \sigma_S^2 \end{bmatrix}. \tag{1.15}$$

There are five parameters to be estimated now. The three fixed effects to be estimated in this model are the productivity intercept, the Level-2 predictor (gmresources), and the Level-1 predictor (gmattitudes). The two random effects at Level 2 are the variability in the intercept and the variability in the attitude-productivity slope. We can confirm this specification (Table 1.8).

If this model converges or reaches a solution without any warnings in the output, we can also try using a completely unstructured (*Unstructured*) covariance matrix of random effects. As suggested in Table 1.9, specifying an unstructured covariance matrix at Level 2 provides an additional random effect in the model, which represents the covariance between the random intercept and random slope; that is, there will be three covariance parameters

TABLE 1.8   Model 4: Covariance Parameters

Covariance parameters	
Residual effect	0
Random effects	2
Design matrix columns	
Fixed effects	3
Random effects	2
Common subjects	122

TABLE 1.9 Model 4a: Covariance Parameters (Unstructured)

Covariance parameters	
Residual effect	0
Random effects	3
Design matrix columns	
Fixed effects	3
Random effects	2
Common subjects	122

to be estimated. We can summarize this unstructured matrix (which is a symmetric matrix) as follows:

$$\begin{bmatrix} \sigma_I^2 & \sigma_{IS} \\ \sigma_{IS} & \sigma_S^2 \end{bmatrix} \tag{1.16}$$

The variances of the intercept and slope are in the diagonals, with their covariance as the off-diagonal element. The extra covariance parameter to be estimated is confirmed below. We note, however, that specifying an unstructured covariance matrix for random effects results in more models that fail to converge, often because the relationship between the two is nearly zero and may not vary sufficiently across units.

Adding Cross-Level Interactions to Explain Variation in the Slope

Finally, we would build a Level-2 model to explain variation in the Level-1 randomly varying slope of interest (i.e., assuming that the slope has significant variance across groups). Our simplified model in Figure 1.4 suggests that organizational resource levels may moderate the within-unit (e.g., attitude-productivity) slope:

$$\beta_{1j} = \gamma_{10} + \gamma_{11} gmresources_j + u_{1j}. \tag{1.17}$$

The cross-level, fixed-effect interaction term, which combines a Level-2 predictor with the Level-1 predictor ($y_{11}X_{ij}W_j$), is built during the model specification phase in IBM SPSS. Substitution of the revised β_{1j} model into Eq. 1.6 results in the following single-equation model:

$$\eta_{ij} = \gamma_{00} + \gamma_{01} gmresources_j + \gamma_{10} gmattitude_{ij} + \gamma_{11} gmresources_j * gmattitude_{ij} + u_{1j} gmattitude_{ij} + u_{0j}. \tag{1.18}$$

The model represented in Equation 1.18 results in seven total parameters to estimate (i.e., if we maintain the unstructured covariance matrix at Level 2). The four fixed effects are the intercept, resources slope, employee attitudes slope, and the cross-level interaction slope (gmattitude*gmresources). The three random covariance parameters at Level 2 are again the intercept (u_{0j}), the attitude-productivity slope (u_{1j}), and the covariance between them. Once again, we can confirm these model specifications in the model dimension table (Table 1.10).

Selecting Level-1 and Level-2 Covariance Structures

Our discussion of these basic modeling steps suggests that there is considerable flexibility in defining covariance structures at Level 2. We suggest working with less complex structures first because they are more likely to converge on a solution. If there is little variance to be estimated between groups, the program will issue a warning.

TABLE 1.10   Model 5: Covariance Parameters

Covariance Parameters	
Residual effect	0
Random effects	3
Design Matrix Columns	
Fixed effects	4
Random effects	2
Common subjects	122

For models that have repeated measures at Level 1, several types of covariance matrices can be used to account for possible temporal dependencies in the data. When there are repeated measures on individuals, it is likely that the error terms within each individual are correlated (IBM SPSS, 2002). Often, a completely unstructured covariance structure with four or more repeated measures requires too many parameters to have to be fit. Another possibility is a diagonal covariance matrix, which provides a separate estimate for each variance at Level 1 but no correlations between time-related observations. This choice can sometimes be more difficult to justify with repeated measures.

Other, more simplified structures will provide estimates of the correlations between successive measurements but require fewer unique parameters to be estimated. An autoregressive error structure, for example, allows the analyst to investigate the extent to which the measurements are correlated. The model assumes that the errors within each subject are correlated but are independent across subjects. Compound symmetry is another possibility if the assumptions hold. We discuss these different choices further in subsequent chapters.

Model Estimation and Other Typical Multilevel Modeling Issues

Model estimation attempts to determine the extent to which the sample covariance matrix representing our model specification is a good approximation of the true population covariance matrix. This determination is made through a formal test of a null hypothesis that the data are consistent with the model we have proposed. In general, confirmation of a proposed model relies on a failure to reject the null hypothesis. Here, one wishes to accept the null hypothesis that the model cannot be rejected on statistical grounds alone. This implies that the model is a plausible representation of those data, although it is important to emphasize that it may not be the only plausible representation.

As we have developed in this chapter, when outcomes are categorical and therefore expected values result from probability distributions other than normal, nonlinear link functions are required. Model estimation requires an iterative computational procedure to estimate the parameters optimally. Maximum likelihood (ML) estimation methods are most often used for generalized linear models and also for multilevel models. ML determines the optimal population values for parameters in a model that maximize the probability or likelihood function—that is, the function that gives the probability of finding the observed sample data, given the current parameter estimates (Hox, 2002). Because the likelihood, or probability, can vary from 0 to 1, minimizing this discrepancy function amounts to maximizing the likelihood of the observed data.

If we consider the sample covariance matrix (S) to represent the population covariance matrix (Σ), then the difference between the observed sample matrix $\hat{S}$ and the model-implied covariance matrix should be small if the proposed model fits the data. The evaluation of the difference between these two matrices depends on the estimation method used to solve for the model's parameters (Marcoulides & Hershberger, 1997). The mathematical relationships implied by the model are solved iteratively until the estimates are optimized. The difference between S and $\hat{S}$ is described as a discrepancy function—that is, the actual difference in the two estimates based on a likelihood. The greater the difference between these two covariance matrices, the larger the discrepancy in the function becomes (Marcoulides & Hershberger, 1997).

For single-level categorical analyses, IBM SPSS GENLIN supports ML estimation using Newton-Raphson and Fisher scoring methods, which are among the most efficient and widely used estimation methods for categorical outcomes (Dobson, 1990). For multilevel models with categorical outcomes, model estimation becomes more difficult. Estimation typically relies on quasilikelihood approaches, which approximate the nonlinear functions used by a nearly linear transformation. This complicates the solving of complex nonlinear mathematical equations representing the relationships among variables in the model (Hox, 2010).

Combining multilevel data structures and GLM, therefore, can lead to much greater complexity in estimating model parameters and require considerably more computer time in providing model estimation. The default estimation method for categorical outcomes used in GENLIN MIXED is referred to as active set method (ASM) with Newton-Raphson estimation—one well-known theoretical way to solve a constrained optimization problem that proceeds by finding a feasible starting point and adjusting iterative solutions until it reaches a final solution that is "optimal enough" according to a set of criteria (Nocedal & Wright, 2006).

Three-level models can be more complicated to estimate than two-level models (e.g., more cross-level interactions, changing sample sizes, etc.), but the strategy is basically the same. As we have already stated at several points, it is best to keep models simplified—for example, by including only random effects that are of strong theoretical or empirical interest. Although it is desirable to include covariances between intercepts and slopes, it is not always possible to achieve model convergence. We will address this more completely in subsequent chapters.

Determining How Well the Model Fits

Maximum likelihood estimation approaches provide a number of ways to determine how well a proposed model fits, assuming that the required assumptions for producing optimal estimates hold (e.g., sufficient sample sizes, proper model specification, convergence in solving the likelihood function is achieved). As we noted, the difference between S and $\hat{S}$ is described as a discrepancy function. ML estimation produces a model deviance statistic, which is defined as $-2*$log likelihood, where likelihood is the value of the likelihood function at convergence and log is the natural logarithm (Hox, 2002). This is often abbreviated as –2LL. The deviance is an indicator of how well the model fits the data. Models with lower deviance (i.e., a smaller discrepancy function) fit better than models with higher deviance. Nested models (i.e., when a more specific model is formed from a more general one) can be compared by examining differences in these deviance coefficients under specified conditions (e.g., changes in deviance between models per differences in degrees of freedom).

For comparing two successive single-level models, the likelihood ratio test is often used. The GENLIN output provides a likelihood ratio chi-square comparison between a current model under consideration (e.g., with one or more predictors) against an intercept-only model. Additionally, successive models can also be compared (when all of the elements of the smaller model are in the larger model) using the change in log likelihood (or the change in deviance). We can also use Akaike's information criterion (AIC) and Bayesian information criterion (BIC) to compare competing models, where we favor models with smaller values of these indices.

Because multilevel estimation procedures with categorical outcomes are approximate only, as Hox (2010) notes, related procedures for comparisons between successive models are more tenuous because of the nature of the quasilikelihood estimation and the rescaling of the variance that takes place each time variables are added to a model. Because solutions are approximate, test statistics based on deviance (–2LL) are not very accurate (Hox, 2010). The GENLIN MIXED output produces an estimate of the –2 log pseudolikelihood, and we can examine the change in

–2 log pseudolikelihood between two successive models. However, caution should be used in interpreting this coefficient because different data transformations may be used across models.

We note also that while information criterion indices are provided (Akiake or Bayesian), they are based on the real likelihood, and Hox (2010) cautions that they may not be accurate when estimated with quasilikelihood estimation procedures. We can also examine individual parameters and residuals to determine where possible sources of model misfit may reside. Readers should be aware that hypothesis testing of individual parameters is typically based on a normal distribution that, with categorical variables, is based on a normal *approximation*. This approximation may not be optimally accurate under certain conditions, such as small sample sizes (Azen & Walker, 2011).

Syntax Versus IBM SPSS Menu Command Formulation

In MIXED, we can formulate models using either syntax statements or menu commands. We suspect that most users are more familiar with the menu framework because they have likely used that to examine single-level models (e.g., analysis of variance, multiple regression, factor analysis). We have chosen the menu-command approach for presenting our model-building procedures, but we also provide examples of syntax for each chapter in Appendix A.

In the "Statistics Display" screen, one can use the "Paste" function (within the GENLIN MIXED dialog box) to generate the underlying syntax before actually running the model. The syntax is sent to its own window within IBM SPSS and can be saved for future use. We note that with more complex models, it is often a benefit to use the syntax statements to formulate and run models for a few reasons. First, the syntax statements provide a nice record of a progression of models (if the syntax files are saved and labeled according to the general modeling steps we have laid out). This is helpful if the IBM SPSS program is closed. The syntax statements provide a record of when one started and left off in building a series of models. This saves time when working with dozens of models over an extended period of time.

Second, we find that pasting the syntax is a good way to check whether the model we are running actually corresponds with the model that we had in mind. Often, doing this helps us see where we have forgotten or incorrectly specified something. This becomes more important when the multilevel models being investigated take a considerable amount of time to produce a solution. We have also found that it is easier to reorganize the syntax statements than to rebuild the whole model from the menu commands if we wish to change the order of the variables in our output tables. Generally, we like to present the output such that the predictors explaining intercepts are organized by levels of the data hierarchy and the predictors that explain variation in a random slope (i.e., cross-level interactions) are similarly organized. This helps readers see that there are two separate components (i.e., the intercept model and the slope model). Using the syntax statements also saves time if we just wish to change the specification of a covariance matrix or include (or exclude) a particular random or fixed effect.

Third, we recommend caution in reflexively adopting the IBM SPSS defaults, as these will sometimes introduce problematic conditions on the models being estimated. We find that printing the syntax statements can help alert analysts to the underlying defaults embedded in the model-building procedures, as well as one's choices to override default settings. Regardless of the type of model being developed, it is always good practice to review these default settings to ensure that they are appropriate for one's purposes.

Sample Size

Under various sampling conditions (e.g., how individuals and units were selected, numbers selected, missing or incomplete responses), there has been considerable debate among methodologists about the efficiency of ML estimation (Goldstein, 1995; Longford, 1993; Morris, 1995). An important conceptual distinction between single-level and multilevel analyses is that sample size considerations are quite different. Multilevel modeling requires sufficient sample size at each level of analysis for optimal estimation. In smaller group samples, typical ML-based estimation may result in a downward bias in the estimation of variance components and standard errors. As we have noted, multilevel modeling of categorical outcomes creates more complex estimation due to the nature of their scales of measurement and the resulting need to use approximate estimation methods (e.g., quasilikelihood). Under certain types of sample conditions (e.g., small samples, several hierarchical levels) and model specification, estimation may become much more challenging. When the estimation does not lead to an acceptable solution (e.g., being able to estimate a variance component), a warning message is issued as part of the output. In other cases, the model may stop estimating and issue a message that an "internal error" has stopped the model estimation process.

It is important to keep in mind that under less than ideal sampling conditions (e.g., small numbers of groups, convenience samples), it may be difficult to determine whether model results might be replicated in other samples. When only a limited number of groups is available, users can apply the Satterthwaite correction to the calculation of standard errors; this provides more conservative (i.e., larger) estimates of standard errors (Loh, 1987). We note that it is often more efficient to add higher level units than to add individuals within groups because this former approach generally reduces the need for sizable samples within the groups and tends to be more efficient in estimating random coefficients.

Power

Power refers to the ability to detect an effect if one exists. In the single-level analysis, most researchers know that the significance level (α), the effect size (i.e., with larger effects easier to detect), and the sample size are determinants of power. Multilevel models raise a number of additional issues involving power. Issues about power typically concern the appropriate (or minimum) sample size needed for various types of multilevel analyses (e.g., determining whether an intercept or slope varies across groups). As we suggested previously, one issue refers to the sample size required to ensure that estimates of fixed effects (e.g., at Level 1 and Level 2) and variances are unbiased (i.e., sampling bias). In most multilevel studies, the estimation of Level-2 effects is generally of greater concern because the number of groups available may be limited. Adding groups (as opposed to individuals) will tend to reduce parameter bias at that level. As Snijders (2005) shows, when fixed effects are the focus, characteristics of the groups themselves have little bearing on the precision of Level-1 estimates.

Another issue refers to the minimum sample size required to ensure that an effect would be detected if, in fact, one existed (i.e., power). In addition to these two determinants of power, in multilevel analyses at least two other considerations inform estimates of power: sample size at each level (i.e., the number of individuals i within each group j and the number of j groups) and the intraclass correlation (see Muthén & Satorra, 1995, for further discussion). With larger intraclass correlations, the power to detect Level-1 effects will be lower (because the groups are more homogeneous), holding sample size constant at all levels. This suggests that the power to detect Level-2 effects is much more sensitive to the number of groups in the sample as opposed to the number of observations within groups.

As designs become more complex, the need for larger samples at both levels increases. For example, in a given sample of individuals within groups, slopes in some units may be less reliably estimated than intercepts because, although intercepts depend only on the average level (mean) of a variable within each group, slope estimates depend on the levels of an outcome and a particular covariate, as well as the variability of their covariance among individuals within each group (Mehta & Neale, 2005). This means that estimating variability in random slopes across units generally requires larger sample sizes for more stable estimates than simply estimating random intercepts.

Complications can also arise due to correlations between random effects and the investigation of cross-level interactions (Raudenbush & Bryk, 2002), as well as the challenges of estimating models with categorical outcomes and complex sets of relationships that we noted previously. As this limited discussion of power suggests, a number of considerations must take place to assess potential bias in parameter estimates and power in various types of multilevel designs. See Snijders and Bosker (1999) or Heck and Thomas (2009) for further discussion of issues related to power.

Missing Data

Users should consider data preparation and data analysis as two separate steps. Missing data can be a problem in multilevel applications, depending upon the extent to which the data are missing and whether or not the data are missing at random. In some modeling situations, there may be considerable missing data. It is often useful to determine the amount of missing data as a first step in a data analysis, as well as the number of missing data patterns. A number of strategies for dealing with missing data are available. Some traditional approaches (e.g., listwise or pairwise deletion, mean substitution, simple imputation, weighting) lead to biased results in various situations (Peugh & Enders, 2004; Raykov & Marcoulides, 2008). It helps to know the defaults for software programs.

Traditionally, the typical way of handling missing data was to use listwise (i.e., eliminating any case with at least one missing value) or pairwise (i.e., eliminating pairs of cases when missing data are present, as in calculating a correlation) deletion, mean substitution, or various regression-based approaches (e.g., estimating outcomes with dummy-coded missing data flags to determine whether there were differences in outcomes associated with individuals with missing versus complete data). Generally, listwise, pairwise, mean substitution, and various regression-based approaches are not considered acceptable solutions because they lead to biased parameter estimation. For example, listwise deletion is only valid when the data are missing completely at random (e.g., as when selecting a random sample from a population), which is seldom the case using real data. Acceptable solutions include multiple imputation (MI) of plausible values and FIML estimation in the presence of missing data (Peugh & Enders, 2004; Raykov & Marcoulides, 2008).

Handling missing data in an appropriate manner depends on one's knowledge of the data set and why particular values may be missing. In general, there are three main types of missing data (see Raykov & Marcoulides, 2008, for further discussion). These include data that are missing completely at random (MCAR), missing at random (MAR), and nonignorable missing (NIM) data. For data to be MCAR, strong assumptions must hold (Raykov & Marcoulides, 2008). The missing data on a given outcome should be unrelated to a subject's standing on the outcome or to other observed data or unobserved (missing) data in the analysis. Typically, this assumption is only met when the data are missing by design, as in the situation where we draw a random sample of the studied population.

For example, we could generate a sample with, say, 25% data missing on an outcome such as student math proficiency scores. If the probability of data being missing on the outcome is related to missing data on a covariate, but not to subjects' standing on the outcome (e.g., if students who have missing data on attendance do not have a greater probability to have missing data on math proficiency outcomes), then the data are MAR (Raykov & Marcoulides, 2008). If the probability of missing on the outcome is related to standing on the outcome, even for individuals with the same value on a covariate (e.g., there are more missing low-math data than missing average- and high-math data among students with the same attendance level), the data are NIM. This latter type of missing data can produce more bias on model estimation than either of the other situations because the missing data on achievement are related to actual values of individual achievement for those subjects who do not take the test.

We note that, as an analytic routine, the MIXED program is limited in its ability to deal with missing data. As a default, the program uses listwise deletion of cases when data are missing. This means that any individual with data missing on any variable will be dropped from the analysis. This can result in a tremendous loss of data and biased parameter estimation. IBM SPSS does provide a number of options for examining missing data in addition to listwise, pairwise, or mean substitution. For example, the Statistics Base program includes a basic program to replace missing values. This is, however, a limited routine that provides the following replacement methods: series mean, mean of nearby points, median of nearby points, linear interpolation, and linear trend at point. It is accessed from the program's toolbar, where the user can select TRANSFORM and REPLACE MISSING VALUES.

IBM SPSS also has a missing-data module that can provide multiple imputation for missing data, but it must be purchased as an add-on program. This allows the user to generate a number of data sets with random values imputed for missing values. Students using the IBM SPSS Graduate Pack may have to upgrade to a regular version of the IBM SPSS Statistics Base and Advanced Statistics software in order to add on the missing data module to use prior to IBM SPSS MIXED. Using this add-on module, patterns of missing data can first be identified and then plausible values can be imputed using the EM (expectation maximization) algorithm.

Expectation maximization is a common method for obtaining ML estimates with incomplete data that has been shown to reduce bias due to missing data (Peugh & Enders, 2004). In this approach, the model parameters are viewed as missing values to be estimated. Obtaining estimates involves an iterative, two-step process where missing values are first imputed and then a covariance matrix and mean vector are estimated. This repeats until the difference between covariance matrices from adjacent iterations differs by a trivial amount (see Peugh & Enders, 2004, for further discussion). The process can be used to create a number of imputed data sets, where each simulates a random draw from the distribution of plausible values for the missing data (Peugh & Enders, 2004). These can be saved as separate data sets and then analyzed. One of the advantages of this approach is that other variables can also be used to supply information about missing data, but they need not be included in the actual model estimation. This approach to missing data is recommended when the assumption that the data are MAR is plausible.

For multilevel data, we note that at present IBM SPSS does not generally support FIML estimation in the presence of observations that may be MCAR or MAR, as is found in some software programs (Raykov & Marcoulides, 2008). What this means is that the cases are simply dropped from the analysis. In contrast, however, when the data are vertically arranged (e.g., when a single individual may have several observations or rows representing different time periods), only that particular occasion will be dropped if there is a missing value present. So, for example, if there are four observations per individual (e.g., four successive math scores) and one observation is missing for a particular individual, the other three observations would be used in calculating that individual's growth over time. This will typically yield unbiased estimates if data are MAR (Raykov & Marcoulides, 2008).

In Table 1.11 we summarize a simple logistic regression model to estimate the effects of gender and socioeconomic status on likelihood to be proficient in reading based on 20 individuals with

TABLE 1.11 Log Odds Estimates ($N = 20$)

Parameter	B	Std. error	95% Wald confidence interval		Hypothesis test		
			Lower	Upper	Wald χ^2	df	Sig.
(Intercept)	1.892	1.1090	−0.282	4.065	2.910	1	0.088
Female	−0.564	1.2336	−2.981	1.854	0.209	1	0.648
Zses	1.945	0.9478	0.087	3.802	4.210	1	0.040
(Scale)	1[a]						

Notes: Dependent variable: readprof; model: (intercept), female, Zses.
[a] Fixed at the displayed value.

TABLE 1.12 Log Odds Estimates (N = 15)

Parameter	B	Std. error	95% Wald confidence interval		Hypothesis test		
			Lower	Upper	Wald χ^2	df	Sig.
(Intercept)	2.504	1.8459	−1.114	6.122	1.840	1	0.175
Female	0.956	1.9963	−2.957	4.868	0.229	1	0.632
Zses	4.774	3.2190	−1.535	11.083	2.200	1	0.138
(Scale)	1[a]						

Notes: Dependent variable: readprof; model: (intercept), female, Zses.
[a] Fixed at the displayed value.

complete data. The log odds coefficients suggest that socioeconomic status (*Zses*) is statistically significant in explaining likelihood of being proficient and that gender is not significant.

In Table 1.12 we estimate the same model but with missing data on five individuals. We will assume that the data are MAR. Listwise deletion, therefore, will result in a loss of 25% of the data. We can see in Table 1.12 that estimating the model with 15 individuals results in a model where socioeconomic status now does not affect the proficiency outcome ($p > .05$). We can see that the estimate for *Zses* (log odds = 4.774) also lies outside the 95% confidence interval in Table 1.11 for the sample with $N = 20$.

In Table 1.13, we provide estimates using multiple imputation of three data sets under MAR. The results indicate that *Zses* is significant in each analysis ($p < .05$) and that female is not significant ($p > .05$). Moreover, the estimates of *Zses* all lie within the original 95% confidence limits in Table 1.11 (i.e., 0.087–3.802), as do the estimates for female (i.e., −2.981 to 1.854). We also present similar results using Mplus, where the model is estimated using full information maximum likelihood (FIML) with the individuals having missing data on *Zses* included. FIML estimation provides unbiased estimation when the missing data are MAR (Asparouhov, 2006). This latter analysis retains the original number of individuals in the study ($N = 20$). No definitive conclusion should be drawn from this simple illustration. Our point is simply to suggest that missing data can have a considerable influence on the credibility of one's findings. It is a problem that should be addressed in preparing the data for analysis.

TABLE 1.13 Log Odds Estimates for Three Imputed Data Sets and FIML Estimates

Parameter	Coefficient	Std. error	Wald χ^2	Sig.
Data set 1				
Female	−0.275	1.30	0.05	0.832
Zses	2.419	1.08	5.01	0.025
Data set 2				
Female	0.572	1.34	0.18	0.670
Zses	2.277	1.16	3.86	0.049
Data set 3				
Female	0.240	1.32	0.03	0.855
Zses	3.444	1.72	4.03	0.045
Mplus FIML estimates				
Female	0.685	1.32	0.27	0.602
Zses	2.463	1.10	5.05	0.024

We note in passing that although user-missing values can be specified in IBM SPSS, this approach is typically used for categorical responses, where some possible responses are coded as missing (e.g., "not applicable" in survey questions). If these user-defined missing values are included in the analysis, however, they will bias parameter estimates. We emphasize that it is therefore incumbent upon the analyst to be aware of how missing data will affect the analysis. Ideally, when using IBM SPSS MIXED, analysts should have close to complete data (e.g., 5% or less missing); however, we caution that even relatively small amounts of missing data can create some parameter bias, so analysts should first use some type of multiple imputation software to generate a series of "complete" data sets that can then be used to estimate the models.

Design Effects, Sample Weights, and the Complex Samples Routine in IBM SPSS

When working with existing data sets (e.g., cross-sectional or longitudinal survey), applying sample weights is important to correct the analyses for features of the sampling design (e.g., probability of selection at multiple levels of the data hierarchy) and data collection problems. Procedures for the selection of Level-2 units and individuals within those units can vary from being simple (e.g., simple random sampling at each level) to relatively complex. Weights may be available at the individual level (Level 1), at the group level (Level 2), or at both levels. Currently, there are no commonly established procedures for applying weights in multilevel analyses, although a considerable number of different approaches have been proposed (e.g., Asparouhov, 2005, 2006; Grilli & Pratesi, 2004; Jia, Stokes, Harris, & Wang, 2011; Pfeffermann, Skinner, Holmes, Goldstein, & Rasbash, 1998; Stapleton, 2002).

The consideration of weights and design effects is vitally important in analyses using disproportionate sampling and multistage cluster samples. Disproportionate sampling will lead to samples that over-represent certain segments of the populations of interest. Typically, this results from the researcher's interest in including a sufficient number of subjects (or objects) from smaller but important subpopulations. Sampling members of such groups proportionally often results in too few sample members to allow meaningful analyses. We therefore oversample many groups to ensure sufficient numbers in the final sample. The result is that the analytical sample is not representative of the true populations because it has too many sample members from the oversampled groups. Sample weights—typically, probability or frequency weights—are used to readjust the sample to be representative of the population from which it was drawn. Failure to use a sample weight in these instances can result in incorrect parameter estimates biased in the direction of the oversampled members of the population.

Disproportionate sampling is often found in multistage cluster samples. Cluster sampling is simply when the researcher first draws a sample at a higher level—organizations, for example—and then within each organization draws a sample of lower level units—employees, for example. The units at each level may or may not be drawn proportionate to their presence in the larger population. To illustrate this point, if one were to sample organizations for some substantive reason related to the research purposes, it might be desirable to oversample rural organizations. If this were to occur, one would want to be sure to adjust for this at the organizational level by using a Level-2 (organizational) weight in the same fashion that the Level-1 individual weight discussed previously was used. Hence, there can be sampling weights for each level of the data, although we note that many currently available data sets do not include weights at multiple levels.

To the degree that the observations within each of the higher order clusters are more similar to each other, there will be a *design effect* present that biases the estimated standard errors downward. Because hypothesis tests are based on the ratio of the estimate to its standard error, having standard errors that are too small will lead to a greater propensity to commit a Type I error (i.e., falsely concluding that an effect is statistically significant when it is not) than if the sample were drawn through a simple random sampling procedure. The design effect is defined as the ratio of the biased standard error to the standard error that would be estimated under a true random sample design. So, for example, if we know that the true standard error was 1.5, but the biased standard error estimated from the data collected through the multistage cluster sample was 1.2, the calculated design effect would be 1.5/1.2 = 1.25.

One standard measure of this within-unit (dis)similarity is the intraclass correlation, or ICC—the proportion of total variance in the outcome due to within-unit differences at higher levels. The higher the ICC is, the larger will be the design effect. Hox (2010) notes that the ICC can be viewed as another way to think about the degree of correlation within clusters. An ICC of .165, which suggests that 16.5% of the variance in the individual-level outcome exists *between* clusters, could also be viewed as an indication that one might expect a within-cluster correlation of .165 between individuals.

This conceptual connection between the ICC and within-cluster correlation is important in understanding design effects. In short, the greater the between-cluster variance in the individual-level outcome is, the more homogenous will be the individual observations within each of the clusters. To the extent that there exist within-cluster similarities, estimates of the Level-1 variances will be smaller than they would be if the sample were collected through a simple random sample design (where such clustering would be irrelevant to the variance structure). The implication central to our interests is that ignoring the clustering that is part of the sample design will yield downwardly biased estimates of the standard error.

The last several versions of SPSS have made available a COMPLEX SAMPLES module (as an add-on) that allows the user to incorporate design information into the model to adjust for the design effects described previously. This module produces results for single-level models incorporating design information and sample weights. As such, the parameter estimates are adjusted for both disproportionate sampling and cluster sampling. In the single-level context, this is the appropriate way to analyze data collected through complex sample designs. In this type of approach, similarities among individuals due to clustering are treated as "noise" that is adjusted out of the analysis, rather than as the focus of the analysis.

Multilevel models, by design, capitalize on the clustered nature of data, and it is quite common to see these models used with large-scale survey data that have been collected through complex sample designs. The same cautions outlined previously apply to estimates produced using various forms of multilevel models. Although multilevel models capitalize on the clustered nature of the data, they do nothing to address disproportionate sampling and, without proper weighting, they will produce incorrect parameter estimates. Sample weights are often essential to generate accurate estimates.

Weighting for unequal selection is relatively well established for single-level analysis. The COMPLEX SAMPLE module allows adjustments to be made for sample design effects (which can include clustering) but maintains a single-level analysis after adjustment for features of the sampling scheme. In this type of approach, similarities among individuals due to clustering are treated as unwanted variance that is adjusted out of the analysis. We are unaware of any documentation that specifically discusses multilevel models as complex sampling models within IBM SPSS. A standard two-level model would be an example of a two-stage cluster sampling design.

In contrast to weighting in single-level analyses, developing weighted analyses in the multilevel context presents a number of more complicated challenges and limitations. Research in this area is ongoing, and important advances have been made over the last 10 years. Most multilevel software programs now include one or more weighting options. Several programs with which we are familiar (HLM 7, LISREL 8.8, Mplus 6.1) incorporate design weights that combine information on clustering, the degree of intraclass correlation, and disproportionate sampling to create a set of scaled weights that will produce accurate estimates at each level of analysis. IBM SPSS allows for the incorporation of simple sample weights (an approximate frequency weight) in the MIXED and GENLIN MIXED routines; however, the current version does not enable a scaling adjustment that accommodates the effects of clustering in the sample design. We believe this is an important limitation of the IBM SPSS program.

A number of factors can influence the estimation of model parameters. These factors include the method of scaling the weights at Level 1 (i.e., how weights are scaled within the clusters), the size of the clusters, the relative invariance of the selection method applied to clusters (referred to as "informativeness"), the presence of missing data, and the intraclass correlation (Asparouhov, 2006). The scaling of Level-1 sample weights is very important in the multilevel context, helping to improve efficiency and decrease bias in estimation (Pfefferman et al., 1998; Skinner, 2005). Asparouhov explains that the scaling of the weights at Level 1 involves multiplying the weights by a scaling constant so that the sum of the weights is equal to some kind of characteristic of the cluster (e.g., cluster size).

Although many of the software programs used for estimating multilevel models enable the appropriate scaling, IBM SPSS does not yet include this feature. If sample weighting is essential to the analysis, it may be better to use another of the available programs or to revert to a single-level formulation within SPSS through its COMPLEX SAMPLES module. More specifically, as Aspouhuov (2006) suggests, if sampling weights present in a secondary data set are designed for a single-level analysis, it may be best to stick with that type of design and conduct a single-level analysis designed for stratified and cluster sampling designs.

Aspouhuov (2006) provides two contrasting situations illustrating this point. First, he notes that when weights are only present at Level 2 (i.e., where clusters have been sampled with unequal probability), we can identify this situation as within the framework of single-level weighted modeling, and methods available for single-level weighted analysis can be applied with consistent estimations regardless of the size of the clusters. Although the model is multilevel, the nature of the weighting is not. Of course, if sample weights are also provided at Level 1, this will change. Second, he suggests that the situation is different when weights are only provided at Level 1, as the unequal probability of selection is applied to dependent units and, therefore, the assumptions of the single-level method of analysis will be violated. The bottom line is that if the single-level sample weights cannot be properly scaled to the multilevel context, it may be better to use a single-level analysis. This threat may be more severe when estimating models with categorical outcomes (Rabe-Hesketh & Skrondal, 2006).

An Example

We provide one simple example of a comparison between results we obtained with Mplus and HLM, which have Level-1 and Level-2 weights available for multilevel analysis, and the single-level weighted analysis available in IBM SPSS. In the example, there are 5,952 individuals nested in 185 units with a dichotomous outcome. For HLM, Level-1 weights were scaled within Level-2 units such that the sum of these weights equals the size of the respective Level-2 unit (Raudenbush et al., 2004). For Mplus, we scaled Level-1 weights so that they also summed to the cluster sample size.

In Table 1.14, we first provide unweighted estimates for the within- and between-group variables using SPSS (i.e., SPSS UN). We next provide the HLM estimates with sample weights included at Levels 1 and 2. Third, we provide the SPSS estimates using only the Level-1 weights (SPSS L1W), and then we provide the SPSS estimates using only Level-2 weights (SPSS L2W). Finally, we also provide the Mplus estimates for the within-groups portion of the model only (i.e., Mplus Level-2 estimates are not log odds coefficients).

We note that the unweighted estimates differ considerably on most variables from the other estimates at both levels. First, there are more findings of significance at Level 2 than for the HLM results (i.e., five versus four, respectively). The Level-1 estimates also seem a bit too large for a couple of the variables. Second, applying only the Level-1 weights in SPSS also results in five significant parameters at Level 2. Third, when we apply only the Level-2 weights in SPSS, the pattern of the Level-2 results is consistent with HLM (but the size of the coefficients differs in some cases), and the estimated Level-1 coefficients and significance levels are consistent with HLM and Mplus. In this model, however, the significance levels of all effects are based on the

TABLE 1.14 Fixed-Effect Estimates for Unweighted and Weighted Analyses

	SPSS(UN)	HLM	SPSS (L1W)	SPSS (L2W)	MPLUS[a]
School					
Intercept	−2.26[d]	−2.41[d]	−2.22[d]	−2.26[d]	
sSES	1.18[d]	1.21[d]	1.29[d]	1.32[d]	
Private	−0.28[b]	−0.30	−0.29[b]	−0.35	
General	0.49[d]	0.47[b]	0.44[c]	0.43[b]	
RankMath	0.01[d]	0.00[b]	0.00[d]	0.00[b]	
City	0.17	0.27	0.14	0.28	
Large City	0.08	0.19	0.12	0.13	
Individual					
SES	0.25[d]	0.19[b]	0.24[d]	0.19[b]	0.20[b]
Math	−0.00[d]	−0.00[d]	−0.00[d]	−0.00[d]	−0.00[d]
Female	−0.24[c]	−0.23[b]	−0.22[c]	−0.22[b]	−0.23[b]
Interaction	0.56[d]	0.52[c]	0.54[d]	0.50[c]	0.51[c]

[a] Mplus between-group estimates are not in a log odds metric.
[b] $p < .05$.
[c] $p < .01$.
[d] $p < .001$.

number of Level-2 units rather than on the number of individuals in the sample. Finally, Mplus and HLM provide consistent estimates at Level 1, as we might expect.

No definitive conclusions should be drawn from this one example data set. We provide these results only to make the point that using sample weights (and using them correctly) does make a difference in the accuracy of the estimates obtained. We are hopeful that the application of multilevel sampling weights will be included in future versions of the software. In the interim, we call attention to recent work by Chantala, Blanchette, and Suchindran (2011) providing SAS and Stata routines to generate scaled weights that could be imported into other multilevel software programs (such as IBM SPSS).

Differences Between Multilevel Software Programs

In preparing this handbook, we compared categorical models estimated using IBM SPSS with models estimated using other multilevel software with which we are familiar (e.g., HLM, Mplus). Different software programs use slightly different algorithms to estimate models—for example, in calculating standard errors, especially in small groups. There are also differences in the means of testing the variance components. For example, IBM SPSS uses a Wald z-test, and HLM uses a chi-square test. In general, however, we have found that the differences in software are generally small; that is, output will carry the same substantive interpretation. We provide a couple of examples in Appendix B.

As we noted in the previous section, HLM and Mplus currently have procedures available for applying sample weights at two levels, but IBM SPSS does not yet have this capability. However, it has taken a number of years for these other two programs to incorporate sample weights at two levels into their analytic routines. The MIXED routine is still relatively new as a multilevel analytic program. We find that it is very compatible with well-established programs in almost every other way (and it exceeds them in some other ways).

Summary

In this chapter, we have developed a context and rationale for the use of multilevel models in the social and behavioral sciences. The use of multilevel analysis can add substantive information about how processes unfold at various levels of a data hierarchy. We suggested that multilevel techniques support the specification of more complex theoretical relationships than is possible using traditional single-level regression analyses. Analytical approaches that can be used to model complex relationships have greatly expanded over the past couple of decades, and these analytical alternatives allow us to investigate social processes in more theoretically and methodologically appropriate ways. Substantive progress in a field is often achieved when headway occurs simultaneously on conceptual and methodological fronts. In the next chapter, we take care of a bit of housekeeping by providing an overview of some important data management techniques. Arranging the data for analysis in IBM SPSS is fairly straightforward. We provide the reader with a few essential steps necessary to put data sets in proper order for analysis using IBM SPSS MIXED.

CHAPTER 2

Preparing and Examining the Data for Multilevel Analyses

Introduction

Essential to any type of analysis is the organization and vetting of the data that will be analyzed. In this chapter we identify a number of practical and substantive issues associated with preparing data for analysis in IBM SPSS MIXED and assessing the adequacy of those data for a variety of multilevel analyses.

Data Requirements

In this workbook we deal exclusively with multilevel models using categorical outcomes. We note that within the past couple of years, IBM SPSS has expanded the capacity of the MIXED program to accommodate nominal, ordinal, and count outcomes. It will now handle both mixed models for continuous outcomes and generalized linear mixed models (GLMMs) for categorical outcomes. Predictors can be continuous, ordinal, or dichotomous.

The sample sizes we employ throughout our presentation are typically large at each level of the data hierarchy. The variables used may come from a variety of different sources, many of which are specific to a particular level of analysis. One might, for example, draw on student-level attitudinal, behavioral, or performance data from national surveys such as the National Educational Longitudinal Study of 1988 (Curtin, Ingels, Wu, & Heuer, 2002). If an objective were to understand the effects of school settings on these individual characteristics, we might assemble school-level data drawing on information from the Common Core of Data (Sable & Noel, 2008). School-level data might include school size; demographic composition; financial characteristics such as state dollars per enrolled student, on-time progression, or graduation rates; teacher and administrative numbers; and the like. To carry this to a third level, we could draw on US Census data to define characteristics of the school districts in which the schools at Level 2 were located (e.g., household income, number of people in the household, their levels of education, etc.).

We will show that there are many variants on this modeling framework. We might, for example, want to understand a specific behavioral change resulting from participating in a treatment program. We might be interested in a student outcome such as the probability of being proficient in reading or math over time. In such an instance, we might conceptualize time points across which we presume change to occur within students who could, in turn, be nested within schools, and so on. However the nesting is conceptualized, each level of analysis will have its own set of variables defining features of the units being measured at that level. Although conceptualizing data at discrete levels of the hierarchy may be relatively straightforward, organizing the data set requires an understanding of how the data need to be arranged to represent that hierarchical conceptualization correctly. In the next section, we outline the main organizational features of data sets that can be used in a multilevel analysis. We return to data requirements in more detail in a subsequent section.

File Layout

We described in the previous chapter an important difference between the single-equation and multiple-equation approaches to estimating multilevel models. The multiple-equation approach (e.g., used in HLM) requires a separate data set for each level of data being analyzed. This can make conceptualization of the different levels clear. If Level 1 consists of 6,871 students, for example, the Level-1 file would contain student data including identifier information about higher order group membership (e.g., the classroom identifier or school identifier for each of the 6,871 students in the file).

Figure 2.1 shows what such a Level-1 file might look like, based on the data set used in Chapter 3. In this particular example (*ch2level-1data.sav*), we have included a student identifying variable (*id*), a school identifying variable (*schcode*), and three variables describing student characteristics—gender (*female*), socioeconomic status on a normalized scale (*ses*), and a dichotomous score indicating whether or not students were proficient in math (*mathprof*).

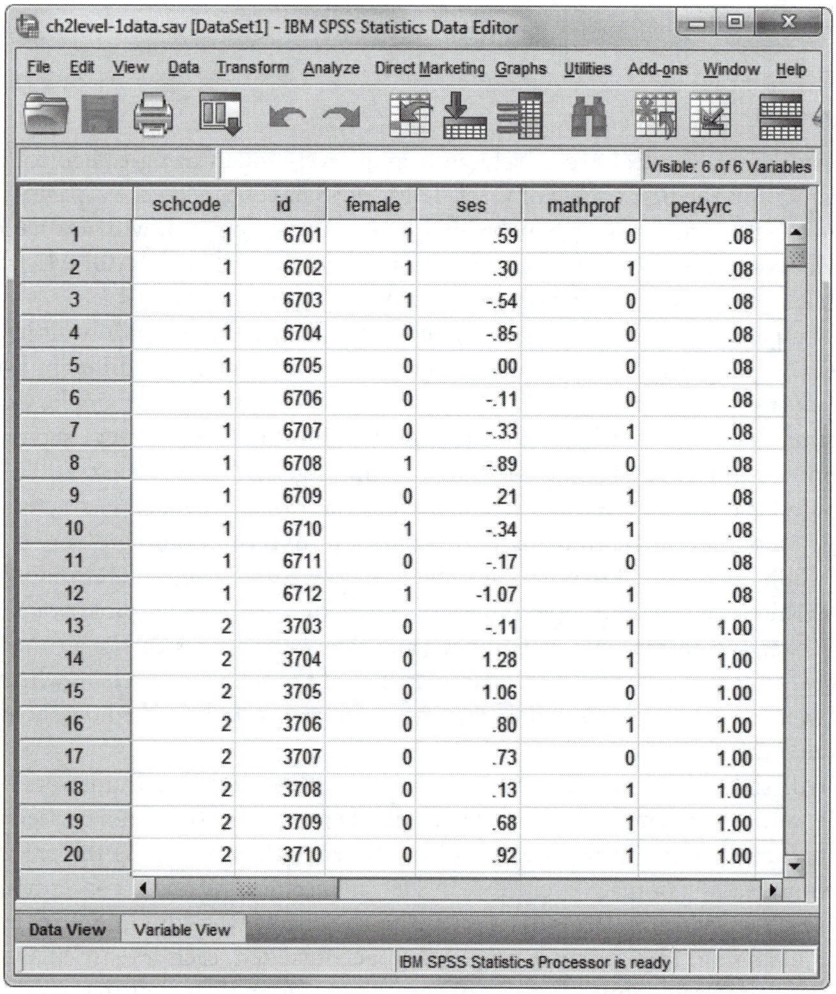

	schcode	id	female	ses	mathprof	per4yrc
1	1	6701	1	.59	0	.08
2	1	6702	1	.30	1	.08
3	1	6703	1	-.54	0	.08
4	1	6704	0	-.85	0	.08
5	1	6705	0	.00	0	.08
6	1	6706	0	-.11	0	.08
7	1	6707	0	-.33	1	.08
8	1	6708	1	-.89	0	.08
9	1	6709	0	.21	1	.08
10	1	6710	1	-.34	1	.08
11	1	6711	0	-.17	0	.08
12	1	6712	1	-1.07	1	.08
13	2	3703	0	-.11	1	1.00
14	2	3704	0	1.28	1	1.00
15	2	3705	0	1.06	0	1.00
16	2	3706	0	.80	1	1.00
17	2	3707	0	.73	0	1.00
18	2	3708	0	.13	1	1.00
19	2	3709	0	.68	1	1.00
20	2	3710	0	.92	1	1.00

FIGURE 2.1 A Level-1 data file (multiple-equation approach, $N = 6,871$).

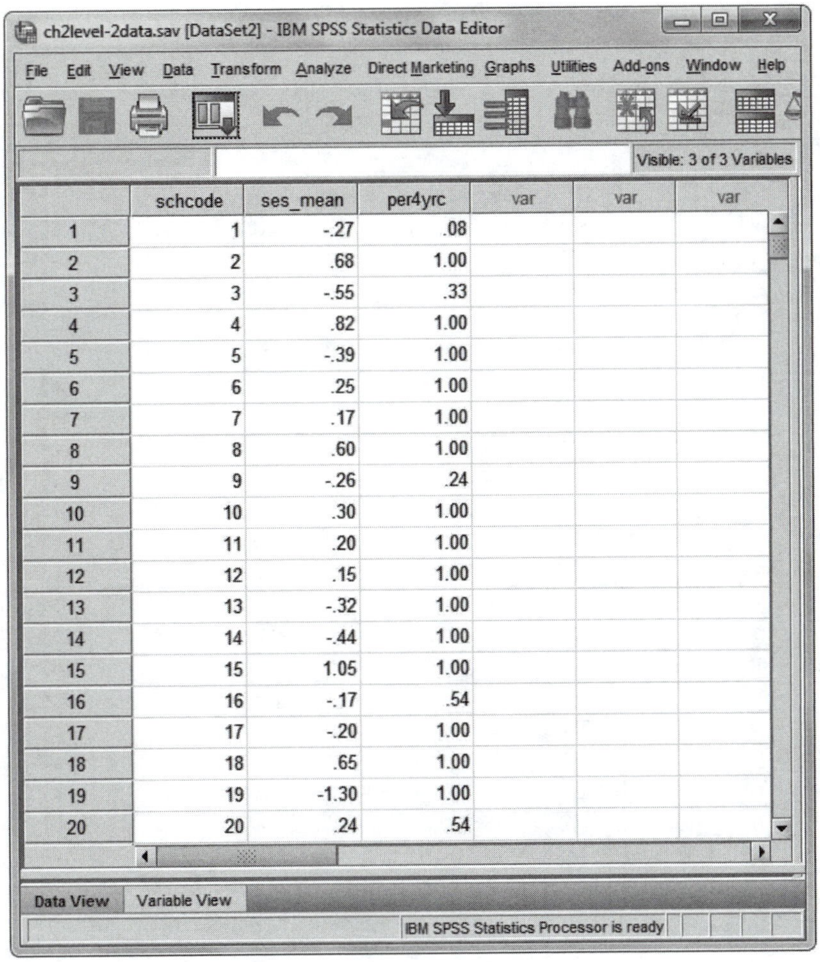

	schcode	ses_mean	per4yrc	var	var	var
1	1	-.27	.08			
2	2	.68	1.00			
3	3	-.55	.33			
4	4	.82	1.00			
5	5	-.39	1.00			
6	6	.25	1.00			
7	7	.17	1.00			
8	8	.60	1.00			
9	9	-.26	.24			
10	10	.30	1.00			
11	11	.20	1.00			
12	12	.15	1.00			
13	13	-.32	1.00			
14	14	-.44	1.00			
15	15	1.05	1.00			
16	16	-.17	.54			
17	17	-.20	1.00			
18	18	.65	1.00			
19	19	-1.30	1.00			
20	20	.24	.54			

FIGURE 2.2 A Level-2 data file (multiple-equation approach, $N = 419$).

If Level 2 in the analysis consisted of schools of which students were members, the Level-2 data set would contain all information about those 419 schools, including a unique school identifier and perhaps aggregated data from the Level-1 student file shown in Figure 2.1.

The Level-2 file (*ch2level-2data.sav*) shown in Figure 2.2 contains a unique school identifier (*schcode*), two variables describing the characteristics of the school, average socioeconomic status (*ses_mean*), and the proportion of students planning to attend a 4-year college (*per4yr*). The data sets are linked through a group-level identifier (in this case, *schcode*) during the multilevel analysis.

In contrast, the single-level approach makes use of one file that combines data from each level (Figure 2.3). In the univariate multilevel model, the file will consist of one record for each Level-1 unit. Values for variables from higher levels will be constant within groups. For example, in a data set (*ch2level-1&2data.sav*) with 6,871 students from 419 schools, there would be a single file of 6,871 records. The values for the student-level variables would vary across all 6,871 students. However, values on the school-level variables would be constant within each of the 419 schools; that is, students within each school would all have the same value on each of the school-level variables.

IBM SPSS MIXED uses the single-equation approach and one omnibus file containing data on each level of the analysis. In the sections that follow, we provide an overview of some of the data management steps within IBM SPSS that will help you organize and prepare your data and files for use within the MIXED routine.

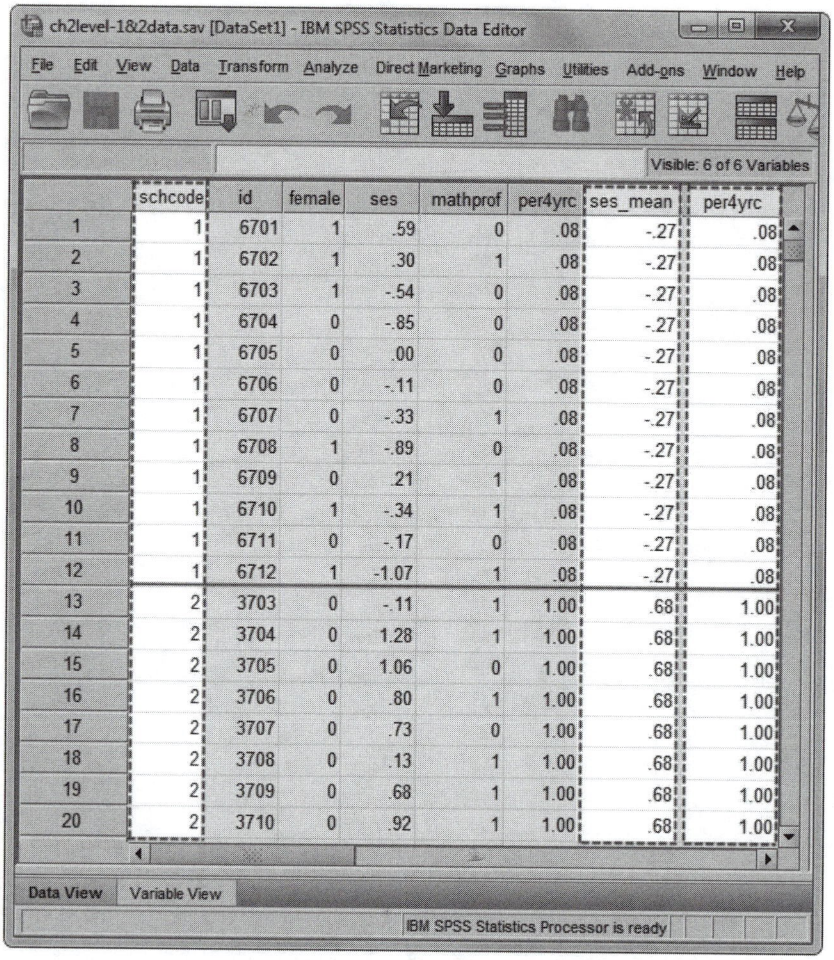

	schcode	id	female	ses	mathprof	per4yrc	ses_mean	per4yrc
1	1	6701	1	.59	0	.08	-.27	.08
2	1	6702	1	.30	1	.08	-.27	.08
3	1	6703	1	-.54	0	.08	-.27	.08
4	1	6704	0	-.85	0	.08	-.27	.08
5	1	6705	0	.00	0	.08	-.27	.08
6	1	6706	0	-.11	0	.08	-.27	.08
7	1	6707	0	-.33	1	.08	-.27	.08
8	1	6708	1	-.89	0	.08	-.27	.08
9	1	6709	0	.21	1	.08	-.27	.08
10	1	6710	1	-.34	1	.08	-.27	.08
11	1	6711	0	-.17	0	.08	-.27	.08
12	1	6712	1	-1.07	1	.08	-.27	.08
13	2	3703	0	-.11	1	1.00	.68	1.00
14	2	3704	0	1.28	1	1.00	.68	1.00
15	2	3705	0	1.06	0	1.00	.68	1.00
16	2	3706	0	.80	1	1.00	.68	1.00
17	2	3707	0	.73	0	1.00	.68	1.00
18	2	3708	0	.13	1	1.00	.68	1.00
19	2	3709	0	.68	1	1.00	.68	1.00
20	2	3710	0	.92	1	1.00	.68	1.00

FIGURE 2.3 Combined multilevel data file (single-equation approach, $N = 6{,}871$).

Getting Familiar With Basic IBM SPSS Data Commands

Organizing and managing the data at various levels is accomplished through five basic IBM SPSS procedures. Of course, many other procedures can be used to modify a data set, but we feel that these are primary to the data management tasks associated with organizing files for multilevel analyses within IBM SPSS. We will have much more to say about the IBM SPSS commands and the menu system itself in the chapters that follow. Here, however, we wish only to introduce these commands and a few principles of data management that we think will prove helpful for getting the most out of the workbook. In order of importance, the five primary procedures are

1. RECODE: changes, rearranges, or consolidates the values of an existing variable.
2. COMPUTE: creates new numeric variables or modifies the values of existing string or numeric variables.
3. MATCH FILES: combines variables from IBM SPSS-format data files.
4. AGGREGATE: aggregates groups of cases in the active data set into single cases and creates a new aggregated file or creates new variables in the active data set that contain aggregated data. The values of one or more variables in the active data set define the case groups.
5. VARSTOCASES: restructures complex data structures (i.e., in which information about a variable is stored in more than one column) into a data file in which those measurements are organized into separate rows of a single column.

TABLE 2.1 Descriptive Statistics

	N	Minimum	Maximum	Mean	Std. dev.
id	8335	1.00	8670.00	4308.3500	2510.57800
nschcode	8335	1.00	525.00	261.9800	152.81700
test1	8335	0.00	1.00	0.5300	0.49900
test2	8335	0.00	1.00	0.5200	0.50000
test3	8335	0.00	1.00	0.4900	0.50000
effective	8335	0.00	1.00	0.5622	0.49614
courses	8335	0.00	4.00	0.7481	0.79013
female	8335	0.00	1.00	0.5100	0.50000
ses	8335	−2.41	1.87	0.0336	0.78390
Valid N (listwise)	8335	0.00			

In this section we will build a multilevel data set using each of these primary commands. It is based on the example used in Chapter 5 but modified to exclude missing data. The data set contains three proficiency test scores taken over time (*test1, test2, test3*), dichotomous indicators of teacher effectiveness (*effective*) and gender (*female*), a continuous variable capturing the number of advanced placement courses a student has taken (*courses*), and a continuous measure of family socioeconomic status (*ses*). There is also an identifier for students (*id*) and for the schools in which they are enrolled (*nschcode*). The descriptive statistics in Table 2.1 show that there are 8,335 records in the data set. Each record represents a single student. Figure 2.4 displays a partial view of the data structure.

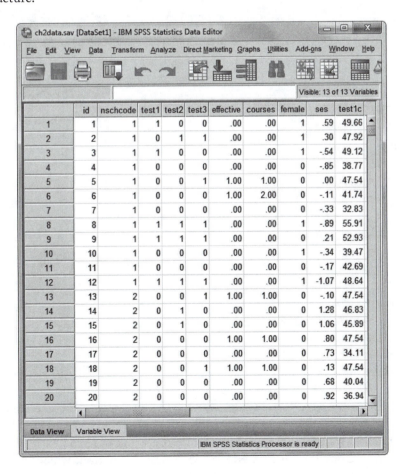

FIGURE 2.4   Horizontal data matrix.

RECODE: Creating a New Variable Through Recoding

Our first task will be to create a new categorical SES variable by recoding *ses* into a variable called *ses4cat* (suggesting that we are going to recode this into a four-category variable). We will use three somewhat arbitrary cut points to create four categories for our recoded variable: −0.5180, 0.0250, and 0.6130 (these actually represent the 25th, 50th, and 75th percentiles, respectively).

Launch the IBM SPSS application program and select the *ch2data.sav* data file.

1. Go to the toolbar and select TRANSFORM, RECODE INTO DIFFERENT VARIABLES. This command will open the *Recode into Different Variables* dialog box.

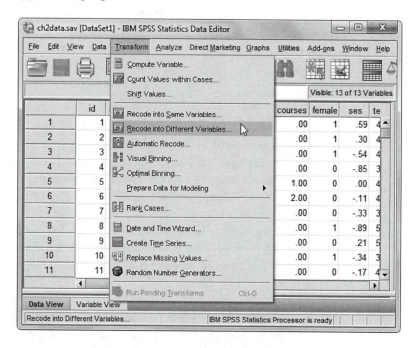

2a. The *Recode into Different Variables* enables creating a new variable (*ses4cat*) using a variable (*ses*) from the current data set. First, click to select *ses* from the left column and then click the right-arrow button to move the variable into the *Input Variable -->Output Variable* box.

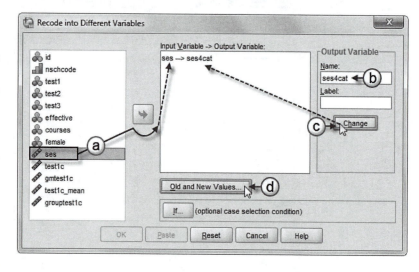

b. Now enter the output (new) variable name for the *Output Variable* by typing *ses4cat* into the *Name* box.

c. Then click the CHANGE button, which will add *ses4cat* and complete the command for *ses -->ses4cat*.

d. We will now transform the single-category "old" value of *ses* to four "new values" categories for *ses4cat*. Click the OLD AND NEW VALUES button, which will then display the *Recode into Different Variables: Old and New Values* screen.

3a. The *Recode into Different Variables: Old and New Values* screen displays multiple options. To define the first category or "cut point" (−.5180), click to select the option *Range, LOWEST through value*.

b. Now enter the value: −.5180.

c. Next, enter "1" as the *New Value*.

d. Then click the ADD button, which will place the first range command *Lowest thru −.5180 --->1* into the *Old --->New* box.

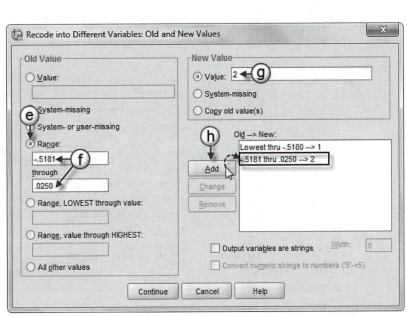

e. Next we will define the second category (cut point), which is a range of values (−.5181 through .0250). First, click the *Range* option.

f. Now enter the low value of −.5181 and then the upper value limit of .0250.

g. Next, enter "2" as the *New Value*.

h. Then click the ADD button, which will place the first range command *−.5181 thru .0250 --->2* into the *Old --->New* box.

i. We will now define the third category which is a range of values (.0251 through .6130). Continue using the *Range* option and enter .0251 and then .6130.

j. Next, enter "3" as the *New Value*.

k. Now click the ADD button to place the range command *.0251 thru .6130 -->3* in the *Old -->New* box.

l. We will define the final category (cut point), which is a single value (.6131). Click to select *Range, value through HIGHEST* option.

m. Now enter the value: .6131.

n. Next, enter "4" as the *New Value*.

o. Then click the ADD button, which will place the range command *.6131 thru Highest -->4* into the *Old-->New* box.

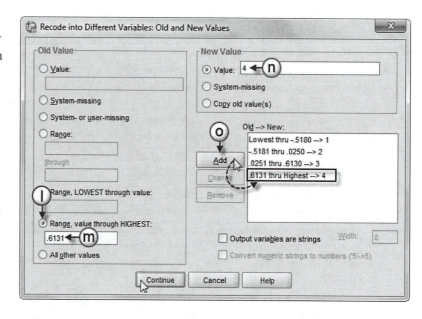

Click the CONTINUE button to return to the *Recode into Different Variables* main dialog box.

4. Click the OK button to generate the recoded variable *ses4cat* and corresponding values.

Note: The icon for *ses4cat* indicates the variable has a nominal measurement but may later be changed to an ordinal setting by accessing the *Measure*

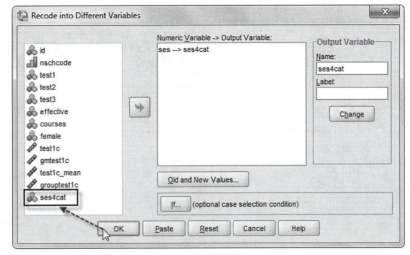

column in the *Variable View* dialog box. Refer to Figure 3.9 in Chapter 3. We will leave *ses4cat* with its nominal measurement setting.

5. Scroll across the *Data View* window; the new variable *ses4cat* with its recoded four categorical scores will be found in the last column. (You may also verify that the raw *ses* scores conform to the categories defined through the *Recode* process.)

Note: The asterisk displayed next to the file name at the top of the display

window is a reminder that the original file has been changed due to adding *ses4cat*. To retain changes made to the data, save the file by going to the toolbar and select FILE, SAVE. You may opt to save the changes under a different file name to retain the original *ch2data.sav* file.

COMPUTE: Creating a New Variable That Is a Function of Some Other Variable

Suppose that we wanted to create a variable to summarize the proportion of occasions (or mean) that students met proficiency standards (i.e., 0.00, 0.33, 0.67, 1.00). Using the TRANSFORM > COMPUTE VARIABLE menu command, we call up the *Compute Variable* dialog box.

Continue using the *ch2data.sav* data file.

1. Go to the tool-bar and select TRANSFORM, COMPUTE VARIABLE. This command will open the *Compute Variable* dialog box.

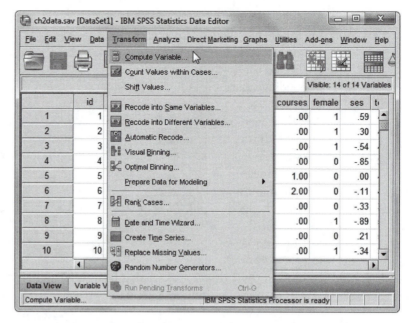

2a. Enter *testmean* as the *Target Variable*.

b. Scroll down the *Function group* list to locate and then click to select the *Statistical* group that displays different assorted statistical *Functions and Special Variables*.

c. Click to select the *Mean* function and then click the up-arrow button, which will place it in the *Numeric Expression* box formatted as: MEAN(?,?). The question marks indi-cate placement of the designated variables.

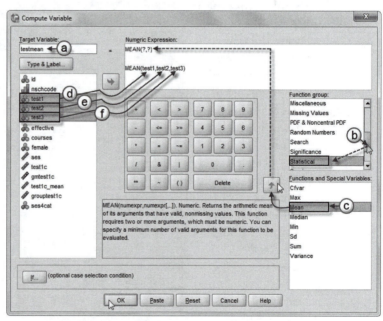

d. The mean will be computed from three variables: *test1*, *test2*, and *test3*. To build the corre-sponding numeric expression, click *test1* and then click the right-arrow button to place the variable in the box. Note that a comma must appear following the variable: MEAN(*test1*,?).

Note: The three variables (*test1*, *test2*, *test3*) are coded as dichotomous (0, 1) in which the mean of the summed total represents a proportion of times a student is proficient at reading.

e. Continue to build the numeric expression by next clicking *test2* and then the right-arrow button to move the variable into the box: MEAN(*test1*,*test2*,?)

f. Complete the numeric expression by clicking *test3* and then the right-arrow button to move the variable into the box: MEAN(*test1*, *test2*, *test3*).

Click the OK button to perform the function.

3. Scroll across the columns; the new variable *testmean* with its computed mean from the three test scores is found in the last column of the data window. You may also verify that the raw *testmean* values represent the average of the three test scores for each individual in the data set.

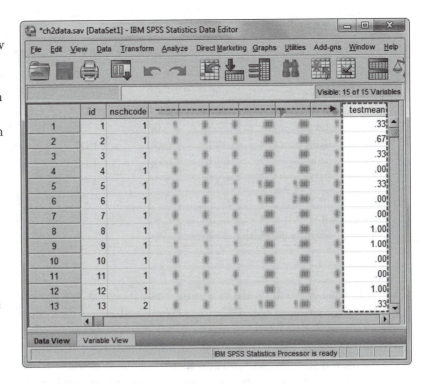

Although we have created two new variables using the RECODE and COMPUTE commands, our data file is still strictly an individual-level (Level-1) file. More specifically, aside from the school id, *nschcode*, we do not have any variables that are school specific and therefore do not have anything to analyze at Level 2. Typically, data from other sources would be brought in and merged with our Level-1 file. If we had information about the schools these students attend (e.g., public or private control, etc.), we could very easily merge those data with the individual-level file with which we have been working.

MATCH FILES: Combining Data From Separate IBM SPSS Files

The MATCH FILES command allows us to combine data from different sources. To demonstrate one use of this command, we will use the existing Level-1 file with which we have been working. MATCH FILES can combine data in two general ways: by adding variables to the existing data set or by adding cases to the data set. We will limit our interest here to the addition of variables that are found in a separate data set.

For our example, we are going to show how to merge a file containing school-level information with our existing file containing student-level data. We want to add information about the frequency of advanced placement testing at each school. The file containing this information (*apexams.sav*) has two variables in it. The first is a school identifier that is the same as the school identifier used in our Level-1 data file (*nschcode*). The second variable is named *apexams*. It is a ratio of the number of AP exams taken by students at the school to the total number of students in the 12th grade at the school. Our student-level (Level-1) file contains 8,335 observations (i.e., the number of students), and the school-level (Level-2) file contains 525 observations representing the schools that the 8,335 students attend. Consider the contents of each file. The

TABLE 2.2   Descriptive Statistics

	N	Minimum	Maximum	Mean	Std. dev.
id	8335	1.00	8670.00	4308.3500	2510.57800
nschcode	8335	1.00	525.00	261.9800	152.81700
test1	8335	0.00	1.00	.5300	.49900
test2	8335	0.00	1.00	.5200	.50000
test3	8335	0.00	1.00	.4900	.50000
effective	8335	0.00	1.00	.5622	.49614
courses	8335	0.00	4.00	.7481	.79013
female	8335	0.00	1.00	.5100	.50000
ses	8335	−2.41	1.87	.0336	.78390
Valid N (listwise)	8335				

Level-1 file will be the target for the data contained in the Level-2 (*apexams.sav*) file; that is, we are going to merge the data from the Level-2 file onto the records in the Level-1 file (see Tables 2.2 and 2.3).

Because this is not a one-to-one match (i.e., there are not the same number of records in each file), we will have to identify a "key" to be used to match the data from the schools to the data from the students. Notice that matching student data to the schools would require aggregating student-level variables because there are fewer schools than students in this case. We will use the single common identifier variable, *nschcode*, which represents the schools in both files. Both data sets will need to be sorted on the key variable. This can be accomplished by opening each data set and choosing from the tool bar the DATA > SORT CASES menu and dialog box. For each file, the sort should be on the variable *nschcode*, and each file needs to be saved after sorting.

TABLE 2.3   Descriptive Statistics

	N	Minimum	Maximum	Mean	Std. dev.
nschcode	525	1.00	525.00	263.00	151.69900
apexams	525	0.00	0.80	0.16	0.13311
Valid N (listwise)	525				

This example will combine data from two files: *ch2data.sav* (primary file) and *apexams.sav* (secondary file). When combining data from separate files, we recommend that the cases are sorted before you begin this procedure to prevent interruption of the workflow.

Continue using the *ch2data.sav* file.

1. Go to the toolbar and select DATA, SORT CASES. This command will open the *Sort Cases* dialog box.

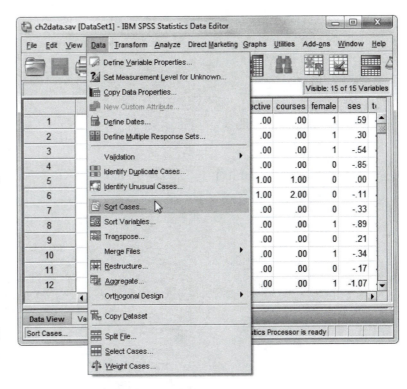

2a. Within the *Sort Cases* dialog box, select *nschcode* (key sort variable) from the left column and then click the right-arrow button to transfer the variable into the *Sort by* box.

b. The default *Sort Order* setting is *Ascending* (low to high), which we will retain. The option *Save file with sorted data* (a new feature available in IBM SPSS v.20) enables saving the sorted data directly to a file and specifying where to save the file. We will not use this option and leave it unchecked.

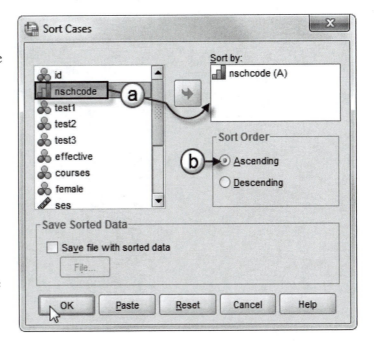

Click the OK button to begin sorting the cases. The next step is to sort the cases for the secondary file, *apexams.sav*.

3. We will now open the secondary file (*apexams.sav*) while keeping the *ch2data.sav* file open. Go to the toolbar and select FILE, OPEN, DATA. This command will then display the *Open Data* screen.

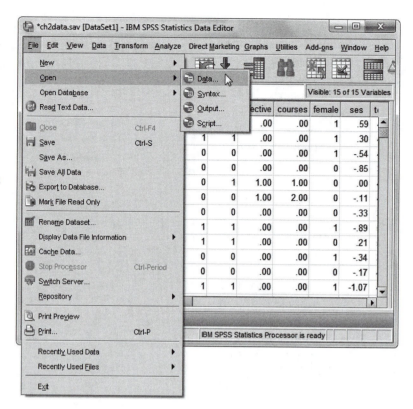

4a. Locate and click to select the data file (*apexams.sav*).

b. Then click the OPEN button.

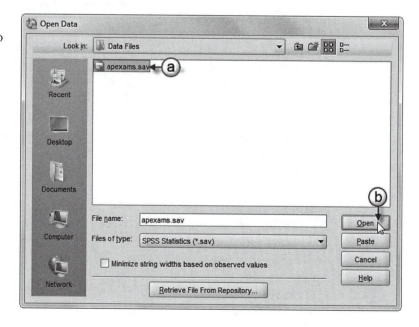

5. With the data file *apexams.sav* opened, go to the toolbar and select DATA, SORT CASES. This command will open the *Sort Cases* dialog box.

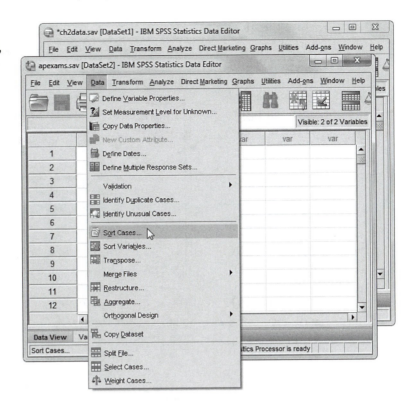

6a. Within the *Sort Cases* dialog box, select *nschcode* from the left column and then click the right-arrow button to transfer the variable into the *Sort by* box.

   Note: We are using the same variable (*nschcode*) in both data sets as the primary or "key" sort variable.

b. The default setting for sorting is the *Ascending* order, which we will retain. The option *Save file with sorted data* (a new feature available in IBM SPSS v.20) enables saving

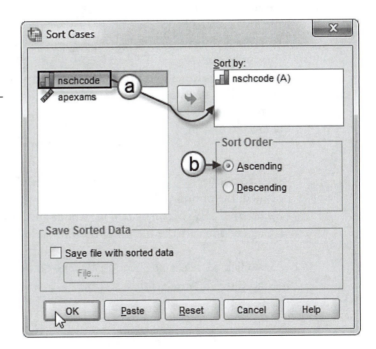

the sorted data directly to a file and specifying where to save the file. We will not use this option and will leave it unchecked.

Click the OK button to begin sorting the cases.

7. Once the files are sorted then merging may begin. Keep the *apexams.sav* data file open but return to the primary data file (*ch2data.sav*) to enable adding variables to the file. Then go to the toolbar and select DATA, MERGE FILES, ADD VARIABLES. This command will open the *Add Variables* dialog box.

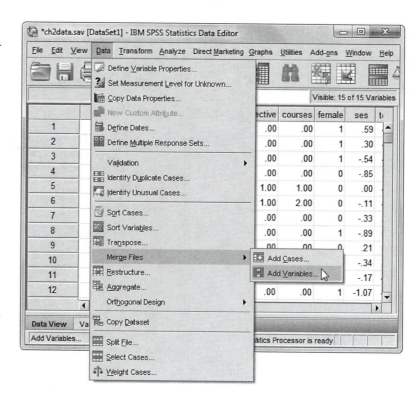

8a. Because the data file *apexams.sav* is an opened file, the option *An open dataset* is preselected and displays the file name in the box below.

Note: If *apexams. sav* had not been opened, the option

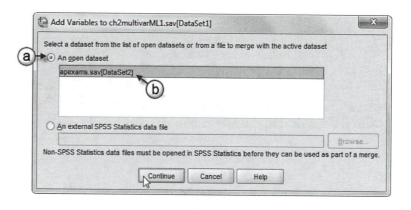

An external SPSS Statistics data file would have been preselected instead, requiring you to locate and identify the file using the *Browse* button.

b. Now click to select *apexams.sav(DataSet2)* and then click the CONTINUE button to access the *Add Variables from* screen.

9a. Within the *Add Variables from* display screen, click to select *nschcode(+)* from the *Excluded Variables* box. This action will activate the *Match cases on key variables in sorted files* option below, enabling the box to be checked.

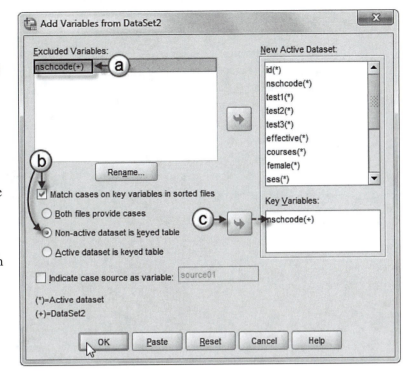

b. Click to select the *Non-active dataset is keyed table* (which is the *apexams.sav* file) option.

Note: A "keyed" table or table lookup file is a file that contains data for each case that can be applied to numerous cases in another data file.

c. Now click the right-arrow button to move *nschcode* into the *Key Variables* box.

10. A warning to presort the data appears but may be disregarded because the data from both files had been sorted using the *nschcode* variable at the outset. Click the OK button.

11. Scroll across the columns; the merged variable *apexams* taken from the *apexams.sav* data file is found in the last column of the *ch2data.sav* data window.

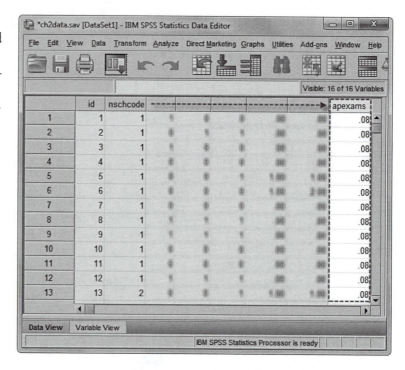

Figure 2.5 illustrates the addition of the new variable, *apexams*, to the *ch2data.sav* file once the merge is completed. Notice that although the individual-level variables vary across all cases within the window, the new school-level variable (*apexams*) is constant within each school (compare *nschcode* 4 with *nschcode* 5 in the screenshot).

Descriptive statistics on the data set will now show a value for *apexams* for each of the 8,335 Level-1 observations. Notice that the descriptive statistics provide no hint of the lack of variance within each *nschcode* for the *apexams* variable. It now looks like an individual-level variable (see Table 2.4).

AGGREGATE: Collapsing Data Within Level-2 Units

Many times the analyst will be interested in creating group-level measures by aggregating the characteristics of individuals within each group. One candidate variable might be student socioeconomic status (*ses*). Another might be gender (*female*) (aggregating this to the mean will

FIGURE 2.5 Data matrix after performing the MATCH FILES function.

TABLE 2.4 Descriptive Statistics

	N	Minimum	Maximum	Mean	Std. dev.
id	8335	1.00	8670.00	4308.3500	2510.57800
nschcode	8335	1.00	525.00	261.9800	152.81700
test1	8335	0.00	1.00	0.5300	0.49900
test2	8335	0.00	1.00	0.5200	0.50000
test3	8335	0.00	1.00	0.4900	0.50000
effective	8335	0.00	1.00	0.5622	0.49614
courses	8335	0.00	4.00	0.7481	0.79013
female	8335	0.00	1.00	0.5100	0.50000
ses	8335	–2.41	1.87	0.0336	0.78390
apexams	8335	0.00	0.80	0.1659	0.12952
Valid N (listwise)	8335				

yield the proportion of students at Level-1 who are female). In this section we will use the AGGREGATE command to create both of these variables.

The objective in this instance is to take the within-school means of the variables *female* and *ses*. This will yield the proportion of females in the sample for each school and the average socio-economic status of students in the sample for each school.

If you are continuing after completing the *Merge Files* tutorial, the *ch2data.sav* file is already open. If not, locate and open the data file.

1. Go to the toolbar and select DATA, AGGREGATE. This command will open the *Aggregate Data* dialog box.

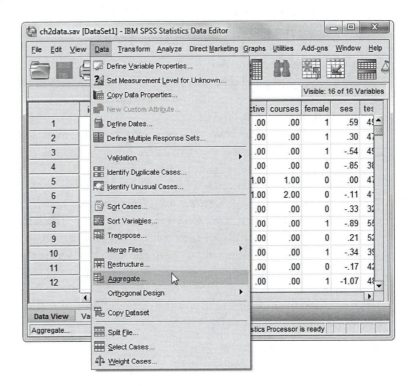

2a. Click to select *nschcode* from the left column of the *Aggregate Data* dialog box and then click the right-arrow button to move the variable into the *Break Variable(s)* box.

b. Now click to select *female* and *ses* from the left column and then click the right-arrow button to move the variables into the *Summaries of Variable(s)* box.

Note: IBM SPSS uses MEAN as the default function, which will be used for this example. Other functions in addition to MEAN are also available via the *Function* button to generate a variety of aggregated data.

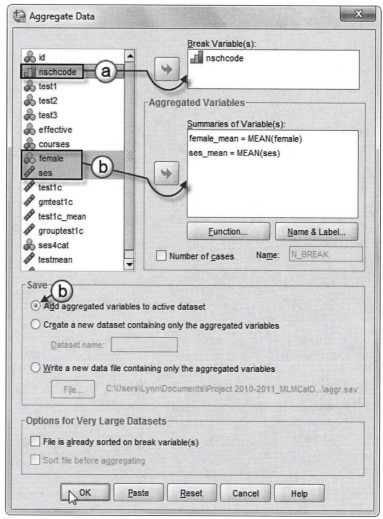

c. Click the option *Add aggregated variables to active dataset,* which will add the two variables *female_mean* and *ses_mean* directly into the active Level-1 data set.

Note: In addition to the *Add aggregated variables to the active dataset* option, we could choose two other options: *Create a new dataset containing only the aggregated variables* or *Write a new data file containing only the aggregated variables.* The latter choices would require us to use the MATCH FILES routine to merge the new aggregated variables back onto the individual-level data set. Choosing to have the variables written straight to the active data set is much more efficient and reduces the risk of errors.

Click the OK button to run the aggregation and merge the new variables to the student level data set.

3. Scroll across the columns; the new *female_mean* and *ses_mean* aggregated variables appear in the last two columns of the *ch2data.sav* data window. Notice that School 4 (*nschcode* = 4) has an all-female sample (*female_mean* = 1.00) but that less than one half of the School 5 sample was female (*female_mean* = 0.47). The SES average for School 4 was also higher than the

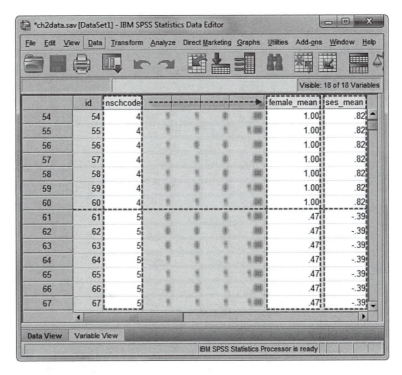

average for School 5 (0.82 vs. −0.39, respectively).

The RECODE, COMPUTE, MATCH FILES, and AGGREGATE commands provide all the tools necessary for structuring our IBM SPSS files for a multilevel analysis within IBM SPSS MIXED. We will expand on some of the commands in subsequent sections to create variables that can be very useful in multilevel analyses. Before turning our attention to those additional examples, however, we introduce a different data structure that enables multivariate analyses and analyses of change over time using the multilevel model through IBM SPSS MIXED.

VARSTOCASES: Vertical Versus Horizontal Data Structures

The data sets we created in the previous sections are arranged horizontally; that is, each observation is contained on a single row with variables arrayed across the columns. In the horizontal data sets, we have the variables arranged in such a way that the lower level units can be seen as nested within high-level units (e.g., students within schools where student values vary within and between schools, but school-level values vary only between schools).

In multivariate and time-varying models, we reconceptualize the nesting and deal with vertical rather than horizontal data structures. Instead of students being our Level-1 unit of analysis, we might nest time periods or indicator variables within students who could, in turn, be nested within schools. So our time periods or indicator variables become Level 1, students become Level 2, and schools become Level 3. In terms of the data structure, what this means is that each individual (i.e., students, to continue our example) will have multiple records. If there are three occasions of interest, each student will have three records, one for each occasion. Similarly, if we are interested in defining a latent variable with five indicators, each student would have five records.

So if we have a data set with 1,000 students and we are interested in looking at change, say, in test scores over three occasions, our data set will have three occasions multiplied by 1,000 students = 3,000 records. Using the latent variable example, if we had 1,000 students and five indicators, we would have 5,000 records in the data set. This is quite in contrast to the univariate outcome models in which individuals are nested in successively higher organizational levels. In those models, the individual defines Level 1 of the analysis, and there are as many records as there are individuals.

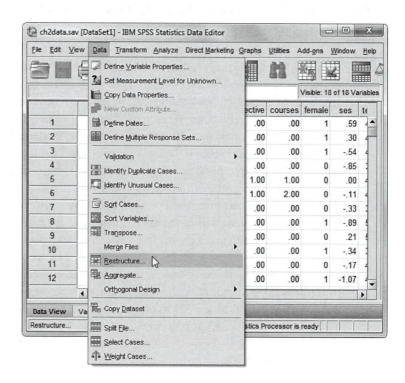

FIGURE 2.6 Horizontal data matrix.

Rearranging the data to accommodate the change and multivariate models is quite straightforward. Let us refresh our memory of the data set we have been using. Note that in the data view in Figure 2.6, there is a single record for each student, and we have three proficiency test scores across successive time periods.

Our objective is to create three records for each student, each representing a distinct time point for each individual in the sample. In other words, we are going to nest these three testing occasions within each student (students will still be nested within their schools). To accomplish this we will use the VARTOCASES routine contained in the IBM SPSS *Restructure Data Wizard*.

Continue using the *ch2data.sav* file.

1. Go to the toolbar and select DATA, RESTRUCTURE. This command will open the *Welcome to the Restructure Data Wizard* dialog box.

2. The Restructure Data Wizard presents three options: (1) restructure selected variables into cases, (2) restructure selected cases into variables, and (3) transpose all data. For this situation we want to treat the three test variables (*test1, test2, test3*) as a single grouped variable, so we will use the default setting: *Restructure selected variables into cases.*

Click the NEXT button to go to the *Variables to Cases* screen.

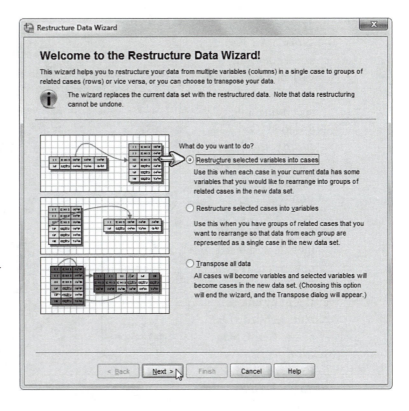

3. The *Variables to Cases: Number of Variable Groups* display screen allows defining the number of variable groups to create in the new data file. In this case, the three tests (*test1, test2, test3*) are to be treated as a single group, so the default setting of *One* will be used.

Click the NEXT button to continue to the next screen.

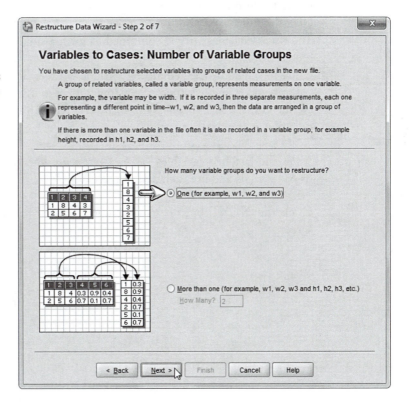

4. The *Variables to Cases: Select Variables* allows defining the variables for the new data. This includes specifying a new group id (*Case Group Identification*), the variables to define the transposition (*Variables to be Deposed*), and the variables to include with each new record (*Fixed Variables*).

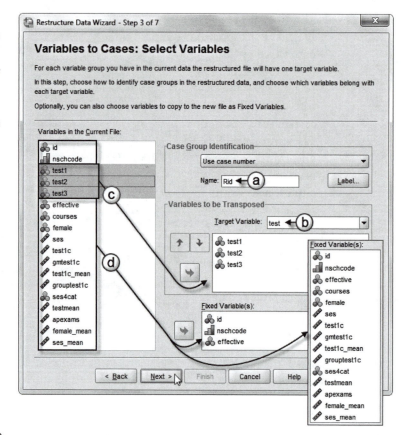

a. Begin by entering *Rid* (recoded ID) as the *Case Group Identification Name*.

b. Enter *test* as the *Target Variable*.

c. Click to select *test1*, *test2*, and *test3* from the left column, then click the right-arrow button to move the variables into the *Variables to be Transposed* box. The three variables will be combined to form a single variable (*test*) that is related to each record (row).

d. Click to select all the remaining variables (excluding *test1*, *test2*, and *test3*) from the left column and then click the right-arrow button to move them into the *Fixed Variable(s)* box. (Refer to the *Fixed Variables* boxed insert.) These variables will then be appended to each new row in the data set.

Note: The last five variables (*ses4cat*, *testmean*, *apexams*, *female_mean*, *ses_mean*) are from prior tutorials and are included in the example.

Click the NEXT button to continue to the *Variables to Cases: Create Index Variables* screen.

5. *The Variables to Cases: Create Index Variables* allows for creating one or more index variables. In this case we want to create one indexed variable encompassing the "timing" of each test. Because *One* is the default option, click the NEXT button to continue to the *Variables to Cases: Create One Index Variable* screen.

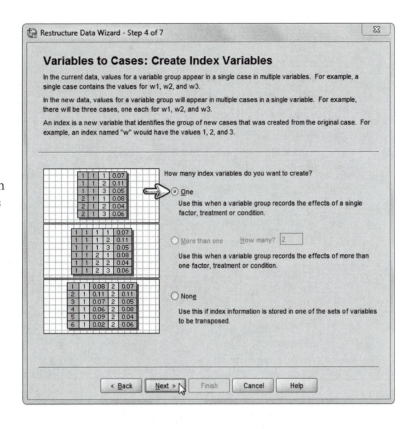

6a. Further information may be specified for the one indexed variable on this page. *Sequential numbers* is the default setting, which will be retained for this example.

b. The variable name may be changed by clicking on *Index1* and then changing it to *time*.

Click the FINISH button to indicate no further changes will be made.

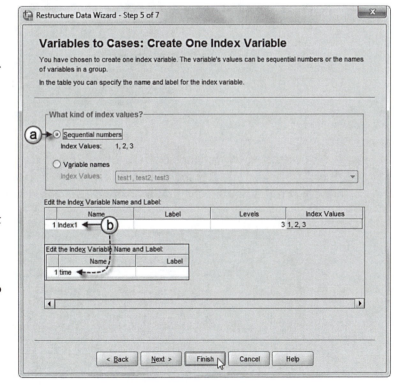

7. After clicking the *Finish* button, a message immediately appears warning that some of the original data will remain in use and that care should be administered to ensure use

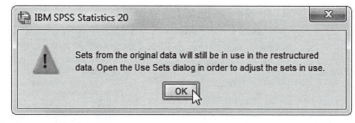

of the appropriate data set for subsequent analyses. Click the OK button to proceed with generating the new data set.

8. The new data set appears and, scrolling across the columns, the *time* and *test* variables are found in the last two columns. A truncated view of the data set shows that each student (as identified by *Rid* and *id*) now has three lines of data—one line for each of the three tests and time periods. A more complete picture is presented in Figure 2.7.

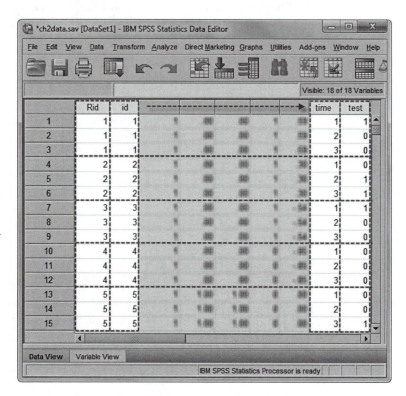

Notice that each student now has three lines of data—one for each test. The test scores (Level 1) vary across all respondents; student characteristics vary across all students (Level 2), but are constant within the group of three test scores; and school characteristics vary across

FIGURE 2.7 Data matrix after performing the VARTOCASES function.

TABLE 2.5 Descriptive Statistics

	N	Minimum	Maximum	Mean	Std. dev.
Rid	25,005	1.00	8335.00	4168.0000	2406.15500
id	25,005	1.00	8670.00	4308.3500	2510.47700
nschcode	25,005	1.00	525.00	261.9800	152.81100
effective	25,005	0.00	1.00	0.5622	0.49613
courses	25,005	0.00	4.00	0.7481	0.79009
female	25,005	0.00	1.00	0.5100	0.50000
ses	25,005	−2.41	1.87	0.0333	0.78386
ses4cat	25,005	1.00	4.00	2.4973	1.11714
testmean	25,005	0.00	1.00	0.5143	0.41044
apexams	25,005	0.00	0.80	0.1659	0.12951
female_mean	25,005	0.00	1.00	0.5052	0.15325
ses_mean	25,005	−1.30	1.42	0.0333	0.50002
time	25,005	1.00	3.00	2.0000	0.81700
test	25,005	0.00	1.00	0.5100	0.50000
Valid N (listwise)	25,005				

all schools (Level 3), but are constant within the group of students attending each school (see Table 2.5).

Notice also that when looking at the descriptive statistics for the new data set, there are 25,005 observations. The original data set had 8,335 observations. From our earlier description, we can see that 8,335 (students) multiplied by 3 (test scores) equals 25,005 records in the new data set.

Using "Rank" to Recode the Level-1 or Level-2 Data for Nested Models

From this, it becomes apparent how quickly data sets can expand as models become more complicated. The computing time necessary for the solution to converge within each level can become substantial as the models become more complicated. This is especially true in the multivariate and change models when variables or time periods are nested within individuals. Leyland (2004) notes that one can save a great deal of the time that it takes to estimate a proposed model by reindexing the Level-1 (and, in some cases, the Level-2) identifiers within each school. Reindexing the individual-level data in these models can yield significant savings in run time, and, in some cases, may make a difference in whether the model can even be estimated within the confines of the computer's available memory. The objective of reindexing is to create a new set of individual identifiers numbered only with reference to each group. So, for example, the 10 individuals in Group 1 would be numbered from 1 to 10 (assuming only 10 in the group), and the 14 individuals in Group 2 would be numbered from 1 through 14, and so forth (1, 2,…, *n*).

Situations in which this may prove beneficial will be identified in later chapters. For now, however, we want to introduce the steps involved. This reindexing can be easily accomplished using the IBM SPSS RANK command. From the TRANSFORM > RANK CASES menu within IBM SPSS, we can call up the *Rank Cases* dialog box.

Creating an Identifier Variable

Creating identification variables is straightforward in IBM SPSS. We will first show how to generate an ID variable for each case. This can be useful in those instances when an identifier is not found in the file being used. For this first example, we will use the *ch2level-1data.sav* we introduced at the beginning of this chapter (see Figure 2.1). You may recall that this data set is single level and cross sectional with 6,871 observations. We have removed the *id* variable that appeared in the version of the data set we used earlier. Our first task will be to recreate an individual level identifier using the COMPUTE command.

Creating an Individual-Level Identifier Using COMPUTE

We will use the *ch2level-1data.sav* for this tutorial.

Begin by deleting the current *id* variable from the data set.

1a. Click on the *Variable View* tab, which displays the variables in the data set.

b. To delete *id*, first locate and click to select the second row (2). Right-click the mouse to display the submenu and select *Clear*, which will delete the variable.

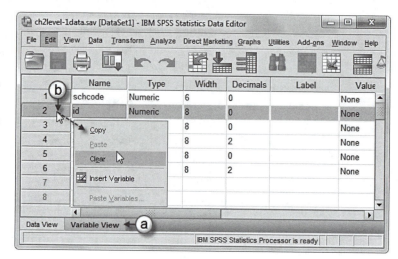

Note: An alternative method to deleting the variable is to go to the toolbar and select EDIT, CLEAR.

2. After removing *id* from the data set, go to the toolbar and select TRANSFORM, COMPUTE VARIABLE. This command will open the *Compute Variable* dialog box.

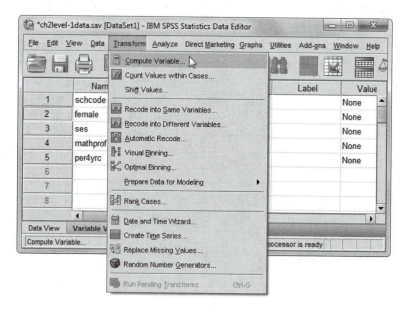

3a. Enter *id* as the *Target Variable.*

b. Within the *Function group* list, click to select *All,* which will display assorted functions in the *Functions and Special Variables* box.

c. Click to select the *$Casenum* function and then click the up-arrow button, which will place the term into the *Numeric Expression* box.

Click the OK button to perform the function.

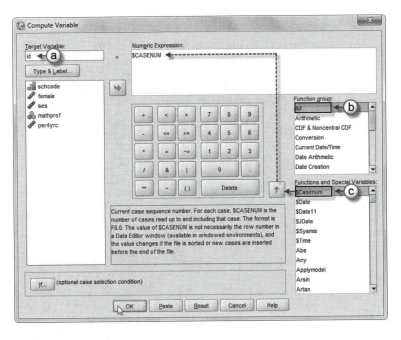

4. The new variable *id* is found in the last column of the data window and corresponds to the case number (or row number) for each case.

Note: The decimal place-setting may be adjusted from the default two places (2) by clicking on the *Variable View* tab and changing the number to the desired setting (i.e., "0").

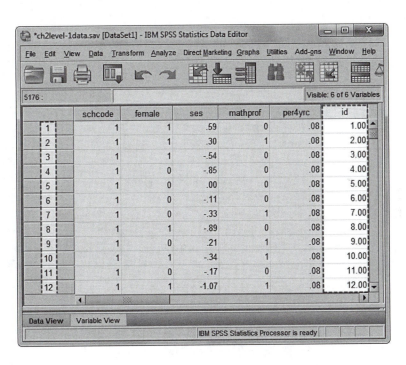

Once a Level-1 identifier exists, it is quite easy to create a variety of different within-group identifiers. These identifiers will become very important in later chapters, where we use more complex models. Using the TRANSFORM > RANK CASES commands from the IBM SPSS menu, we will now create another identifier assigning a sequential id variable (*Rid*) within each group. The *Rid* variable will range from 1 through *n* within each Level-2 unit.

Creating a Group-Level Identifier Using Rank Cases

We will use the original single-level *ch2data.sav* (having 13 variables) for this tutorial. This data set does not contain the *Rid* variable.

1. Go to the IBM SPSS toolbar and select TRANSFORM, RANK CASES. This command will open the *Rank Cases* dialog box.

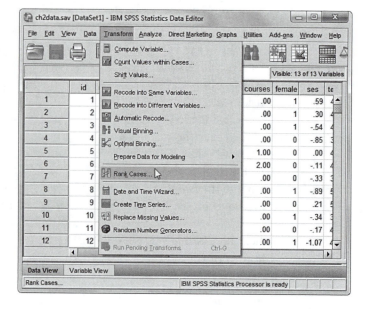

2a. Click to select the Level-1 identifier (*id*) from the left column and then click the right-arrow button to move it into the *Variable(s)* box.

b. Next click to select group identifier (*nschcode*) from the left column and then click the right-arrow button to move the variable into the *By* box.

c. In the *Assign Rank 1 to* section (bottom left), confirm that *Smallest value*, which happens to be the default setting.

d. Click the RANK TYPES button, which will open the *Rank Cases: Type* dialog box. *Rank* is the default setting, so click the CONTINUE button to close the dialog box.

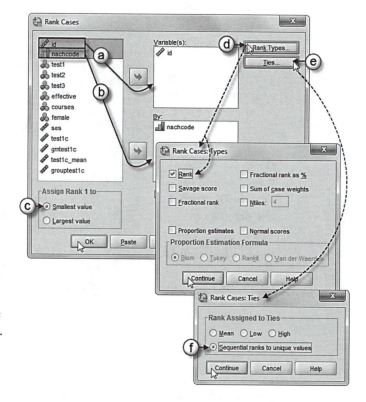

e. Now click the TIES button, which will open the *Rank Cases: Ties* dialog box.

f. Among the options listed for *Rank Assigned to ties*, click to select *Sequential ranks to unique values*. The option must be selected to ensure that each person is assigned an identifier within each unit. Then click the CONTINUE button to close the box.

Now click the OK button to create the *Rid* Level-1 group identifier.

3. Scroll across the columns; the *Rid* Level-1 group identifier variable is located in the last column of the data window. Notice that each Level-2 unit (*nschcode*) will have its own sequence of *Rid* beginning with 1 and continuing to 12.

Note: The decimal place-setting may be adjusted by clicking on the *Variable View* tab and changing the number to the desired setting (i.e., "0").

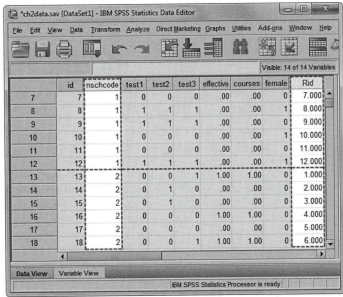

Finally, we will take this one step further by creating a new within-group id (*Rid*) variable for the multivariate data set shown in *ch2data.sav* (see Figure 2.7).

Creating a Within-Group-Level Identifier Using Rank Cases

We will use the *ch2data–vert.sav* file for this tutorial.

Begin by deleting the current *Rid* variable from the data set.

1a. Click on the *Variable View* tab at the bottom of the screen, which displays the variables in the data set.

b. To delete *Rid*, first locate the variable's row and then click to select the first row (1). Right-click your mouse to display the submenu and select *Clear*, which will delete the variable.

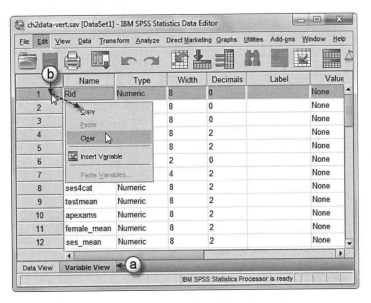

Note: An alternative method to deleting the variable is to go to the toolbar and select EDIT, CLEAR.

2. After removing *Rid* from the data set, go to the toolbar and select TRANSFORM, RANK CASES. This command will open the *Rank Cases* dialog box.

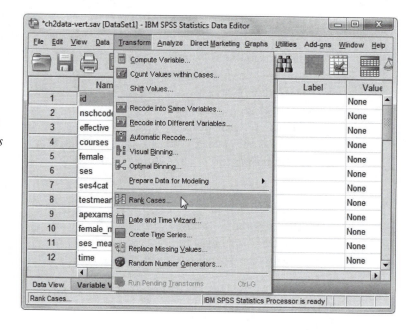

3. Because the data in the multivariate example are vertically arranged (that is, there are multiple records for each individual), we need to generate sequential, within-group identifiers for each person, constant across the three time points (recall that we had three time points nested within each person).

a. Click to select *id* and then click the right-arrow button to move it into the *Variable(s)* box.

b. Now click to select *nschcode* and then click the right-arrow button to place it into the *By* box.

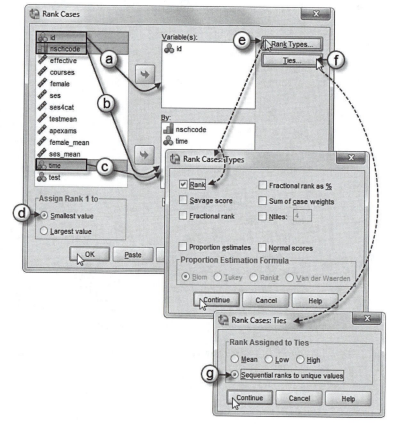

c. In a prior tutorial we selected only one group identifier, *nschcode*. Because the data in the multivariate example are vertically arranged (i.e., there are multiple records for each individual), we need to generate sequential, within-group identifiers for each person, constant across the three time points (recall that we had three time points nested within each person).

We will now add the other grouping variable (*time*) to the *By* box. This will generate a sequential identifier for each person across the three time periods within each group.

d. In the *Assign Rank 1 to* section (bottom left), confirm that *Smallest value*, which happens to be the default, is selected.

e. In the *Rank Types* option, we will retain the default *Rank* setting. To view the options, click the *Rank Types* button, which causes the *Rank Cases: Types* menu to appear. Now click the CONTINUE button to close the dialog box.

f. Click the *Ties* button to change the setting.

g. Instead of the default (*Mean*) we will click to select: *Sequential ranks to unique values* and then click the CONTINUE button to close the dialog box.

Click the OK button.

4. Scroll across the columns; the *Rid* variable is shown in the last column of the data viewer. Notice that the *Rid* variable reflects the number of observations in the data set but not the number of individuals.

Note: The decimal place setting may be adjusted by clicking on the *Variable View* tab and changing the number to the desired setting (i.e., "0").

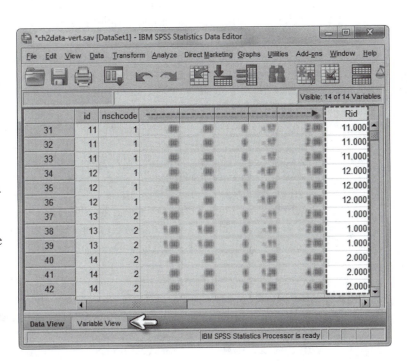

In this section we have shown the primary commands necessary to create new variables, recode existing variables, merge new data onto the existing data set, create group-level aggregates from individual-level variables, and how to restructure the data set for multivariate analyses using IBM SPSS MIXED. In the sections that follow, we use these data management procedures to offer instruction on the creation of new variables that will be needed for the analyses presented in subsequent chapters.

Centering

The basic multilevel model (i.e., a random intercept model) treats the Level-1 intercept as an outcome with variance that can be explained using variables from a higher level. Although we develop this model in some detail in the following chapter, we want to convey the importance of the intercept here. From our second data set (*ch2data2.sav*), consider as a starting

TABLE 2.6   Descriptive Statistics

	N	Minimum	Maximum	Mean	Std. dev.
test3	8335	0.00	1.00	.4900	.50000
test1c	8335	24.35	69.25	47.6439	6.32466
Valid N (listwise)	8335				

point the traditional fixed-effect logistic regression model using a continuous outcome math test (*test1c*) as a predictor for the student's probability of being proficient in math at Time 3:

$$\eta_i = \log\left(\frac{\pi_i}{1 - \pi_i}\right) = \beta_0 + \beta_1 test1c. \tag{2.1}$$

We can see in Table 2.6 that the values on the continuous math test range from 24.35 to 69.25 with a mean of 47.64.

Recall that the intercept in a model such as that in Equation 2.1 is the value of the outcome *Y* when the predictor (*test1c* in this example) is equal to 0. When additional terms are added to the model, the interpretation generalizes to the value of the *Y* when each of the predictors in the model is equal to 0. In instances where the predictor cannot be zero, such as this example when the lowest value of the math test score is 24.35, the intercept will have little substantive use. This is fine when the emphasis is on the interpretation of slopes that are constant across groups that may exist in the sample.

But now imagine a scenario in which, for some reason, we have an interest in an interpretable intercept. One way to ensure a meaningful and interpretable intercept is to alter the predictor in a way that makes zero a meaningful value. This is often accomplished by "centering" the predictor on zero (or some other value). Consider the results of the logistic regression model specified in Equation 2.1. We will use scores on *test1c* to predict probability of being proficient in math (*test3*), which has a mean of 0.49 in Table 2.6. We start with the intercept only. In Table 2.7, the log odds coefficient is −0.036, which is equal to an odds ratio of .965. Converting the log odds to the probability that *Y* = 1 [exp(β)/(1 + exp(β)) or .965/1.965], we obtain the proportion mean of 0.49.

Now we add the predictor. From the results in Table 2.8, we can see that a one-point increase in performance on *test1c* is associated with a 0.226 increase in log odds of being proficient in math on *test3*. The intercept value is −10.860, which represents a small probability of being proficient if one had a *test1c* score of 0.

With these two values in Table 2.8, one could calculate a predicted probability for being proficient on *test3*, given knowledge of performance on *test1c*. The intercept is used only to generate that predicted value for *test3* and has no real interpretative use because 0 is not a valid option for *test1c* performance.

TABLE 2.7   Parameter Estimates

Parameter	β	Std. error	95% Wald confidence interval		Hypothesis test		
			Lower	Upper	Wald χ^2	df	Sig.
(Intercept)	−0.036	0.0219	−0.079	0.007	2.735	1	0.098
(Scale)	1[a]						

Note: Dependent variable: test3; model: (intercept).
[a] Fixed at the displayed value.

TABLE 2.8   Parameter Estimates

Parameter	β	Std. error	95% Wald confidence interval		Hypothesis test		
			Lower	Upper	Wald χ^2	df	Sig.
(Intercept)	−10.860	0.3106	−11.469	−10.252	1222.680	1	0.000
test1c	0.226	0.0064	0.214	0.239	1232.261	1	0.000
(Scale)	1[a]						

Note: Dependent variable: test3; model: (intercept), test1c.
[a] Fixed at the displayed value.

By centering the *test1c* variable on 0, we can make the intercept more interpretable. To do this, we simply subtract the mean of *test1c* (47.64) from the score for each student in the data set on *test1c*; that is,

$$X_{ij} - \bar{X}. \tag{2.2}$$

We will show how to do this in a moment. For now let us just consider the change in the logistic regression results in Table 2.9.

Notice that the *test1c* slope coefficient (and standard error) remains the same as in the previous model, but the intercept is now considerably different (−0.08) because it now represents the adjusted probability of being proficient for *test3* (see preceding discussion) when the score on *test1c* is at the grand mean of the sample. So when *test1c* is equal to 0 (and 0 is now the overall mean for *test1c*), *test3* is equal to its adjusted overall mean. The intercept now has a useful interpretation.

Because the multilevel model treats the intercept as an outcome, as we will show in the next chapter, it is very important that the Level-1 model yield an interpretable value for β_0. Centering makes this possible and therefore is an important feature of the multilevel model. We are primarily concerned with grand-mean centering in this workbook, as shown in the previous example. However, there are occasions when group-mean centering, which centers the variable on the mean of each higher level group, may be of theoretical interest. In the sections that follow, we demonstrate how to create these two types of centered predictors within IBM SPSS MIXED.

Grand-Mean Centering

As in our previous example, variables in a multilevel model are most frequently grand-mean centered. For example, the grand mean for *test1c* is 47.64. Using *Compute* in IBM SPSS, one can name the new variable *gmtest1c* and then compute the new variable by subtracting the grand mean of *test1c* from students' scores on that variable (i.e., *test1c* − 47.6439). This will transform the scores in terms of the grand mean of the sample. So, if a student has a *test1c* score of 50.6439, her new score will be 3 (50.6439 − 47.6439 = 3), which carries the meaning that its relative position is three points above the grand mean with respect to other students in the sample.

TABLE 2.9   Parameter Estimates

Parameter	β	Std. error	95% wald confidence interval		Hypothesis test		
			Lower	Upper	Wald χ^2	df	Sig.
(Intercept)	−0.080	0.0250	−0.129	−0.031	10.249	1	0.001
gmtest1c	0.226	0.0064	0.214	0.239	1232.261	1	0.000
(Scale)	1[a]						

Note: Dependent variable: test3; model: (intercept), gmtest1c.
[a] Fixed at the displayed value.

Grand-mean centering results in unit-level means that have been adjusted for differences among individuals within the units. Notice that the distribution remains exactly the same when we center. The only thing that shifts is the scale itself.

Use the original *ch2data.sav* (with 13 variables) for this tutorial.

1. Go to the tool-bar and select TRANSFORM, COMPUTE VARIABLE. This command will open the *Compute Variable* dialog box.

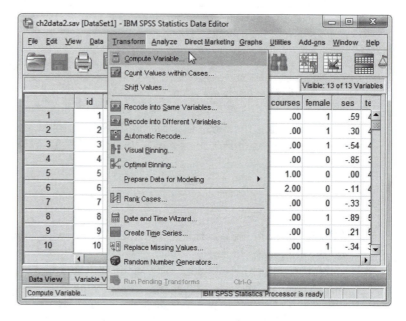

2a. In the *Compute Variable* display screen enter *gmtest1c* as the *Target Variable*.

Note: We will use the same name for the same grand-mean centered variable already in the data set. You may elect to rename this variable if you prefer not to overwrite the original *gmtest1c* variable.

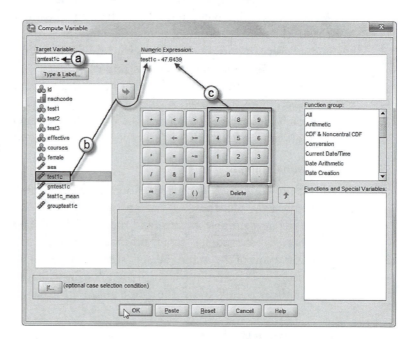

b. Now select *test1c* and click the right-arrow button to move the variable into the *Numeric Expression* box.

c. Use the numeric keypad (or enter from your keyboard) to complete the equation: *test1c − 47.6439* (the mean of *test1c*).

Click the OK button to create the computed variable *gmtest1c*.

3. IBM SPSS will issue a warning prompt because the target variable (*gmtest1c*) has the same name as another variable in the data set. We are aware of the variable being over-written, so click OK.

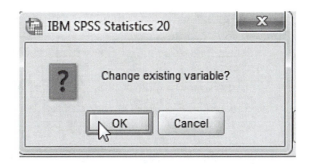

4. Scroll across the columns to locate the computed variable *gmtest1c*.

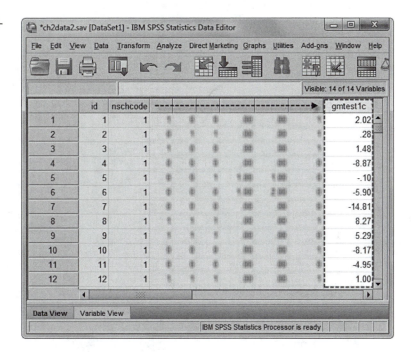

A quick look at the descriptive statistics of the original *test1c* variable and its grand-mean centered counterpart shows that although the standard deviation remains the same, the mean and range values have changed. Again, this simply demonstrates that only the scale has changed, while the distribution (or variability around the mean) remains the same. Note that we have not standardized the variable but, rather, have simply readjusted it so that its mean is equal to zero (see Table 2.10).

Group-Mean Centering

Group-mean centering of variables yields values that represent the unadjusted mean for group *j*. So rather than using the overall mean as the reference, the group mean is used instead:

$$X_{ij} - \bar{X}_{.j}. \tag{2.3}$$

Group-mean centering a variable is a two-step process. The first part involves aggregating the focal variable to the group level. The second part then is similar to the procedure used previously

TABLE 2.10 Descriptive Statistics

	N	Minimum	Maximum	Mean	Std. dev.
test1c	8335	24.35	69.25	47.6439	6.32466
gmtest1c	8335	−23.29	21.61	0.0000	6.32466
Valid N (listwise)	8335				

for grand-mean centering. We will use the AGGREGATE and COMPUTE commands to accomplish this.

Use the *ch2data2.sav* file (with 13 variables) for this tutorial.

1. Go to the toolbar and select DATA, AGGREGATE. This command opens the *Aggregate Data* dialog box.

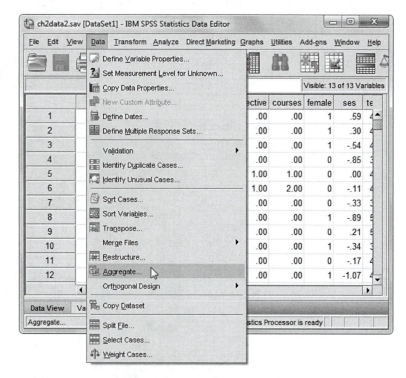

2a. Within the *Aggregate Data* dialog box, click to select *nschcode* from the left column and then click the right-arrow button to move the variable into the *Break Variable(s)* box.

b. Next, click to select *test1c* from the left column and then click the right-arrow button to move the variable into the *Summaries of Variable(s)* box. (IBM SPSS uses MEAN as the default function, which will be used for this example.)

c. Change the output variable name by clicking on the NAME & LABEL

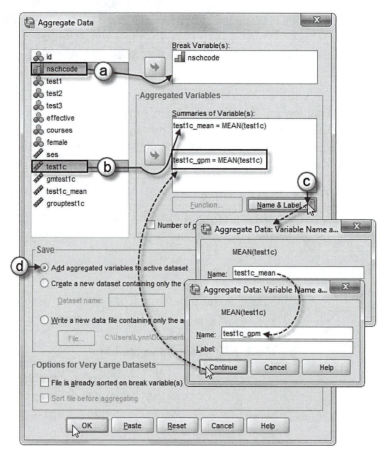

button, which will open the *Aggregate Data: Variable Name and Label* box. Then replace the initial variable name from *test1c_mean* to the one representing the group mean: *test1c_gpm*.

d. Click the option *Add aggregated variables to active data set* which will add the two variable *test1c_gpm* directly into the active level-1 dataset.

Click the CONTINUE button to close the box when completed to return to the *Aggregate Data* main dialog box. Click the OK button to perform the aggregation and create the new *test1c_gpm*.

3. Scroll across the columns; the new aggregated variable *test1c_gpm* appears in the last column of the *ch2data2.sav* data window.

Notice that this new variable is constant within each Level-2 unit (schools). That is, the value represents the mean *test1c* score within each school. These values provide the reference value for the group centering of *test1c* (just as the overall mean provided the reference value for grand mean centering).

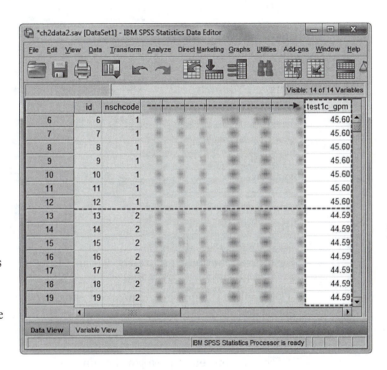

With the group mean value known, it is easy to calculate the group centered *test1c* variable which we will call *grouptest1c*.

Next, we will specify the target variable we wish to create (*grouptest1c*) and create the numeric expression: *test1c – test1c_gpm*.

4. Go to the toolbar and select TRANSFORM, COMPUTE VARIABLE. The command opens the *Compute Variable* dialog box.

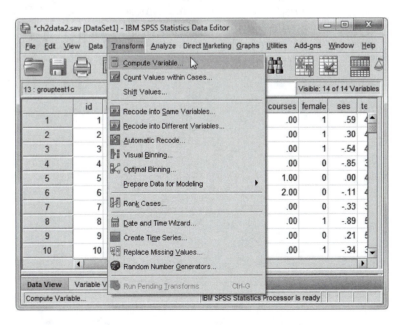

Note: The formula [*test1* − 47.643788] used in the preceding exercise may appear in the *Numeric Expression* box. To remove the formula, click the RESET button at the bottom of the screen.

5a. Now enter *grouptest1c* as the *Target Variable.*

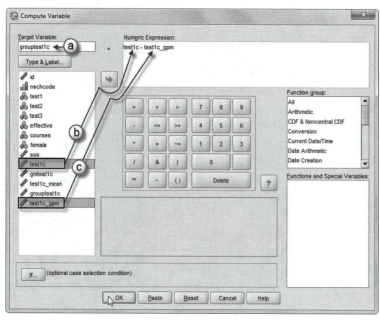

Note: We will use the same name for the same group-mean centered variable already in the data set. You may elect to rename this variable if you prefer not to overwrite the original *grouptest1c* variable.

b. Click to select *test1c* from the left column and then click the right-arrow button to move the variable into the *Numeric Expression* box. Insert a minus sign (−) by clicking on the key.

c. To complete the numeric expression click *test1c_gpm* from the left column and then click the right-arrow button to move the variable into the box. This completes the *Numeric Expression* of subtracting *test1c_gpm* from *test1c.*

Click the OK button to perform the function.

6. IBM SPSS will issue a warning prompt because the target variable (*grouptest1c*) has the same name as another variable in the data set. We are aware of the variable being over-written, so click OK.

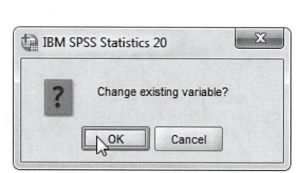

7. Scroll across the columns to locate the new *grouptest1c* variable in the data window.

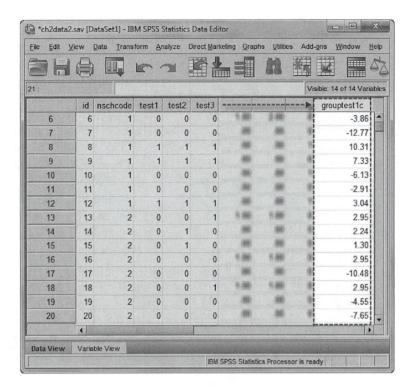

We can now compare the three variables: *test1c, gmtest1c,* and *grouptest1c* (see Table 2.11). Notice that although the uncentered *test1c* and the grand-mean centered *test1c* variables share the same distribution (but different means), the group-mean centered *test1c* variable has a mean of 0 (like the grand-mean version) but a different standard deviation. This results from the variance in *test1c* means across the schools in the sample—an artifact that will become very important in subsequent analyses.

For reasons we will discuss in Chapter 4, researchers often enter dummy variables in the model as uncentered (although there are occasions when analysts may also wish to grand-mean center them), and they grand-mean center the continuous variables at Level 1 and at Level 2. We will also show how this logic generalizes to situations in which there are more than two levels of analysis.

TABLE 2.11 **Descriptive Statistics**

	N	Minimum	Maximum	Mean	Std. dev.
test1c	8335	24.35	69.25	47.6439	6.32466
gmtest1c	8335	–23.29	21.61	0.0000	6.32466
grouptest1c	8335	–22.90	22.90	0.0000	5.87943
Valid N (listwise)	8335				

Checking the Data

The diligent analyst always takes great care to examine the data thoroughly. As data sets for multilevel analyses can become quite complex in terms of structure and content, great attention to detail should be given when reviewing their contents. IBM SPSS has a rich set of tools for the exploratory analysis of the data. The IBM SPSS EXPLORE routine, for example, can provide rich detail on data coding, distributions, missing data, and the like (this can be accessed through the ANALYZE > DESCRIPTIVE STATISTICS > EXPLORE menu). We reiterate that missing data can prove to be a real headache when using IBM SPSS MIXED or any other multilevel modeling routine. We highly recommend that missing data be carefully assessed to determine any patterns that may exist and to find remedies where at all possible.

Beyond a careful examination of the data, thought should be given to model specification and the distribution of residuals. Raudenbush and Bryk (2002, p. 253) point out that model specification assumptions apply at each level and that misspecification at one level can impact the results at other levels. Moreover, as we have noted earlier, with categorical outcomes the probability distributions are different from the typical normal distribution at Level 1. There are assumptions about the distribution of the residuals that apply at each level in the multilevel model. Although we do not spend a great amount of time addressing the investigation of residuals in this workbook, we do recommend that readers familiarize themselves with the possibilities that exist for model checking in the multilevel framework

A Note About Model Building

As will become clear in the chapters that follow, developing and testing multilevel models requires a great deal of thought as well as trial and error. Even with a basic two-level model there are many intermediate steps over which the model evolves. Keeping track of this evolution is essential for understanding the way the model is behaving and for replicating the models in subsequent steps.

Over the years we have developed a fairly simple naming scheme for our models and take care to document each model as fully as possible. Moreover, although we may use the IBM SPSS graphical user interface to develop the model, we always have IBM SPSS export the syntax so that we can save it for future reference. Our naming scheme is applied to the syntax files themselves. At a glance we can determine the type of model specified through the syntax (e.g., from a simple analysis of variance model to a fully specified three-level, random-slopes, and intercept model). With the syntax for each model saved and annotated, we can always document the evolution of model specification and easily modify models at any point in the future. We will have more to say about this throughout the workbook.

Summary

This chapter has provided an overview of the data management tools necessary for understanding and working with the hierarchical data files used in multilevel modeling. We have introduced five primary commands for manipulating data files to suit the needs of univariate and multivariate analyses using IBM SPSS MIXED. There is, of course, a great deal more than could be presented. Our main purpose in this workbook, however, is the modeling techniques themselves, rather than developing the more universal data management skills used to structure data within IBM SPSS. The treatment provided in this chapter is designed to highlight the elementary skills associated with data management relating to the specification of the multilevel model within IBM SPSS.

CHAPTER 3

Specification of Generalized Linear Models

Introduction

In this chapter we provide an overview of the generalized linear modeling (GLM) framework. As we noted in Chapter 1, techniques like multiple regression and analysis of variance (ANOVA) are actually special cases of this larger class of quantitative models. The term *generalized linear model* is from Nelder and Wedderbern (1972), who unified several of these linear regression-type models for examining continuous and categorical outcomes under one framework. The GLM framework facilitates the examination of different probability distributions for a dependent variable using a link function—that is, a mathematical model that specifies the relationship between the linear predictors and the mean of the distribution function. Our intent is to develop the rationale behind the underpinnings and specification of GLM in a relatively nontechnical manner. Further discussions can be found in Agresti (2007), Aitkin, Anderson, Francis, and Hinde (1989), Azen and Walker (2011), Long (1997), and McCullagh and Nelder (1989).

The single-level GLM that we develop in this chapter using the GENLIN program in IBM SPSS assumes that the observations are uncorrelated. Extensions of the GLM have been developed that allow for correlation between observations, as occurs in clustered designs and repeated measures designs. We address these extensions in the subsequent chapters.

Describing Outcomes

Statistical modeling depends on a family of probability distributions for outcome variables (Agresti, 2007). We often use the term *random variable* to describe the possible values that an outcome may have. A probability distribution is a mathematical function that links a particular observed outcome obtained in a sample to the probability of its occurrence in a specific population. The most common example involves the sampling distribution of the mean from which the probability of obtaining particular samples with a particular mean can be estimated (Azen & Walker, 2011). Most commonly, the observed mean is assumed to result from a normal distribution of possible values around the population mean, which has some variance. We assume that the general shape of this distribution resembles a bell-shaped curve; that is, the majority of individuals are closer to the mean and there are fewer individuals as distance from the mean increases.

Some Differences in Describing a Continuous or Categorical Outcome

An example may help to illustrate this point. Suppose we have a set of fifth grade students' reading scores in a particular elementary school. The population is 200 students. We draw a random sample of 50 students and obtain reading scores ranging from 0 to 10 on a test to estimate student proficiency in reading. We summarize this distribution in Figure 3.1. The sample mean is about 6.0 on the 10-point test, with a standard deviation of a little less than 2.5. Given

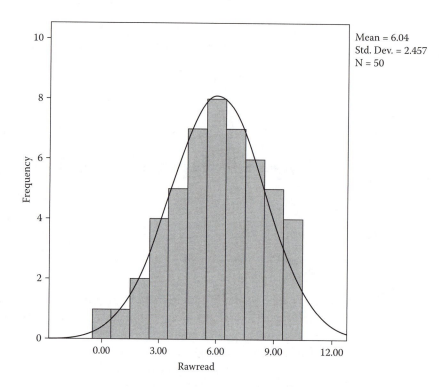

FIGURE 3.1 Raw reading scores.

properties of the normal distribution, we can expect about 68% of the sample to lie within a standard deviation above or below the sample mean (i.e., roughly between 3.5 and 8.5). We can see in Figure 3.1 that we have 33 students (66%) within a standard deviation (SD) of the mean (i.e., with scores ranging from 4 to 8). We also see that there are 17 students (34%) farther than one standard deviation from the mean, with 9 of them having scores at least one standard deviation above the mean (i.e., scores of 9 or 10) and 8 students having scores at least one standard deviation below the mean (i.e., scores of 3 or lower). Although these data do not form a perfect bell curve, for a small sample they represent the general shape of a bell curve quite well.

This conclusion is also supported by the descriptive statistics in Table 3.1. The negative skewness (–.306) indicates that there are a few more higher scores than lower scores, as shown in Figure 3.1. The negative kurtosis (–.405) suggests that the distribution is a bit flat compared to the theoretical normal distribution; however, both coefficients suggest that the data are relatively normally distributed for this small sample. Importantly, with a continuous outcome, we could select many other samples with a mean of approximately 6.0, but with different variances and corresponding standard deviations (i.e., the square root of the variance), which describe how dispersed the set of individuals is from the mean. For example, we could imagine one sample where everyone scores on the mean (i.e., variance = 0). In another, perhaps everyone lies between 5 and 7 (e.g., with variance = 0.25 and standard deviation = 0.5). The point is that with a normal distribution, the mean and the variance represent independent estimates. This is not the case for non-normal distributions such as the logistic distribution, for example.

TABLE 3.1 **Descriptive Statistics**

	N statistic	Mean statistic	Std. dev. statistic	Variance statistic	Skewness		Kurtosis	
					Statistic	Std. error	Statistic	Std. error
Rawread	50	6.0400	2.45748	6.039	–0.306	0.337	–0.405	0.662
Valid *N* (listwise)	50							

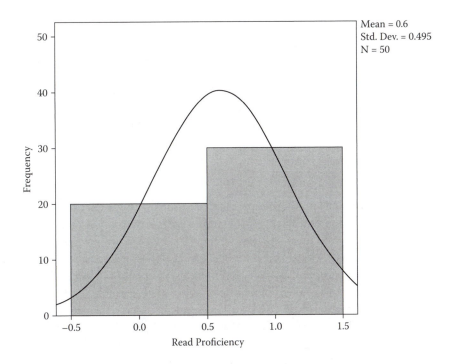

FIGURE 3.2 Distribution of reading proficiency scores.

When the outcome is categorical, we can no longer assume that the random value obtained (e.g., a sample mean) came from a normal distribution. Categorical variables are almost always discrete, which means that they take on a finite (or countable infinite) number of values. Examples include being proficient or not proficient in reading, purchasing a sports car (among several possible types of cars), expressing a level of agreement or disagreement on a political issue, or receiving a number of traffic tickets during a specified interval of time. Random values of these different types of categorical outcomes are obtained from several probability distributions (i.e., binomial, multinomial, Poisson) other than the normal distribution.

Suppose now that we wish to summarize the number of fifth grade students who are proficient in reading within the previously identified school. For demonstration purposes, we will assume that the cut score is a score of 6.0. Students who obtain a score of 6 or above are considered to be proficient in reading, and students whose scores fall below the cut point are considered not to be proficient. Our sample consists of 30 students who are proficient and 20 who are not. The observed mean now is simply the proportion of individuals who are proficient (30) divided by the total number (50) in the sample, or 0.60, but the distribution of values around the mean in Figure 3.2 now looks very different from a normal distribution.

Categorical variables that follow a binomial distribution can take on only one of two outcomes over a given number of n trials (such as flipping a coin). In the case of a dichotomous outcome (where there are only two possible outcomes), we encounter a type of probability distribution known as a Bernoulli distribution, where the number of trials is equal to 1.0. In this case, the proportion of proficient students (0.60) and students who are not proficient (0.40) must add up to 1.0. The probability of selecting a proficient student is then simply the proportion of students who are proficient (often denoted by π to refer to a population and p within a sample), and the probability of selecting a student who is not proficient is $1 - \pi$. The expected value, or mean (μ), is then the population proportion ($\mu = \pi$). We can obtain the variance in a binomial distribution as $\sigma^2 = n\pi(\pi - 1)$, so in the case of a Bernoulli distribution where $n = 1$, the variance is simply the proportion of students who are proficient multiplied by the proportion of students who are not proficient ($\sigma^2 = .60*.40 = .24$).

In Table 3.2, we consider the proportion of proficient readers that might result from five random samples of students drawn from the population of 200 students. One can quickly see that,

TABLE 3.2 Descriptive Statistics

	N	Minimum	Maximum	Mean	Variance
Read1	50	0.00	1.00	0.20	0.16
Read2	50	0.00	1.00	0.40	0.24
Read3	50	0.00	1.00	0.50	0.25
Read4	50	0.00	1.00	0.60	0.24
Read5	50	0.00	1.00	0.80	0.16
Valid N (listwise)	50				

unlike the normal distribution, the variance is tied to the mean proportions of proficient students. For example, we see the same variance in a sample with 20% proficient in reading and a sample with 80% proficient, as well as in samples where 40% or 60% are proficient. We will encounter this idea again in estimating categorical models because the variance also depends on the level of the mean in some other probability distributions we will encounter (e.g., Poisson, multinomial).

Because categorical outcomes are almost always discrete variables, other types of probability distributions may better describe the specific populations from which they are obtained (see Azen & Walker, 2011, for further discussion of probability distributions encountered in categorical data analysis). Discrete variables are often best summarized as frequency and percentage distributions rather than as means and standard deviations. A frequency distribution provides a way of summarizing the *probability* that particular responses occur. The cumulative percentages can be divided by 100 to obtain the probability (or proportion) of each event. In this case, the proportion that is proficient is 0.60, which is also the mean of the sample. The proportion can also be obtained by dividing the frequency in each category by the total (*N*) (see Table 3.3).

Although we can calculate various descriptive statistics for categorical variables (e.g., mean, variance, standard deviation), we can no longer assume that a normal distribution is the mechanism that produces the observed response patterns. In particular, although the mean describes the proportion of sample members that are proficient, we can see that the responses are not normally distributed (Table 3.4).

TABLE 3.3 Frequency Distribution for Reading Proficiency

		Frequency	Percent	Valid percent	Cumulative percent
Valid	0	20	40.0	40.0	40.0
	1	30	60.0	60.0	100.0
	Total	50	100.0	100.0	

TABLE 3.4 Descriptive Statistics for Reading Proficiency

N	Valid	50
	Missing	0
Mean		0.60
Std. deviation		0.49
Variance		0.24
Skewness		−0.42
Std. error of skewness		0.34
Kurtosis		−1.90
Std. error of kurtosis		0.66

Similarly, the measures of association between two categorical variables will be different from the typical Pearson correlation (*r*) because of the restricted range of observed values present in calculating the association and the particular measurement scale of each variable—that is, whether the two variables are both nominal, both ordinal, or one of each type.

Over the past few decades, analyses for statistical models with categorical outcomes have grown in sophistication. The GLM consists of three components needed to specify the relationship between the categorical dependent variable *Y* and a set of independent *X* variables (McCullagh & Nelder, 1989):

- An outcome variable with an underlying *random component* that describes the sampling distribution of the dependent variable. Because the dependent variable of interest has a discrete probability distribution of random values, its predicted values should also follow the respective distribution. The assumption is that the expected value (or mean) of the dependent variable can follow one of several known probability distributions (including the normal distribution for a continuous outcome).
- A *structural component* that refers to a linear combination of independent variables that predicts values of the outcome.
- A *link function*. Because the relationship between the effects of the predictors on the dependent variable may not be linear in nature, a link function is required to transform the expected value of the outcome so that a linear model can be used to model the relationship between the predictors and the transformed outcome (Hox, 2010). For the continuous outcome that is approximately normally distributed (as in Table 3.1), the link function is a simple identity function, which means that the linear combination of values predicting the dependent variable does not have to be transformed.

Measurement Properties of Outcome Variables

The type of probability distribution to employ in a particular analysis is primarily a function of characteristics of the dependent, or response, variable. In this section we outline some very basic issues associated with the measurement properties of variables. Stevens (1951) proposed a set of broad measurement scales that are widely known (i.e., nominal, ordinal, interval, ratio). The conceptual dimensions of these measurement properties are essential to an understanding of the methods and models we describe throughout the workbook. But beyond a desire to connect with these basic definitional properties, we also call attention here to the considerable debate that exists around the meaning and definition of levels of measurement and the enduring criticism that the Stevens classification has faced over the decades since its introduction (see Chrisman, 1998; Luce, 1997; Michell, 1986; Velleman & Wilkinson, 1993, for examples).

Variables are defined by the presence or absence of several measurement characteristics, including distinctiveness, magnitude (or ordering), equal intervals, and the presence of absolute zero (Azen & Walker, 2011). For continuous outcomes, ratio variables (e.g., income, height, weight) have all four measurement characteristics, and interval variables (e.g., math achievement scores, temperature) have all except the presence of an absolute zero. For example, receiving a score of zero on a math test generally does not imply that the student has no knowledge of mathematics; however, if we can assume that assigning a score of zero to an individual indicates the absence of the attribute being measured, then the measurement scale has the characteristic of an absolute zero (Azen & Walker, 2011). An example would be if an individual were assigned a zero for having no typing errors on a 2-minute timed test. As we have noted, continuous variables typically follow a normal distribution.

If the variable is nominal, we assign numbers to observations based on their distinctive characteristics. Common examples include race/ethnicity, sex, residential region in the United States, and home language. Suppose we wish to study students' immediate plans after graduating from high school. In this latter case, we might have five categories: attending a postsecondary educational institution (coded 1), attending a vocationally oriented school (coded 2), joining the military (coded 3), entering the work world (coded 4), or traveling (coded 5). In this case, there is not any particular ordered set of responses implied in these choices. Because there is no ordering of responses, the number we assign to any particular category will not affect the results (i.e., we could select any category to code 1).

Ordinal data have both distinctiveness and magnitude. For example, we may be interested in examining students' likelihood to persist in graduating from high school. For a particular student cohort, we might code dropped out as 0, still enrolled in school but behind peers (i.e., through not having earned enough credits to graduate) as 1, and graduated as 2. Alternatively, we could also treat this ordinal outcome as a nominal variable and code the three categories in any fashion. If we assume only distinctiveness between categories (i.e., by assigning any number to any category), however, we lose information about possible differences in the rank order of individuals between the categories of the outcome. We note in passing that some dichotomous variables may imply a simple type of ordering, such that passing a course (coded 1) may be assumed to have greater magnitude (or value) than failing (coded 0), although, strictly speaking, it could be coded in either way.

Certain assumptions about the distribution of individuals on predictors across the ordered outcome categories are required for such ordinal models to be estimated optimally. More specifically, because ordinal variables imply not only distinctiveness but also differences in magnitude, we expect that predictors in our model, such as student motivation, academic background, or attendance, should either increase or decrease systematically across the ordered outcome categories. In this case, we might expect higher levels of these predictors to be associated with greater likelihood to graduate (coded 2) versus still attending school (coded 1) versus dropped out (coded 0). If these assumptions do not hold in our preliminary analyses of the data, we may have to explore alternative modeling options.

As we noted, one possibility is to consider the outcome as nominal (i.e., having only distinctiveness between categories) rather than ordinal (i.e., having a clear ordering that can be explained by different values of the set of relevant predictors). In this case, the same probability distribution (multinomial) would underlie the nominal or ordinal outcome; however, a different link function would be required for each (e.g., one choice being the generalized logit and cumulative logit, respectively). Such decisions are often left to the researcher and can be made on the basis of the characteristics of the outcome data in each particular instance.

Similarly, researchers sometimes decide to treat ordinal response patterns as if they were continuous (e.g., a scale from "unimportant" to "very important" or from "strongly disagree" to "strongly agree"), but it is important to keep in mind that there can be consequences resulting from assuming that such ordered categorical responses actually constitute interval data (Olsson, 1979). As points on the scale are added (e.g., 5-point scale, 7-point scale), the ordinal outcome may behave more like an interval outcome, but it should be noted that it is always prudent to investigate the skewness associated with an ordinal variable to evaluate whether it is appropriate to consider it as interval.

The existence of possible values on a scale is oftentimes very different from the actual distribution of responses across those values. If, for example, a five-category ordinal scale has responses limited largely to two of those categories (e.g., strongly agree and agree), the analyst may need to rethink the actual level of measurement that such a variable represents. We encourage the reader to consider carefully both the scale and the actual distribution of responses across that scale when deciding which level of measurement is appropriate for a given variable.

Counts are another type of outcome that generally does not have a normal distribution. Count data refer to the frequency of occurrence of an event such as passing or failing a course, having a traffic accident, or seeking medical attention during some specified period of time. By definition, counts cannot be below zero, so they always have zero as their lower bound. For example, if we examine students' course taking during their ninth grade year in high school, we might expect the majority of students not to fail any courses; however, there will be a subset of students who fails one course to perhaps the maximum number of courses an individual could take during that year. What makes count data non-normal is the tendency for many more responses of zero than would be expected in a normal distribution, as well as an upward limit on the number of courses that a student can fail in a given time period specified for the study.

Events-in-trials data are often used in predicting the occurrence (or the time to recurrence) of an event. This refers to a response variable that represents the frequency of a particular event occurring in a set of trials (or different levels of exposure to the event). For example, the event might be the frequency of contracting a particular illness at different ages (or its frequency of recurrence) or the frequency of having car accidents with respect to differing levels of driving experience. The response variable includes the number of events that occur, and an additional variable (referred to as an offset) can be used to contain the number of trials, or level of exposure. If the number of trials is the same across individuals in the study, the trials may be specified as a fixed value, as previously described. Events should be non-negative integers and trials should be positive integers (IBM SPSS, 2010).

Explanatory Models for Categorical Outcomes

As we have noted, in contrast to their continuous outcome counterparts, models with categorical outcomes typically have a restricted range of possible values. As a reference point, we begin with the familiar multiple regression model to explain individual i's outcome on Y, given her or his levels of X_1 and X_2:

$$Y_i = \beta_0 + \beta_1 X_1 + \beta_2 X_2 + \varepsilon, \tag{3.1}$$

where β_0 is the intercept, β_1 and β_2 are slope parameters, and ε represents errors in predicting individual outcomes from the equation.

We note that in this chapter (and subsequent chapters), when we discuss single-level models we will not include the subscript i, as it is understood that all variables (and the residual) are defined at the individual level. An important assumption of the linear model is that the residuals (or errors) in predicting Y from X will be normally distributed, have a mean of 0, have constant variance, and will be uncorrelated with each other (Hamilton, 1992). This type of model is appropriate to explain relationships between a set of predictors and the outcome when Y is a continuous variable, such as level of reading achievement on a standardized test, but not when the outcome is categorical.

If the outcome variable is dichotomous (coded proficient = 1; not proficient = 0), responses can only take on two values. In this case, if we imposed the typical linear model on the data, we would eventually observe estimates of Y that fall outside the allowable range of values for the dependent variable. Because the dependent variable Y can only be 0 or 1, the model should not predict values of Y beyond the lower or upper limits of this constraint. When there are only a few possible values for Y, the residuals (ε) will be tied to levels of X; therefore, the assumption of constant error variance will be untenable. More specifically, when Y is dichotomous, for any level of X (e.g., a motivation score of 5.0), there will be only two residuals, which will therefore not be normally distributed. In this case, the assumption is that the probability distribution is binomial rather than normal.

As a simple illustration, consider the probability of a student being proficient in reading (i.e., 1 = proficient; 0 = not proficient). We might hypothesize a relationship between student motivation and probability of being proficient, with greater levels of motivation increasing the likelihood that a student is proficient in reading. For convenience, we will say that motivation (X) is measured on a scale from 0 to 10. When we estimate the model, we obtain the following equation to predict $\hat{Y}$ (where Y hat, or $\hat{Y}$, indicates a predicted value):

$$\hat{Y} = 0.1 + 0.2X$$

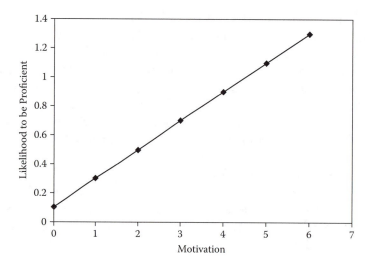

FIGURE 3.3 Linear relationship between motivation and likelihood to be proficient.

As long as motivation is 4 or less, our model returns plausible predicted values of Y (i.e., between 0 and 1). For example, if motivation is 0, $Y = 0.1$ [$0.1 + 0.2(0) = 0.1$]. If motivation is 4, $Y = 0.9$. If motivation is 5 or 6, however, our predicted values for Y become 1.1 and 1.3, respectively, which would fall outside the boundary of 1.0 for predicted values of Y. This linear relationship is represented in Figure 3.3.

We can next illustrate that another assumption of the multiple regression model is also violated when the outcome is not continuous; that is, prediction errors should be independent and normally distributed with a mean of zero and some variance (Heck & Thomas, 2009). Let us consider again our reading example. Residuals evaluate how much observed values deviate from model estimated values ($Y - \hat{Y}$). If we use the raw reading scores (from 0 to 10 in Figure 3.1), we get the normal distribution of residuals with mean of 0.00 and some variance (i.e., SD = 0.95) from predicting students' raw reading scores from a predictor such as motivation (Figure 3.4). Remember that the scores range from 0 to 10, so for any level of motivation, the figure suggests that the residuals can take on a considerable number of different values. Moreover, even in this small sample, Figure 3.4 suggests that the residuals are relatively normally distributed for levels of predicted reading scores (i.e., skewness = 0.02; kurtosis = −0.40). Larger standardized residuals (e.g., ±2) indicate outliers.

In contrast, Figure 3.5 represents the distribution of residuals in predicting whether an individual is proficient in reading (coded 1) or not (coded 0) for given levels of the predictor. We see that when the outcome is dichotomous, residuals are not normally distributed around the sample mean (0.60). For each level of X, there are now only two possible residuals. A different type of statistical model is required that can extend the basic linear regression model by transforming the dichotomous responses in an appropriate manner so that they can be modeled as a linear function of the predictors (Azen & Walker, 2011).

In the past, the approach to non-normally distributed variables and heteroscedastic errors (i.e., errors having different variances) was to apply some type of empirical transformation directly to the outcome in order to achieve normality and reduce the heteroscedasticity (Hox, 2002). This would be followed by an analysis using a linear modeling technique such as multiple regression. A number of different empirical transformations can be used as possible remedies for certain types of outcomes (e.g., square root transformation, logarithmic transformation). For example, suppose we have the following set of numbers: 1, 4, 9, 25, 36. The mean is 15.0 and the standard deviation is 14.95 (with skewness of 0.74). The positive skew suggests more low scores than high scores. If we use a square root transformation, the new data set (1, 2, 3, 4, 5) has a mean of 3.0 and a standard deviation of 1.58 (with skewness of 0.0).

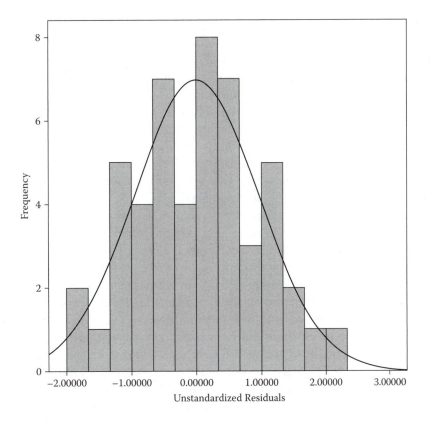

FIGURE 3.4 Unstandardized residuals of predicting values of raw reading scores

When the outcome is dichotomous, however, trying to transform it is problematic. Suppose, for example, we have five individuals: three who are proficient in reading and two who are not proficient. If we code not proficient versus proficient in different ways (e.g., 0, 1 or 0, 4), when we take the square root, we find that the skewness and kurtosis coefficients are unaffected by the transformation. Instead, if we try to take the natural logarithm of the outcome directly, the

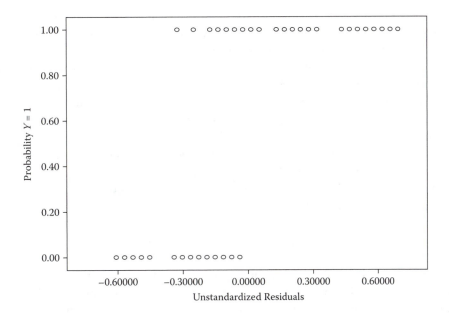

FIGURE 3.5 Unstandardized residuals for predicting the probability of proficiency in reading.

transformation breaks down because the function is undefined for 0, so we will have missing data for our two individuals who are not proficient (coded 0). Because the predicted value can only take on one of two values (as shown in Figure 3.2), no empirical transformation will make a dichotomous outcome become normally distributed. Hence, use of these outcomes will always lead to a violation of the main assumptions of the GLM.

Similarly, count data can also present problems in terms of trying to transform the outcome responses to make them appear more normally distributed. For example, we may wish to examine student course failures over an academic year, where the possible values of Y are non-negative integers (0, 1, 2,..., N). Such data are usually positively skewed; that is, there are more responses at the low end of the range of possible values for Y. If there were only a few zeros in the data, a logarithmic transformation might be acceptable [$\log(Y+1)$] so that the model could then be formulated as a typical multiple regression with continuous outcome (Raudenbush, Bryk, Cheong, & Congdon, 2004). It is likely, however, that at least 50%–60% of a sample of students would not fail a course during a typical academic year. Therefore, the assumption of normality would still be violated after transforming the outcome. In this case, we might assume that the underlying probability distribution is Poisson rather than normal.

As these situations illustrate, when the dependent variable is categorical, the response function is not optimally described by a linear relationship as specified in multiple regression because the response variable is not measured on a continuous scale; therefore, individuals' standing on outcome is not normally distributed. Rather than trying to transform the outcome directly, GLM incorporates the necessary transformation and choice of an appropriate error distribution directly into the statistical model (Hedeker, 2007; Hox, 2002). Nonlinear link functions make it possible to extend standard regression models to include categorical outcomes having different non-normal error distributions. These link functions provide a means of linking *expected* values of the categorical outcome Y to an underlying (latent) variable that represents predicted values of the outcome (Hox, 2002). Therefore, generalized linear models avoid trying to transform the *observed* values of a binary or ordinal variable directly and instead facilitate the application of a transformation to the *expected* values of Y.

Components for Generalized Linear Model

We reiterate that the GLM consists of the following three common components:

- An outcome variable Y with a specific *random distribution* with expected value μ_i and variance σ^2. The random component underlying the outcome (Y) is necessary in computing its expected value or mean (represented by the Greek letter mu, μ); that is, $E(Y) = \mu$.
- A *link function*, often represented as $g(\cdot)$, which connects the expected values (μ) of Y to the transformed predicted values of $\eta[\eta = f(\mu)]$. The Greek letter η (eta) is typically used to denote a linear predictor. The link function converts the expected value Y (i.e., $\mu = E[Y]$) to the linear predictor η. The link function therefore provides the relationship between the linear predictor and the mean of the distribution function. In the case of a normal distribution, no transformation is needed because when the canonical link function (identity) is applied, the same value results; that is, $\eta = \mu$.
- A *linear structural model* that relates the transformed predictor η of Y to a set of Level-1 predictors ($q = 1, 2,..., Q$):

$$\eta = \beta_0 + \beta_1 X_1 + \beta_2 X_2 + \cdots + \beta_Q X_Q. \tag{3.2}$$

The transformed linear predictor therefore incorporates the information about the independent variables into the model. Readers will note that, in Equation 3.2, there is no residual variance (ε) describing errors in predicting the outcomes. For a continuous outcome, residuals are assumed to have a mean of 0 and a constant variance to be estimated. As we have noted, however, in the case of a dichotomous outcome, the variance is determined when the expected value (μ) is known; that is, its variance is not independent of μ. Because the Level-1 variance depends on the expected value of Y, it is not entered in the model as a separate term.

When the outcome is continuous, combining the sampling model, identity link function, and the structural model will result in the familiar model for continuous outcomes expressed in Equation 3.1 because there is no need to transform the expected value of the outcome. In this case, there is a residual variance because the underlying probability distribution associated with Y is normal. One of the attractive properties of the GLM, therefore, is that it allows for continuous outcomes and discrete outcomes to be subsumed under a single framework. It is possible to estimate models where the underlying data are normal, inverse Gaussian, gamma, Poisson, binomial, negative binomial, and geometric by choosing an appropriate link function (Hilbe, 1994; Horton & Lipsitz, 1999).

Outcome Probability Distributions and Link Functions

As this discussion suggests, the GLM facilitates the separation of the probability (or sampling) distribution from the link function, which provides flexibility in describing how expected values of Y are distributed and in choosing appropriate nonlinear functions to represent expected responses of Y. We further assume that the transformed mean follows a linear model.

In practice, however, certain link functions are typically used with particular probability distributions for which sufficient statistics exist. These are often referred to as *canonical* (or natural form) link functions (see McCullagh & Nelder, 1989, for further discussion). A probability density function defines the exponential family of probability distributions. Because the link function is needed to express the relationship between the linear predictor η and the response mean μ, for each probability distribution in the exponential family there exists a special link function, the canonical link, for which θ (the canonical scale parameter in the probability density function) equals η.

The canonical link function, therefore, has some desirable mathematical properties for each particular distribution; one is the use of its inverse link function to transform values of η back to μ $[g^{-1}(E(Y)) = E(Y) = \mu]$. The canonical function tends to simplify the calculations in estimating the model's parameters. The canonical links expressed in terms of the mean parameter of the probability distribution are provided in the next section.

We note that the canonical link functions for each distribution are not the only possible choices, but they are common choices. By making it possible to select several different appropriate probability distributions with different link functions, both the GENLIN and GENLIN MIXED routines in IBM SPSS provide a flexible analytic framework for defining appropriate single-level and multilevel models, respectively, for various categorical (and continuous) outcomes. This type of user-specified link function and error distribution choice can be obtained through using a "custom" specification procedure, which provides several more options rather than the commonly used canonical link function for a particular sampling distribution. We note that some link functions are not appropriate for particular distributions (e.g., because the expected value of μ may be restricted to a particular range, such as 0 to 1).

Continuous Scale Outcome

When the outcome is a continuous (scale) variable, the error distribution is specified as *normal*. The normal distribution describes continuous variables whose values are symmetrically distributed (i.e., bell-shaped distribution) about a central (mean) value with constant variance (σ^2). The structural model linking predicted values of η to the set of predictors is a linear regression equation similar to Equation 3.2, and the link function that brings together the expected values of $Y(\mu)$ and the predicted values of η is referred to as *identity* (i.e., $\eta = \mu$), so no transformation is necessary.

Positive Scale Outcome

When a response variable is continuous, but values are all positive and skewed toward larger positive values, a second type of scale outcome specifies the sampling distribution as *gamma*. The canonical link function is referred to as *inverse* or reciprocal ($\eta = \mu^{-1}$). We note that a canonical link function for a particular distribution is a link function expressed in a form where the location of the observations of Y is related to the mean of the distribution. In the case of the gamma distribution (which ranges from 0 to ∞), however, because the linear predictor η may be negative, using the inverse link function could provide a negative value for the mean (which would lie outside the permitted range).

Therefore, a noncanonical link function is used. The default link function, then, is log [$\eta = \log(\mu)$] for the gamma distribution. The identity and power [$f(x) = x^a$, if $a \neq 0$; if $a = 0$, f(x) = $\log(x)$] link functions can also be used with the gamma distribution. The identity link function is the same as a power link function where the exponent = 1. One example of this type of outcome might be the amount of money paid out in bodily injury insurance claims resulting from car accidents (IBM SPSS, 2010). Because, for the gamma distribution, the domain of the canonical link function is not the same as the permitted range of the mean, if the data value on the dependent variable is 0 or less (or is missing), the case will not be used in the analysis. This is because if the linear predictor were to take on these values, it might result in an impossible negative mean.

Dichotomous Outcome or Proportion

When the outcome Y is dichotomous or a proportion, the probability distribution is specified as *binomial* (μ, n), with mean μ and the number of trials represented as n. The mean can be interpreted as the probability of success. For this outcome, the focus is on determining the proportion of successes (π) in a sequence of n independent trials or events. In other words, in a binomial distribution, the probability of success does not change over successive trials. The expected value of Y will then be

$$E(Y|\pi) = n\pi, \tag{3.3}$$

and the variance will be

$$Var(Y \mid \pi) = n\pi(1 - \pi). \tag{3.4}$$

A special case of the binomial distribution is the Bernoulli trial—that is, an experiment with only two possible outcomes and one event (i.e., all $n = 1$), which represents the most simplified probability distribution for categorical outcomes. The probability of success—that is, the event coded 1[$P(Y = 1)$] occurs—is typically denoted by π in the population (and p in a sample). The probability that the event coded 0[$P(Y = 0)$] occurs is then $1 - \pi$. For example, if the probability of being proficient is 0.60 and the probability of being not proficient is 0.40 $(1 - 0.60)$, then the odds of success (or being proficient) are defined as 0.60/0.40 or 1.5 to 1. The expected value for the distribution (from Equation 3.4) is simply $\mu = \pi$, and the variance (from Equation 3.4) will be $\sigma^2 = \pi(1 - \pi)$. We can quickly see that when the mean proportion is 0.40 (as in Table 3.2), the variance will then simply be 0.40(0.60) = 0.24.

For proportions and dichotomous data, the logit link function (i.e., the canonical link function for the binomial distribution) is most often used because it represents a fairly easy transformation of the expected scores of Y by taking the log of the probabilities of each event occurring. More specifically, the expected values of Y, for a given X, can be transformed so that they are constrained to lie within a particular interval. The logit link function assumes a dichotomous dependent variable *Y* with probability π, where, for individual *i*, we have the ratio of the probability of the two events occurring:

$$\eta = \log\left(\frac{\pi}{1-\pi}\right), \tag{3.5}$$

where log refers to the natural logarithm.[1]

The logit coefficient (η) is the log of the odds of the event coded *Y* = 1 as opposed to *Y* = 0. For example, if the expected probability (π) of being proficient is 0.5, then the odds of success are 1 [0.5/(1 − 0.5) = 1], and the corresponding natural logarithm is 0; that is, log(1) = 0. If the probability of being proficient is greater than 0.5, the odds of success are greater than 1 and the log odds will be positive. Consider a case in which the probability of being proficient is 0.9. The odds of success are 9 [0.9/(1 − 0.9) = 9], and log(9) = 2.197. Conversely, if the probability of being proficient is less than 0.5, the odds of success will be less than 1.0 and the log odds will be negative. In this case, if the probability of being proficient is 0.2, the odds of success are 0.25 [0.2/1 − 0.2) = 0.25], and log(0.25) = −1.383. As this discussion suggests, logits take on any negative value when the probability is less than 0.5 and any positive value when the probability is greater than 0.5. Therefore, although the predicted value for the transformed outcome η can take on any real value, the probability *Y* = 1 will always vary between 0 and 1.

Taking the log of the odds of success provides a means of representing the additive effects of the set of predictors on the outcome. The structural model can then be expressed as a linear logit equation, where the log odds for the likelihood of individual *i* being proficient in reading can be expressed in terms of the underlying variate (η), which is defined by the logit transformation of $\pi(1 − \pi)$ as follows:

$$\eta = \log\left[\frac{\pi}{1-\pi}\right] = \beta_0 + \beta_1 X_1 + \beta_2 X_2, \tag{3.6}$$

or simply the following (Hox, 2010):

$$\pi = \text{logistic}(\beta_0 + \beta_1 X_1 + \beta_2 X_2). \tag{3.7}$$

The intercept β_0 is the value of η when the value of all independent variables is zero. In a model with one predictor, the slope coefficients (β) can be interpreted as the predicted change in the natural log of the odds that *Y* = 1 (versus *Y* = 0) for a one-unit increase in *X* (Azen & Walker, 2011).

We illustrate this in Figure 3.6 using a linear (identity link function) and logistic (logit link function) regression model to predict students' likelihood to persist (coded 1) versus not persist (coded 0), given their socioeconomic status (SES). One can see the predicted values of *Y* in the linear regression model, given different levels of student SES, extend beyond the range of allowable values, and in fact the intercept would cross the *Y* axis at some negative value. In contrast, the logit link transforms the expected values of *Y* such that their predicted probabilities, given SES, will lie within the boundaries of 0 and 1. Because student SES is a z-score, we note in the figure that the probability of persisting is about 0.5 for an individual who is −0.8 of a standard deviation below the sample mean.

We can also notice that a characteristic of a logistic function is that the slope is the steepest halfway between 0 and 1 (π = 0.5), indicating the maximum effect of a unit change in SES on the probability *Y* = 1. This value of *X* is sometimes referred to as the *median effective* level (Azen & Walker, 2011). Because the relationship between the probability that *Y* = 1 and SES is nonlinear, the change in the predicted probability for a unit increase in *X* is not constant,

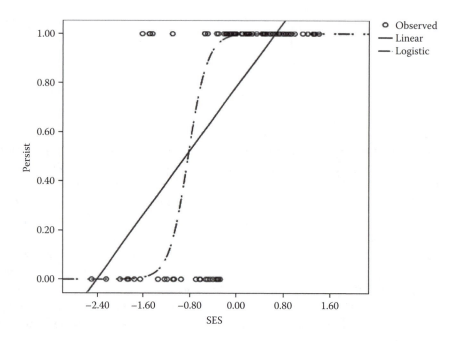

FIGURE 3.6 Student probability to persist as a function of SES.

but instead varies depending on the level of X (Azen & Walker, 2011). Notice, for example, in the figure that the effect of a unit change in SES will lessen as the probability is closer to 1 or 0.

Again, in Equation 3.6 we call readers' attention to the fact that there is no residual variance term. Because the variance in a Bernoulli distribution (where trials = 1) is a function of the population proportion π (or mean), the Level-1 variance cannot be estimated separately. In most software programs (including IBM SPSS), a scale factor of 1.0 is used to define the scale of the continuous variate η. As Hox (2010) notes, setting this scale factor to 1.0 makes the assumption that the Level-1 errors follow the binomial error distribution exactly.

Overdispersion (i.e., when the observed variance is larger than would be expected) or underdispersion, however, can result in situations where there are extreme outliers, a complete level left out of the multilevel analysis, or extremely small groups (Hox, 2010; McCullagh & Nelder, 1989). In these cases, specifying a different scale factor coefficient or allowing the program to estimate the scale factor can correct the standard errors and improve the model's fit (Hox, 2010).

It is also important to keep in mind that the scale factor for the variance is just used to establish a metric for η. The variance of a logistic distribution with a scale factor of 1.0 is $\pi^2/3$, or approximately 3.29; in this case we are using $\pi = 3.14$ (Evans, Hastings, & Peacock, 2000). We note that this within-group variance term can be used in estimating the proportion of variance in a dichotomous or ordinal outcome that lies between groups [i.e., referred to as the intraclass correlation (ρ)]:

$$\rho = \frac{\sigma^2_{Between}}{\sigma^2_{Between} + 3.29} \tag{3.8}$$

When models are built from a simple intercept (or no predictors) model to more complex models containing several added predictors, it is challenging to determine how much variance is being accounted for in successive models because the variance in the underlying outcome η is rescaled to 1.0 each time. In two-level models, this affects variances at Level 2 as well, making it difficult to examine changes in variance accounted for between successive models in a manner similar to multiple regression. To do so requires adjusting the series of models to be on a constant scale of measurement (see Hox, 2010, or Snijders & Bosker, 1999, for further discussion of this issue).

We also note that the natural log changes linearly with a unit increase in X, but the log odds is not an easy metric to interpret. One advantage of the logistic regression model is that the log odds coefficients can be easily transformed into odds ratios (e^{β}), where $e = 2.71828$. Odds ratios express the change in the odds of persisting or not persisting associated with a unit change in the predictor X, while holding the other variables constant. Odds ratios, which should not be confused with the odds of success, can therefore be used to estimate how strongly a predictor is associated with the outcome.

Log odds coefficients can also be converted back to a predicted probability through exponential (exp) transformation. In particular, the inverse transformation of the logit link function is a logistic function of the form

$$\pi = \frac{\exp(\eta)}{1 + \exp(\eta)}. \tag{3.9}$$

For example, if the log odds for η are 0.8, the predicted probability will be $e^{0.8}/(1 + e^{0.8})$, which is $2.23/(1 + 2.23)$ or 0.69. The predicted probability of $Y = 1$ would then be 0.69. Moreover, because of the mathematical relationship, $e^{\eta}/(1 + e^{\eta}) = 1/(1 + e^{-\eta})$, the predicted probability can also be expressed as

$$\pi = \frac{1}{1 + e^{-(\beta_0 + \beta_1 x_1 + \beta_2 x_2)}}, \tag{3.10}$$

where the βs are logistic regression coefficients for two covariates predicting η from Equation 3.6.

In the case where the value of the regression equation predicting η is again 0.8, the predicted probability will be $1/(1 + e^{-(0.8)})$, which is $1/1.445$ or 0.69. We will also make use of variations of Equations 3.9 and 3.10 when predicting probabilities for multinomial and ordinal outcomes. As we suggested previously, applying the logit transformation (and its inverse link function) assures that whether the predicted value for η is positive, negative, or zero, the resulting predicted probability π will lie between 0 and 1. We illustrate this relationship in Figure 3.7, where we summarize predicted values for η on the Y axis and predicted probabilities that $Y = 1$ on the X axis.

We note that the nonlinear relationship between X and the expected value of Y is summarized in Figure 3.8, where we provide predicted probabilities $\pi = 1$ for values of student SES. Readers will notice that the predicted values have been constrained to be larger than 0 but less than 1. We can observe that as student SES increases, the probability that an individual will persist increases. Although the resulting logit coefficient is a linear function of X, the probability $Y = 1$ is a nonlinear S-shaped function (cumulative logistic distribution), as shown in Figure 3.8. The predicted probabilities approach but never reach or exceed the boundaries. As the figure illustrates, the logit regression model provides a more realistic model for the predicted probabilities than a linear regression model would provide. Usually, however, one does not get a full S-curve across a complete range of X, as shown in the figure. Typically, there are only partial S-curves, which represent the range of values of X in a given data set.

The probit (probability unit) link function [$\eta = \Phi^{-1}(\pi)$], where Φ^{-1} is the inverse standard normal cumulative distribution function, is another common S-shaped function favored by some researchers because it is related to the normal distribution. The probit link function is based on the theoretical assumption that a *continuous* scale underlies a dichotomous variable. Theoretically, this probability distribution makes sense when we assume that a threshold divides an underlying latent variable η into the observed dichotomies (e.g., passing versus failing a course), although for practical purposes, there is often little difference between the logit and probit link functions.

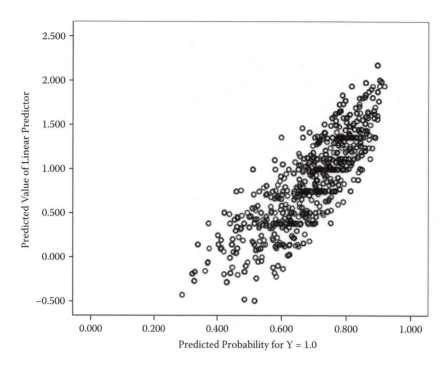

FIGURE 3.7 Scatterplot of predicted linear values of η and predicted probability *Y* = 1.

With population proportions between 0.1 and 0.9, the two link functions will yield virtually identical results, but in a different metric, when the random component is dichotomous and assumed to follow a binomial distribution (Hox, 2010). The *p* values for associated significance tests will also be very similar. For more extreme proportions, the logit transformation may work better because it spreads the population proportions close to 0 or 1 over a wider range of the transformed scale (Hox, 2002). We illustrate the probit and logistic links in more detail in Chapter 4.

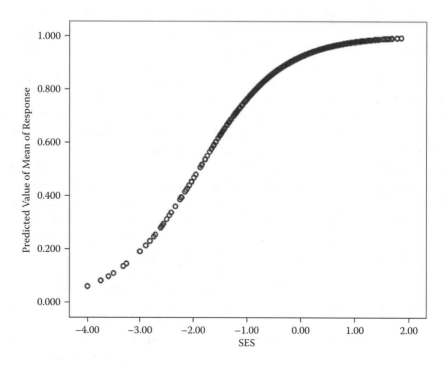

FIGURE 3.8 Predicted probabilities of student persistence for levels of SES.

The logistic regression and probit regression models primarily differ in the scaling of the underlying latent outcome variable η. In a probit model, the observed probabilities (π) are transformed into a z-score that corresponds to the value of the standard normal curve below which the observed proportion of the area is found (Azen & Walker, 2011). The variance in a probit model is standardized at 1.0. In contrast, the variance of the logistic distribution is larger ($\pi^2/3$, or approximately 3.29). This makes the standard deviation about 1.8 for the logistic distribution and 1.0 for the probit distribution. This difference in scaling makes the regression coefficients and standard errors correspondingly larger in the logit model by about 1.6 to 1.8 times (Hox, 2010).

Some other possible link functions that can also be considered with the binomial distribution include the negative log–log [$f(x) = -\log(-\log(x))$]; complementary log–log [$f(x) = -\log(-\log(x))$], which can be used for interval censored survival models; and the log complement [$f(x) = \log(1 - x)$]. In addition, the log [$f(x) = \log(x)$] and power [$f(x) = x^a$, if $a \neq 0$; if $a = 0$, $f(x) = \log(x)$] link functions can be used with the binomial distribution.

Nominal Outcome

For nominal outcomes with C unordered categories ($c = 1, 2,\ldots, C$), the sampling distribution is referred to as multinomial, which represents an extension of the binomial distribution and assumes a set of counts in three or more categories (Agresti, 2007). In a multinomial distribution, the cumulative probabilities of each possible outcome ($\pi_1, \pi_2,\ldots, \pi_c$) can be expressed such that their sum is 1. The probability that an individual falls in a particular category c is π for categories $c = 1,\ldots, C$, shown as

$$P(Y_c = 1) = \pi_c. \tag{3.11}$$

A set of dummy variables is constructed such that $Y_c = 1$ if $\pi = c$ and $Y_c = 0$ otherwise. For a nominal variable with three categories, for example, the predictors have different associations with the probabilities of individuals being in the three different categories of the response variable (Y):

$$
\begin{aligned}
P(Y_1 = 1) &= \pi_1 \\
P(Y_2 = 1) &= \pi_2 \\
P(Y_3 = 1) = \pi_3 &= 1 - \pi_1 - \pi_2.
\end{aligned}
\tag{3.12}
$$

Because $\pi_3 = 1 - \pi_1 - \pi_2$, the last category is redundant. The last category typically becomes the reference category in multinomial analyses. Therefore, there are actually only $C - 1$ probabilities required to specify the multinomial outcome. In the case with three categories, then, there will be two probabilities required to specify the possible outcomes. The expected value of Y_c is

$$E(Y_c | \pi_c) = \pi_c, \tag{3.13}$$

and the variance is then

$$\mathrm{Var}(Y_c | \pi_c) = \pi_c(1 - \pi_c). \tag{3.14}$$

Once again, because the variance depends on the predicted value of π, it cannot be modeled as a separate term. Similar to the binomial distribution, the variance is set to a scale factor of 1.0, which suggests that it does not need to be interpreted (Hox, 2002).

The most common link function for a multinomial outcome is a generalized logit link, where η_c is then the predicted log odds of being in category c ($c = 1,\ldots, C-1$) versus a reference category C:

$$\eta_c = \log\left(\frac{\pi_c}{\pi_C}\right), \tag{3.15}$$

where

$$\pi_C = 1 - \sum_{c=1}^{C-1} \pi_c. \tag{3.16}$$

As we noted, when the outcome is nominal, one category is chosen as the reference category, with the default in SPSS being the last category. Separate log odds coefficients are estimated for each response category of the dependent variable with the exception of the reference category, which is omitted from the analysis. In the case where Y has three response categories, this amounts to equations for the probability that Y is in Category 1 versus Category 3 (i.e., the reference group C), and the probability Y is in Category 2 versus Category 3.

The structural model can then be specified to predict the log of the odds of outcome c occurring relative to outcome C (the reference category) occurring for individual i as follows:

$$\eta_{ic} = \log\left(\frac{\pi_c}{\pi_C}\right) = \beta_{0(c)} + \sum_{q=1}^{Q} \beta_{q(c)} X_q \tag{3.17}$$

for $c = 1,\ldots, C-1$ outcome categories.

Beta coefficients β_q ($q = 1, 2,\ldots, Q$) can be interpreted as the increase in log odds of falling into category c versus category C resulting from a one-unit increase in the qth covariate, holding the other covariates constant. For three unordered categories, therefore, there will be two Level-1 equations for η_{1i} and η_{2i} as follows:

$$\eta_{1i} = \beta_{0(1)} + \sum_{q=1}^{Q_1} \beta_{q(1)} X_q \tag{3.18}$$

$$\eta_{2i} = \beta_{0(2)} + \sum_{q=1}^{Q_2} \beta_{q(2)} X_q. \tag{3.19}$$

As noted previously, other possible link functions that can be used with nominal outcomes include log and power.

Ordinal Outcome

Ordinal outcomes accommodate an ordered set of categories defining the dependent variables. The model is similar to the multinomial outcome except that because the outcome categories are assumed to be ordered, the relationships can be captured in one set of estimates. For example, a Likert scale implies the concept of "greater than" or "less than" but does not actually constitute continuous data. As we noted previously, ordinal response patterns are often treated as if they were continuous, but there can be consequences resulting from assuming that such ordered categorical responses actually constitute interval data (Olsson, 1979).

Because ordinal data are not continuous, using a linear modeling technique such as multiple regression can create bias in estimating the model's parameters, standard errors, and fit indices because the necessary assumptions may not be met when the observed data are discrete. The parameters may be underestimated and standard errors can be downwardly biased (Muthén & Kaplan, 1985, 1992). The major issue to overcome with ordinal data is the calculation of a proper set of correlations for variables measured on ordinal scales because the Pearson product-moment (r) correlation is not an optimal measure of association when there are only a few ordered categories. We note that, currently, there are often better options for ordinal outcomes than assuming that they are continuous.

Similarly to the multinomial distribution, the individual falls into category c and there are C possible categories ($c = 1, 2,..., C$). The difference, however, for an ordinal outcome is that the categories are ordered. Ordinal response models predict the probability of a response being at or *below* any given outcome category. We can predict the probability of persisting or being below; that is, the outcome is generally defined as the probability of being at or below the cth category, as denoted by

$$P(Y \leq c) = \pi_1 + \pi_2 + \cdots + \pi_c. \tag{3.20}$$

Because this is a *cumulative probability* formulation (also referred to as a proportional odds model), we actually only fit the first $C - 1$ categories ($Y_1,..., Y_{C-1}$) because the cumulative probability must always be 1.0 for the set of all possible outcomes (Azen & Walker, 2011). For example, for an ordinal outcome with three ordered response categories, we obtain the following cumulative probabilities:

$$P(Y_1 = 1) = \pi_1$$

$$P(Y_2 = 1) = \pi_1 + \pi_2 \tag{3.21}$$

$$P(Y_3 = 1) = \pi_1 + \pi_2 + \pi_3 = 1$$

As Equation 3.21 suggests, the probability $Y_3 = 1$ is redundant because once the probabilities of the first and second categories are estimated, the third probability will follow by subtracting them from 1. As Azen and Walker (2011) note, the definition of the cumulative probability model allows us to compute the probabilities for a specific category as the difference between two adjacent probabilities:

$$P(y = c) = \pi_c = (\pi_1 + \pi_2 + \cdots + \pi_c) - (\pi_1 + \pi_2 + \cdots + \pi_{c-1}) = P(Y \leq c) - P(Y \leq c - 1). \tag{3.22}$$

The link function links the expected values of the cumulative probabilities (π_c) to the predicted values of η_c. The logit link function for the cumulative logits is defined as follows:

$$\eta_c = \log\left(\frac{P(Y \leq c)}{P(Y > c)}\right) = \log\left(\frac{\pi_c}{1 - \pi_c}\right), \tag{3.23}$$

where $c = 1, 2,..., C - 1$. The structural model assuming a proportional odds formulation may be defined as follows:

$$\eta_c = \beta_0 + \sum_{q=1}^{Q} \beta_q X_q + \sum_{c=2}^{C} \theta_c, \tag{3.24}$$

where $\eta_c = \log(\pi_c/(1 - \pi_c))$.

Thresholds (similarly to intercepts) specify the relationship between the underlying latent variable and the categories of the outcome, with one less threshold than the number of ordered categories of the outcome (Hox, 2010). Equation 3.25 indicates that a series of thresholds (θ_c), beginning with the second threshold ($c = 2$), separate the ordered categories of the outcome. The lowest threshold for η_c is just the intercept β_0. Once again, there is no residual variance term. In the case of three ordered categories, therefore, we would have the following two models:

$$\eta_1 = \beta_0 + \sum_{q=1}^{Q} \beta_q X_q$$

$$\eta_2 = \beta_0 + \sum_{q=1}^{Q} \beta_q X_q + \theta_2.$$

(3.25)

The characteristic feature of the cumulative odds model is that the two expressions share the same slope for X_q. This means that the corresponding conditional probability curves expressed as functions of any predictor variable are parallel and only differ due to the thresholds (and not c). We note that this assumption may not always hold empirically. Various alternative formulations are well known for single-level analyses, but they have not been widely implemented in multi-level software (see Azen & Walker, 2011, and Hox, 2010, for further discussion).

For three ordered categories, the predicted log odds can also be converted to predicted probabilities as follows:

$$P(Y = 2 \mid x) = \frac{1}{1 + e^{-(\beta_0 + \beta x + \theta_2)}}$$

$$P(Y = 1 \ or \ 2 \mid x) = \frac{1}{1 + e^{-(\beta_0 + \beta x)}}$$

(3.26)

$P(Y = 2 \mid x)$ denotes the probability that y equals 2 (the highest category) for a given value of x. The other event $P(Y = 1 \ or \ 2)$ denotes the probability that y is 1 or 2 (the highest two categories) for a given value of x.

We can also use the cumulative probit link function, which is defined as

$$\eta_c = \Phi^{-1}(\pi_c),$$

(3.27)

where Φ^{-1} is the inverse standard normal distribution and c is a set of categories ($c = 1, 2,..., C - 1$).

In a probit ordinal regression model, the observed probabilities are replaced with the value of the standard normal curve below which the observed proportion of the area is found. The probit regression model can be written as

$$\pi_c = \Phi(\beta_0 + \beta_1 X_1 + \beta_2 X_2 + \cdots + \beta_q X_q).$$

(3.28)

This formulation, however, specifies the model in the form of the inverse link because it is specified in terms of the probability π_c. Using the inverse normal probability function Φ^{-1}, the model can be written as follows:

$$\eta_c = \Phi^{-1}(\pi_c) = \beta_0 + \beta_1 X + \beta_2 X + \cdots + \beta_q X_q,$$

(3.29)

where $\Phi^{-1}(\pi_c)$ is the value of η_c such that the area under the curve less than η_c is π_c.

TABLE 3.5 **Summary of Distributions and Link Functions**

	Distributional assumption	Normal	Binomial	Multinomial	Gamma	Inverse Gaussian	Negative binomial	Poisson
Link Function	Identity	X	X		X	X	X	X
	Log	X	X		X	X	X	X
	Power	X	X		X	X	X	X
	Log-complement		X					
	Logit		X	X				
	Negative log–log		X	X				
	Probit		X	X				
	Complementary log–log		X	X				
	Cumulative cauchit			X				

A probit coefficient (β) describes a one-unit increase in a predictor (x) corresponding with a β standard deviation increase in the probit score. The probit model also makes the same assumption of parallel regression lines separated by the thresholds of the underlying latent variable.

The GLM user's guide (IBM SPSS, 2010) suggests that the complementary log–log link is often a good choice when the cumulative probabilities increase from 0 fairly slowly and then rapidly approach 1. In contrast, if the cumulative probability for lower scores is high and the approach to 1 is slow, the negative log–log link may more appropriately describe the data. A number of other link functions can also be used with ordinal data (see Table 3.5). We note that there is no specific criterion that one can use to decide among appropriate link functions. Sometimes, a little trial and error is necessary. The decision to use canonical link functions is often based on their desirable properties for explaining the outcomes, but, given particular data situations, the analyst may wish to consider possible alternatives that appear to fit the data better.

Count Outcome

Count outcomes represent situations where the values of Y are non-negative integers (0, 1, 2,…, N). Data values that are less than 0, or missing, will not be used in the analysis. When the counts represent relatively rare events, a Poisson distribution is often used. The Poisson distribution can be thought of as an approximation to the binomial distribution for rare events where the probability of π is very small and the number of observations is relatively large. Counts in such distributions are likely to be positively skewed because they may have many zeros and considerably more low values than high values. The period of time for which the counts are observed must be specified to occur within a fixed interval (or exposure), with the events occurring at a constant rate (Hox, 2010).

An example might be how many courses students fail during their ninth grade year in high school. A considerable number of students might not fail any courses during this period. In one example, we might assume that Y is the number of courses individual i fails during ninth grade year, such that the time interval $n = 1$. The exposure, however, does not have to be time related. We refer to Y as having a Poisson distribution with exposure n and event rate λ:

$$Y \mid \lambda \sim P(n, \lambda). \tag{3.30}$$

The expected number of events for an individual is then the following:

$$E(Y \mid \lambda) = n\lambda. \tag{3.31}$$

The variance is then the mean of the $n\lambda$ parameter:

$$Var(Y|\lambda) = n\lambda. \tag{3.32}$$

Because the exposure is the same for every individual in a Poisson distribution (i.e., the number of courses failed during a 1-year period for each person in each group), n is fixed to 1. The expected value of Y then is just the event rate λ. The Poisson distribution therefore has only one parameter, the event rate λ. This suggests that as the expected value of Y increases, so does the expected variability of the event occurring.

The link function is the *log* link [$(\eta = \log(\mu)$]. We can represent this as follows:

$$\eta = \log(\lambda). \tag{3.33}$$

Because the log of $\lambda = 1$ is 0, when the event rate is less than 1, the log will be negative, and when the event rate is greater than 1, the log will be positive. The structural model that produces a predictor η of the outcome Y is the same as Equation 3.2. Using its inverse link function, the predicted log rate can be converted to the expected mean event rate as follows:

$$\hat{\lambda} = e^\eta. \tag{3.34}$$

Similar to the dichotomous outcome, then, we can describe the log of the event rate in terms of a "rate ratio" or "relative rate." Therefore, whether η is positive, zero, or negative, λ will be non-negative. For example, if males are observed to have a higher likelihood of failing a course ($\eta = 0.30$), then they are 1.35 times ($e^{0.30} = 1.35$) more likely to fail a course than females are.

One assumption of the Poisson model (from Equation 3.32) is that the mean and variance of the outcome are approximately equal. When the conditional variance exceeds the conditional mean, it is referred to as *overdispersion* (Azen & Walker, 2011). As we have noted previously, the variance is scaled to 1.0, which assumes no overdispersion. What this means is that the observed count data are assumed to follow the theoretical Poisson distribution very closely. In practice, however, the Poisson model often underestimates the variance, which results in overdispersion and a likelihood to underestimate standard errors. This can lead to more significant findings than would be expected because significance tests are based on the ratio of the parameter estimate to its standard error. Another important assumption of the Poisson distribution is that the events are independent and have a constant mean rate (Hox, 2010). In our course failure example, therefore, we would have to assume that failing a math class is independent from failing an English or social studies class. If there is some tendency for events not to be independent for subjects, they may not follow the theoretical distribution very closely, which results in overdispersion.

Negative Binomial Distribution for Count Data

When assumptions of the Poisson distribution are violated (e.g., limited range of count values, overdispersion is present at Level 1), the *negative binomial* (NB) distribution with *log* link can also be used to estimate the model. The NB model provides an additional random variance parameter (ε), which is used in estimating the mean and variance as a correction term for testing the parameter estimates under the Poisson model:

$$\lambda = \exp(\eta + \varepsilon) = \exp(\eta) + \exp(\varepsilon). \tag{3.35}$$

The presence of the residual error term increases the variance compared to the variance implied in the Poisson distribution (Hox, 2010). This dispersion parameter can be either estimated or set to a fixed value (often 1.0).

The NB model has the advantage, therefore, of always having a variance that is greater than the Poisson model, unless the parameter is fixed at the same value as the Poisson distribution. The parameter estimates in the model are not affected by the value of the added dispersion parameter, but the estimated covariance matrix is inflated by this factor, which tends to correct the bias present in the standard errors. Ignoring overdispersion that is present can lead to underestimation of standard errors. Because adjusting for overdispersion tends to increase the standard errors, this makes significance tests more conservative.

For single-level models, one can examine the difference between Poisson and negative binomial models based on the difference in log likelihoods and information indices (AIC and BIC) to determine which might be more advantageous in a particular situation if overdispersion may be present. When there is no overdispersion, the two models are the same. We note that, with repeated-measures data, it is important to take into consideration the correlation between consecutive counts. This necessitates using generalized estimating equations (GEE) or GENLIN MIXED to adjust for the correlated events. We discuss this further in Chapter 5.

Events-in-Trial Outcome

In many situations, models with count data need some sort of mechanism to deal with the fact that counts can be made over different periods of observation. The Poisson regression model can be further extended by considering varied exposure or trials (n). As we noted previously, an example would be a situation in which the response variable was the number of accidents an individual had over a 3-year period and the exposure could be the individual's years of driving experience (IBM SPSS, 2010). A second example might be the number of course failures a student accumulates during college, with the level of exposure being the number of units attempted or the number of semesters enrolled in an institution.

Count models account for these differences by including the log of the exposure variable in the model (the log is used so it is in the same metric as η). Variable exposure for different subjects is specified as an offset variable, and usually the coefficient is constrained to equal 1.0 (Hox, 2010; McCullagh & Nelder, 1989). The use of exposure is superior in many instances because the length of exposure is used to adjust counts on the response variable. It allows the analyst to include various kinds of rates, indexes, or per-capita measures as predictors in examining count outcomes.

When there is interest in modeling varied exposure, the NB distribution can be used. As we noted previously, this distribution can be thought of as the number of trials it takes to observe Y successes. It should be noted that the value of the NB distribution's ancillary parameter can be a number greater than or equal to 0 (i.e., it is set to 1.0 as the default). As noted in Equation 3.35, this has the effect of adding an error term to the model, which can be useful in dealing with overdispersion. As Hox (2010) suggests, the NB model is similar to adding a dispersion parameter to a Poisson distribution (i.e., when the ancillary parameter is set to 0, this distribution is the same as a Poisson distribution). The log link function is used in the NB regression model $[f(x) = \log(x/(x + k^{-1}))]$, where k is the ancillary parameter of the NB distribution. One possible application is in multilevel models where very small group sizes (under five) can lead to overdispersion (Wright, 1997). For single-level models, in GENLIN the analyst can choose to have the program estimate the value of the ancillary parameter. In GENLIN MIXED, however, the analyst currently does not have the option to specify different ancillary values when using a negative binomial distribution. We illustrate this type of model in Chapter 7.

Other Types of Outcomes

Other possible statistical models can be examined, such as those that combine properties of continuous (non-negative real values) and discrete distributions (e.g., a probability mass at a particular value such as zero, referred to as Mixtures in IBM SPSS) and other custom models (i.e., when the analyst can specify her or his own combination of error distribution and link function). The *Custom* command allows the analyst to specify her or his own combination of distribution and link function. Interested readers can consult the user's guide for further information about handling categorical outcomes in IBM SPSS.

In Table 3.5 we summarize the range of distributions and associated link functions that are available within the GLM and GLMM modules in IBM SPSS. We note that several link functions change slightly when used with the binomial or multinomial distributions. More specifically, for the multinomial distribution, the link functions are referred to as *cumulative,* which reflects the cumulative (or proportional) odds assumption. For example, the logit link is used with the binomial distribution, and the cumulative logit link is used with the multinomial distribution. We also note that for multinomial models, which can be specified using GENLIN MIXED (i.e., where the outcome is specified as nominal rather than ordinal), only the logit link (which is referred to as *generalized* for the nominal outcome) is available. Finally, we note in passing that the Tweedle distribution (which is appropriate for variables that are mixtures but is not currently available in GENLIN MIXED) supports the same link functions as the Poisson distribution for single-level analyses.

Estimating Categorical Models With GENLIN

Model estimation attempts to determine the extent to which a sample covariance (or correlation) matrix is a good approximation of the true population matrix. In general, confirmation of a proposed model relies on the retention of the null hypothesis—that is, that the data are consistent with the model (Marcoulides & Hershberger, 1997). Failure to reject this null hypothesis implies that the model is a plausible representation of the data, although it is important to note that it may not be the only plausible representation of the data. Because GLMs have outcomes that are discrete rather than continuous, they cannot be optimally estimated with the ordinary least squares (OLS) estimation procedures used in multiple regression. Instead, nonlinear link functions are required and therefore model estimation requires an iterative computational procedure to estimate the parameters optimally.

Maximum likelihood (ML) estimation determines the optimal population values for parameters in a model that maximize the probability or likelihood function; that is, the function that gives the probability of finding the observed sample data, given the current parameter estimates (Hox, 2002). Likelihood functions allow the parameters of the model (e.g., π) to vary while holding everything else constant (Azen & Walker, 2011). The likelihood function is obtained by multiplying together the probabilities of each pattern of covariate responses, assuming a specific probability distribution (e.g., binomial, multinomial) of responses. This involves a series of iterative *guesses* that determines an optimal set of weights for random parameters in the model that minimizes the negative of the natural logarithm multiplied by the likelihood of the data.

If one considers that the sample covariance matrix represents the population covariance matrix, then the difference between the two should be small if the model fits the data. The mathematical relationships implied in the model are solved iteratively until a set of estimates is optimized. Arriving at a set of final estimates is known as model convergence (i.e., where the estimates no longer change and the likelihood is at its maximum value). It is important that the model actually reaches convergence, as the resulting parameter estimates will not be trustworthy if it has not. A warning message is provided in the output when a model has failed to converge. Sometimes, increasing the number of iterations will result in a model that converges, but often the failure of the model to converge on a unique solution is an indication that it needs to be changed and reestimated.

As we noted in Chapter 1, the difference in the model-implied matrix and sample matrix is described as a discrepancy function—that is, the actual difference in the two estimates based on a likelihood function. The goal of the analysis is to minimize this function by taking partial derivatives of it by the model parameters with respect to the elements of $S - \hat{S}$. The greater the difference is, the larger the discrepancy function becomes, which implies less similarity between the elements in the two matrices (Marcoulides & Hershberger, 1997). Users can request the log likelihood iteration history as part of the output. The first iteration provides the log likelihood of the unconditional model. The last iteration is the final log likelihood result for the current model when the program reaches convergence. As part of this output, users can also see successive estimates of the independent variables' effects on the outcome. The model log likelihood is calculated based on these estimates of the model's parameters.

The output also includes information about the gradient vector (i.e., the vector of partial derivatives of the log likelihood function with respect to the parameter estimates) and Hessian matrix (i.e., the square matrix of second derivatives of the log likelihood function with respect to the parameter estimates). The variance-covariance matrix of the model parameters is the negative of the inverse of the Hessian (i.e., a type of matrix used in solving large-scale optimization problems within interative estimation methods). The iteration history and related information about the variance-covariance matrix of model parameters can provide clues about problems that occur during the estimation process. Because the actual matrix can be difficult to compute, algorithms that approximate the Hessian matrix have been developed. If the model estimation fails to converge, a warning is issued in the output.

ML estimation produces a model deviance statistic, which is an indicator of how well the model fits the data. It is usually defined as the log likelihood of the current model multiplied by –2 (i.e., –2*log likelihood). This is sometimes abbreviated as –2LL. Successive models, which are referred to as nested if all of the elements of the smaller model are contained in the bigger model, can be evaluated by comparing changes in their log likelihoods. Models with lower deviance (i.e., a smaller discrepancy function) fit the data better than models with larger deviance. As part of the model's iteration history output, users can request either the full log likelihood (with an extra term added that does not affect parameter estimation) or the kernel log likelihood (which is typically used in calculating –2LL). The kernel log likelihood is used in calculating the model deviance with the result multiplied by –2, so the difference of two –2 log likelihood values will have a chi-square distribution. This allows a likelihood-ratio test to be constructed.

Because the likelihood, or probability, of the data can vary from 0.0 to 1.0, minimizing this discrepancy function amounts to finding the estimates of the parameters at which the likelihood (or probability) is maximized (Azen & Walker, 2011). It represents a measure of the likelihood or probability of observing the data without assuming the validity of the hypothesis. We emphasize that each general approach to model estimation rests on a somewhat different set of assumptions; the statistical theory underlying the estimation of categorical models, and multilevel categorical models in particular, is relatively more complex than OLS regression for continuous outcomes because it depends on the type of analysis being conducted, the measurement scale of the dependent variable, and characteristics of the data being analyzed within and between groups.

Categorical models with nested data structures require more complex estimation of model parameters that rely on quasilikelihood approximation or numerical integration. Integration over the random-effects distribution is necessary, which requires approximation. A number of different approaches are described in the literature (e.g., Hedeker, 2005; Hox, 2010; McCullagh & Nelder, 1989; Raudenbush & Bryk, 2002). Because the estimates are only approximate, this makes comparison of successive multilevel models with categorical outcomes more tenuous compared with the single-level case.

GENLIN supports several estimation methods and a considerable amount of output can be requested about specific aspects of the model estimation process for each model. There are many iterative methods for ML estimation in the GLM, of which the Newton-Raphson and Fisher scoring methods are among the most efficient and widely used (Dobson, 1990). The analyst can choose from Newton-Raphson, Fisher scoring (iterative reweighted least squares), or a hybrid approach that uses Fisher scoring iterations before switching to the Newton-Raphson method. If convergence is reached during the Fisher scoring phase of estimation before the maximum number of Fisher iterations is reached, the algorithm continues with the Newton-Raphson method. We note that if the canonical link function is used, then these two methods are the same (McCullagh & Nelder, 1989). Users can obtain this information for a specific model by requesting the model iteration history.

In some situations, the scale parameter method can be used instead. This approach employs ML estimation of the scale parameter with the model effects. A fixed value for the scale parameter can also be specified. The IBM SPSS manual cautions that this option may not be valid for outcomes for some distributions commonly encountered with categorical outcomes (i.e., negative binomial, Poisson, binomial, or multinomial distributions). In one example with count data (Poisson distribution) and overdispersion, however, McCullagh and Nelder (1989) relax the assumption that the scale parameter is usually 1.0 and use the scale parameter method's Pearson chi-square estimate to obtain more conservative estimates of variances and significance levels.

GENLIN Model-Building Features

In this section, we introduce some of the essential features for defining single-level models using GLM (GENLIN). We relied on the IBM SPSS user's GLM guide to describe some of the program options. Readers can consult the guide for further information (IBM SPSS, 2010). The GENLIN routine provides a unified analytic framework for an extended range of cross-sectional categorical outcomes (e.g., count, multinomial, ordinal), and the GEE option provides a framework for examining longitudinal models with categorical outcomes (where occasions of measurement are nested within individuals).

These approaches have in common that the individuals are selected at random, which means they are independent observations. This assumption will, of course, be violated if the individuals are clustered in groups. The degree of bias in estimating relevant model parameters (e.g., standard errors) will depend on the degree of similarity between individuals within groups. In Chapter 4, we provide the similar model specification steps using the GLMM routine in IBM SPSS. This is referred to as GENLIN MIXED. In Chapters 5–7, we also provide illustrations of setting up single-level and multilevel models using the two different routines for various categorical outcomes. Users will generally find that programs are quite similar in terms of setting up models.

As a first step, it is important to make sure that scales of measurement are properly assigned to the variables to be included in the model. If dependent variables are not properly defined within IBM SPSS, the program may not be able to run the proposed model. It is important to note that the coding of the variables and their scale of measurement within the data set are important in terms of understanding the meaning of the output produced. For example, similar to regression modeling, in terms of their measurement scale, dichotomous variables (coded 0, 1) can be referred to as either nominal (unordered categories) or continuous (referred to as *scale*). If they are referred to as continuous, the reference group will be the category coded 0, and the estimates produced will be for the category coded 1. This may or may not be the manner in which the researcher wants the output estimates organized.

As we noted previously, if the dichotomous variable is defined as a factor, by referring to it as nominal, the default reference group will be the category coded 1, so the estimates produced will be for the category coded 0. The logic of specifying the reference category extends to categorical variables with three or more categories. This default "last category" organization is referred to as *ascending* in the program's syntax language. As we noted, variables may be defined as scale (continuous), ordinal, or nominal, as shown in Figure 3.9.

Once the variables in the data set have been defined, single-level models may be constructed using the GENLIN procedure, which encompasses a series of command "tabbed" screen displays. We will present an overview of the key command screens and options; however, for further information, readers may press the "Help" button on the lower right-hand corner of

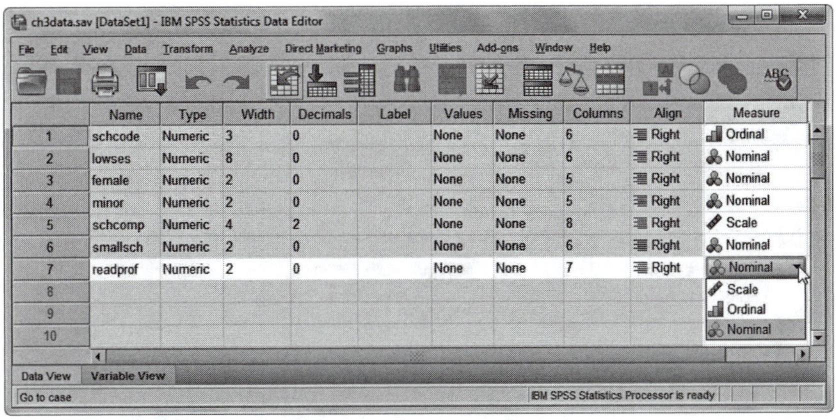

FIGURE 3.9 Defining the measurement setting as nominal for selected variable (*readprof*).

each command screen or at the top right-hand corner of the IBM SPSS Data or Variable View screens. Information may also be found under Generalized Linear Mixed Modeling in the user's guide of the Advanced Module (IBM SPSS, 2010).

Type of Model Command Tab

After defining variables, the generalized linear models option in GENLIN requires the user first to specify the model's distribution and log link. To access the GENLIN feature, go to the IBM SPSS toolbar and select ANALYZE > GENERALIZED LINEAR MODELS > GENERALIZED LINEAR MODELS. The command will cause the *Type of Model* screen to appear (Figure 3.10), which is the default screen when creating a model for the first time.

Distribution and Log Link Function

Specifying the distribution and log link for a model is simplified by options for several common models based on the response type (i.e., scale, ordinal, counts, binary, mixture). Readers may not be as familiar with "mixture" variables, which combine properties of continuous (non-negative real values) and discrete distributions (i.e., they must be numeric, with data values greater than or equal to zero). We note that the measurement level of the dependent variable restricts which distributions and link functions are appropriate for the given analysis and emphasize that the analyst should ensure that the dependent variable is defined as continuous (scale), ordinal, or nominal as intended so that the analysis will be performed properly (refer to Figure 3.9).

GENLIN offers several common distribution and link function models that can be selected. For example, with a dichotomous outcome (reading proficiency), the selection *Binary logistic regression* automatically couples a binomial error distribution with the logit link function.

Custom Distribution and Link Function

If the user selects *Custom*, as shown in Figure 3.10, the program will allow selecting other possible distribution–link function combinations that may be appropriate for a particular modeling situation. The user should select a combination, however, that can be justified by theoretical, methodological, and empirical considerations.

The Response Command Tab
Dependent Variable

The *Response* command tab (Figure 3.11) is used to define the model's dependent (outcome) variable. The outcome should be a single categorical variable that can have any measurement level or an events/trials specification, in which case the events and trials must be continuous.

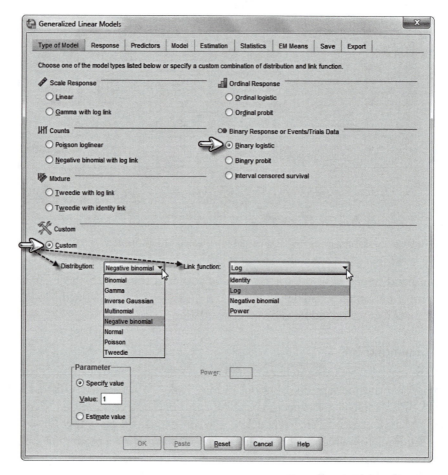

FIGURE 3.10 Specifying a distribution and log link function for a GENLIN model (image modified to display all options).

Reference Category

In the case where the dependent variable has only two values (0, 1), the analyst may specify a reference category (Figure 3.11). The default makes the last (highest valued) category the default, so parameter estimates should be interpreted as relating to the likelihood of the category coded 0. For ordinal multinomial models, the analyst may specify the category order of the response to be ascending, descending, or data (i.e., the first value in the data defines the first category; last value defines the last category).

Number of Events Occurring in a Set of Trials

When the response is a number of events occurring in a set of trials, the dependent variable contains the number of events and the analyst can select an additional variable that indicates the number of trials. If the number of trials is the same for all subjects, then the trials may be specified as a fixed value. It is important to note that the number of trials should be greater than or equal to the number of events for each case. Events should be non-negative integers, and trials should be positive integers.

The Predictors Command Tab

The *Predictors* command tab allows specifying "predictors" to build model effects and to specify an optional offset (Figure 3.12).

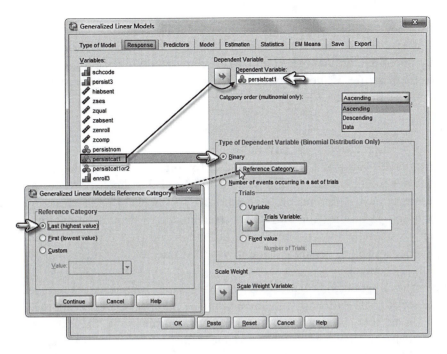

FIGURE 3.11 Defining the dependent variable (*persistcat1*) and default reference category (first, lowest value) for a GENLIN binary logistic regression model (image modified to display all options).

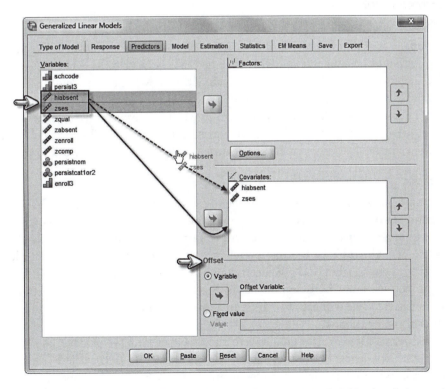

FIGURE 3.12 Defining model predictors and the offset by the "select-and-click" (solid line) or "select-and-drag" (dashed line) method.

Predictors

The predictors refer to the independent variables in explaining the outcome. Categorical variables (e.g., nominal, ordinal measurement) are represented in the program as *factors,* which may be numeric or string. Continuous variables (scale measurement) are referred to as *covariates,* which must be numeric.

Offset

The offset term is a structural parameter. Its coefficient is not estimated, but it is assumed to have a default value of 1 (i.e., the value of the offset is simply added to the linear predictor of the dependent variable). The offset can be useful in Poisson regression models, where each case may have different levels of exposure to the event of interest. For example, when modeling course failures for individual students, there would be an important difference between a student who had one course failure in the first few units attempted in college (a first-time freshman) versus a student who had one course failure in 120 units attempted (a graduating senior).

When the target outcome variable is a count—that is, it represents a number of events occurring in a set of trials—the target field contains the number of events and the user can select an additional field that specifies the number of trials. For example, the number of course failures can be modeled as a Poisson response variable if the number of units the student has attempted is included as an offset term. Alternatively, if the number of trials is the same across all subjects, then trials can be specified to be fixed at some value. It is important to note that the number of trials should be greater than or equal to the number of events for each record. Events should be non-negative integers (i.e., including 0) and trials should be positive integers (1, 2,..., N).

The Model Command Tab

Main Effects

Model effects are defined in the *Main Effects* command tab (Figure 3.13, upper half of the figure). Although the default model is specified as intercept only, a number of different kinds of fixed effects can be defined (i.e., main, interactions). Main effects refer to the direct effect of each variable alone in terms of the outcome. Variables may be added by selecting them and then either (1) clicking the right-arrow button to move them into the *Model* box, or (2) "dragging" them into the *Model* box. For illustration clarity and instructional consistency, we will use the "click the right-arrow button" method in subsequent chapters of the workbook.

Interactions

Interactions between variables are another type of fixed effect (Figure 3.13, bottom half of the figure). Two-way interactions refer to possible pairs of variable effects (e.g., *lowses*female*); three-way interactions refer to possible triplets of variables (e.g., *lowses*female*minor*). Custom terms can also be added to the model by either dragging and dropping variable combinations or building them separately using the *Build Nested Term* dialog window and then adding them to the model. We show both approaches in Figure 3.13, but for illustration clarity and instructional consistency, we will use the "drag-and-drop" method for GENLIN models in subsequent chapters of the workbook.

Nested terms refer to situations where the effect of a factor or covariate is not assumed to interact with the levels of another factor. This is shown as A(B). There are some limitations. First, factors within nested effects must be unique, so if A is a factor, specifying A(A) is invalid. Second, no effect can be nested within a covariate, so if A is a factor and X is a covariate, specifying A(X) is invalid. Interactions, such as higher order polynomial terms for the same covariate (e.g., *age*age*), may also be specified. It should be noted that all factors within an interaction must be unique, such that if A is a factor, A*A would be invalid.

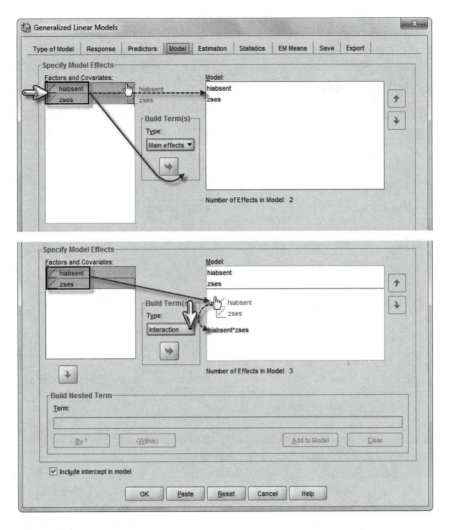

FIGURE 3.13 Adding main effects (*hiabsent, zses*) and cross-level interaction (*hiabsent*zses*) into a model.

The Estimation Command Tab

The *Estimation* command tab (Figure 3.14) allows specifying GENLIN parameter estimation methods and iterations options.

Parameter Estimation

Model estimation methods and initial values for the parameter estimates may be controlled by the following settings:

Method. This option allows choosing from Newton-Raphson, Fisher scoring, or a hybrid method in which Fisher scoring iterations are performed before switching to the Newton-Raphson method. If convergence is achieved during the Fisher scoring phase of the hybrid method before the maximum number of Fisher iterations is reached, the algorithm continues with the Newton-Raphson method.

Scale parameter method. The scale parameter estimation method allows the analyst to choose a fixed value, deviance, or Pearson chi-square option. Maximum likelihood jointly estimates the scale parameter with the model effects. We caution, however, that this option is not valid if the response has a negative binomial, Poisson, binomial, or multinomial distribution. The deviance and Pearson chi-square options estimate the scale parameter from the value of those statistics. Alternatively, a fixed value for the scale parameter may be specified.

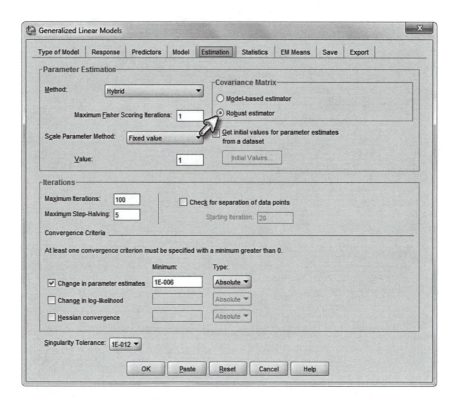

FIGURE 3.14 The *Estimation* command tab and default settings.

Initial values. The procedure will automatically compute initial values for parameters or initial values for the parameter may be specified instead.

Covariance matrix. Traditional methods for estimating variance parameters in logistic regression models are based on maximum likelihood from independent observations. When there may be clustering (e.g., from hierarchical data structures or repeated observations), approaches are needed that can take into consideration the possible clustering effects. Robust estimates of standard errors take into consideration clustering effects. Researchers can choose a model-based or a robust estimator of the standard errors. The default model-based standard errors setting for estimating variance parameters is often chosen when data are assumed to be normal (and there is little reason to believe that clustering is present). The model-based (or naïve, because it does not incorporate possible violation of independence of observations) estimator is the negative of the generalized inverse of the Hessian matrix. The model-based approach provides consistent estimates when both the mean model and the covariance model are correctly specified. If the model assumptions are not met (e.g., dependence of the observations), however, standard errors can be underestimated. The robust standard error setting is favored when there may be departures from normality.

The robust (or the Huber/White/sandwich) estimator is a "corrected" model-based estimator that provides a consistent estimate of the covariance, even when the specification of the variance and link functions is incorrect (Horton & Lipsitz, 1999). In the presence of clustering, the approach amounts to inflating the standard errors to provide more conservative tests of the significance of model parameters. Because the user will typically not know the correct covariance structure ahead of time, the empirical variance, or robust, estimate will be preferred when there is reason to assume departures from normality. Robust standard errors, however, are not a solution to all situations of model misspecification. For example, with small samples, the robust variance estimator may actually yield highly biased estimates. In this case, it may be better to use model-based estimation of variance parameters or to compare results across both approaches in the presence or absence of particular covariates.

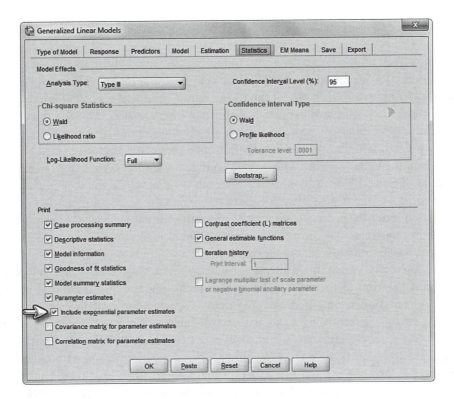

FIGURE 3.15 Model effects and print options in the GENLIN *Statistics* command tab.

The Statistics Command Tab

The *Statistics* command tab (Figure 3.15) allows specifying various model effects and print options.

Model Effects

Model effects include analysis type, confidence intervals, and log likelihood function.

Analysis type. The user may select among three types of analysis: Type I (appropriate when the focus is on a particular ordering of predictors in the proposed model), Type III (generally more applicable), and a combination of Type I and Type III. Model testing statistics (e.g., Wald or likelihood-ratio statistics) are computed based upon the selection in the chi-square statistics group. Goodness-of-fit statistics include deviance and scaled deviance, Pearson chi-square and scaled Pearson chi-square, log likelihood, Akaike's information criterion (AIC), corrected AIC (AICC), Bayesian information criterion (BIC), and consistent AIC (CAIC).

Chi-square statistics. We note that tests of the significance of model parameters are typically constructed using the standard normal distribution. In some situations (e.g., small sample sizes, probability distributions other than normal), however, the test statistics may not approximate a normal distribution very accurately because it may not be possible to achieve the desired significance level exactly. For test statistics having a continuous distribution, the p-value has a uniform null distribution over the interval 0, 1; that is, over the null hypothesis, the p-value is equally likely to fall anywhere between 0 and 1 (i.e., its probability value is 0.50). As Agresti (2007) cautions, however, for a test statistic with a discrete distribution, the null distribution of the p-value is discrete and has an expected value greater than 0.50, leading to more conservative hypothesis testing under small samples.

For single-level analyses, GENLIN uses the Wald chi-square to test the significance of fixed effects. The Wald chi-square is similar to the t-test for a continuous outcome because it is designed to test the statistical significance of a single slope parameter at a time. The Wald test statistic (X^2) is the squared form of the z-statistic (where z refers to the standard normal distribution) and will follow a χ^2 distribution with one degree of freedom for large samples (Azen &

Walker, 2011). Because the Wald test relies on the population parameter (i.e., the ML estimate) to compute the standard error, however, it can be unreliable in small samples. In such cases, constructing a likelihood ratio statistic (by comparing the ratios of the likelihoods of a model with and a model without the parameter of interest) is generally more reliable (Hox, 2002).

Confidence intervals. This allows the user to set a confidence interval band around the estimates produced. Commonly reported confidence intervals are 95% (the default as shown) or 99%. Confidence intervals provide a range of possible values of a parameter that would fall within two standard deviations of the result obtained, given the other variables in the model. For a given predictor variable with a 95% confidence interval, we can say that upon drawing repeated samples, 95% of the confidence intervals would include the true population parameter. This provides a range of possible values between which the true parameter lies and thus provides information about the precision of its estimation. In some situations (e.g., small sample size), the user might request a 90% confidence interval (which increases the range of possible values in which the true population parameter might lie).

Log likelihood function. This controls the display format of the log likelihood function. Users can obtain iteration history about the model, including the log likelihood at each step. The likelihood function is obtained by multiplying together the probabilities of each pattern of covariate responses, assuming a particular distribution of responses (e.g., binomial, multinomial, Poisson). These are then summed over all covariate patterns (using the covariate and parameter values to calculate the probability for each) as the parameter estimates which minimize the sum are sought. The full function includes an additional term that is constant with respect to the parameter estimates but has no effect on parameter estimation (IBM SPSS, 2010). GENLIN calculates the model using aggregated data and the kernel of the negative log likelihood and adds these constants to obtain the full log likelihood. Users can also choose the kernel of the log likelihood function. This option estimates the log likelihood without the extra term used in the full log likelihood. This facilitates simplified (faster) calculation of model parameters. The kernel will differ from the full value when there are multiple cases with the same covariate (and/or factor) values. The kernel log likelihood estimate is half of the −2 log likelihood provided by the binary logistic regression (REGRESSION). The estimate is multiplied by 2 so that the difference of two −2 log likelihood values between nested models will have a chi-square distribution, which allows likelihood ratio tests of model fit to be constructed.

Print. The *Print* options allow the user to specify the model information to be displayed in the output. The default settings preselect six options, which provide model information such as the number and percentage of cases included and excluded from the analysis; descriptive statistics and summary information about the dependent variable, covariates, and factors; model information; goodness-of-fit statistics; model fit tests and summary statistics; parameter estimates with corresponding test statistics; and confidence intervals. Users may also designate additional print options as needed. For example, the GENLIN models shown in subsequent chapters call for displaying the exponentiated parameter estimates (odds ratios) in addition to the raw parameter estimates, so users are instructed to include this option in the print selection.

Additional GENLIN Command Tabs

The six GENLIN command tabs discussed so far are the primary ones that we use to illustrate single-level model construction; however, we will briefly describe three remaining GENLIN screens. For further information about each command tab and the available options, readers may press the *Help* button on the lower right-hand corner of the command screen or at the top right-hand corner of the IBM SPSS *Data* or *Variable View* screens. Information may also be found in the Generalized Linear Mixed Modeling in the user's guide of the Advanced Module (IBM SPSS, 2010).

Estimated Marginal (EM) Means

The *EM Means* command tab allows the user to display the estimated means for levels of factors and factor interactions, as well as the overall estimated mean. Estimated marginal means are not available for ordinal or multinomial models.

Save

The *Save* command tab enables saving selected variable items in several formats, such as saving them with a specified name. The analyst can choose to overwrite existing variables with the same name as the new variables or avoid name conflicts by appendix suffixes to make the new variable names unique.

Export

The *Export* command tab permits writing a data set SPSS format containing the parameter correlation or covariance matrix with parameter estimates, standard errors, significance values, and degrees of freedom.

Building a Single-Level Model

After discussing some of the modeling-building features, we can turn our attention to building a simple analysis. Consider a situation where a researcher wishes to determine whether there is a relationship between elementary-aged students' background variables (e.g., gender, socioeconomic status, language background) and their probability of being proficient in reading.

Research Questions

Our initial research question might simply be the following: Is there a relationship between student background and probability of being proficient in reading? In this case, we use three background predictors: female (coded 1), low socioeconomic status (coded 1), and minority by race/ethnicity (coded 1). When we examine this type of model with main effects and an interaction with a single-level analysis, we make the assumption that students are randomly sampled and that there is not a higher level grouping of students present. This assumption is likely to be violated in most instances, however, in conducting educational analyses. For purposes of demonstration, we will assume that this is true. Our subsequent analysis will allow us to introduce some important concepts about modeling categorical outcomes and to consider the meaning of simplified output before moving on to more complex multilevel formulations in subsequent chapters.

The Data

The data in this extended example consist of 6,528 students (see Table 3.6). When there is no assumption of individuals nested within groups, the coefficients that describe the prediction equation in a single-level analysis (i.e., the intercept and slope coefficients for each predictor in the model) are generally considered as *fixed* population values, which are estimated from the sample data. In some situations, there may be little similarity among individuals within their groups. However, in most cases, we have found that multilevel models result in more accurate estimates of model parameters because they adjust for the presence of clustering effects, or the similarity among individuals within a group setting on particular types of outcomes measured. One advantage of multilevel models is that the analyst does not have to choose between conducting the analysis at the individual level (where organizational settings are ignored) or aggregating the individual data to create group-level analyses (which eliminates the variability among individuals within the group).

TABLE 3.6 Data Definition of ch3data.sav (N = 6,528)

Variable	Level[a]	Description	Values	Measurement
lowses	Individual	Predictor dichotomous variable measuring students' social economic status (SES)	0 = not low socioeconomic status 1 = low socioeconomic status	Nominal
female	Individual	Demographic predictor variable representing student's gender	0 = male 1 = female	Nominal
minor	Individual	Demographic predictor variable representing student's ethnicity	0 = nonminority 1 = minority	Nominal
readprof	Individual	Two-category dependent variable referring to students' reading proficiency	0 = not proficient 1 = proficient	Nominal

[a] Individual = assuming no apparent clustering of individuals within groups.

Specifying the Model

We develop our example analysis using the logit link function, which provides log odds coefficients and corresponding odds ratios and is estimated through an iterative algorithm. It is most commonly used to examine a dichotomous outcome. In Chapter 4, we contrast the logit link function with the probit link function for examining dichotomous data. Readers should see that the two models are very similar, differing primarily in the way in which the coefficients are scaled. Although they provide similar results, it is important to note exactly what the coefficients mean in either approach. We have found that some fields have a preference for one approach over the other.

We might hypothesize that socioeconomic status (coded low = 1 vs. other = 0), ethnicity (coded minority = 1 vs. other = 0), and gender (female coded 1 vs. male coded 0) are related to the subject's likelihood to be proficient in reading. Because the outcome is not continuous, we focus on the probability that an individual is proficient in reading (coded 1) versus not proficient. Recall from earlier in the chapter that we noted that the expected value that an individual i is proficient in reading can be expressed as

$$E(Y_i) = \pi, \tag{3.36}$$

where π is the probability that $Y_i = 1$, and $1 - \pi$ will be the probability of not being proficient.

We begin with a simple model to relate the transformed predicted values to an intercept parameter:

$$\eta_i = \log\left(\frac{\pi}{1-\pi}\right) = \beta_0. \tag{3.37}$$

The initial output confirms that we have defined a model with binomial probability distribution and logit link function (see Table 3.7).

TABLE 3.7 Model 1.1: Model Information

Dependent variable	readprof[a]
Probability distribution	Binomial
Link function	Logit

[a] The procedure models 1 as the response, treating 0 as the reference category.

Defining Model 1.1 With IBM SPSS Menu Commands

Note: Refer to previous sections in this chapter for discussion concerning settings and modeling options available in the IBM SPSS GENLIN routine.

Launch the IBM SPSS application program and select the *ch3data.sav* data file.

1. Go to the toolbar and select ANALYZE, GENERALIZED LINEAR MODELS, GENERALIZED LINEAR MODELS. This command opens the *Generalized Linear Models* (GENLIN) main dialog box.

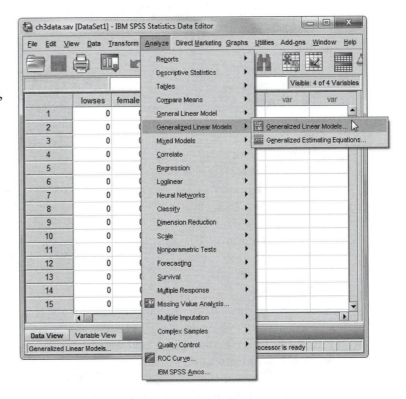

2a. The *Generalized Linear Models* displays screen nine command tabs: *Type of Model, Response, Predictors, Model, Estimation, Statistics, EM Means, Save, Export.* The default command tab when creating a model for the first time is *Type of Model* and it enables specifying a model's distribution of the dependent variable and the link function

b. For this first model, we will designate a binary logistic analy-

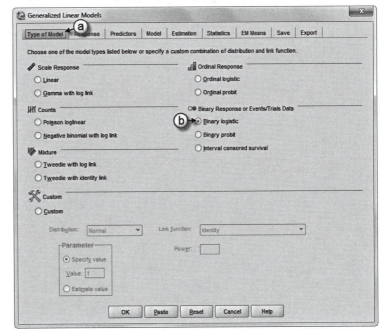

sis due to the dichotomous aspect of the dependent variable (*readprof*) to be used for this model. Click to select *Binary logistic.*

3a. Click the *Response* command tab, which enables specifying a dependent variable for the model.

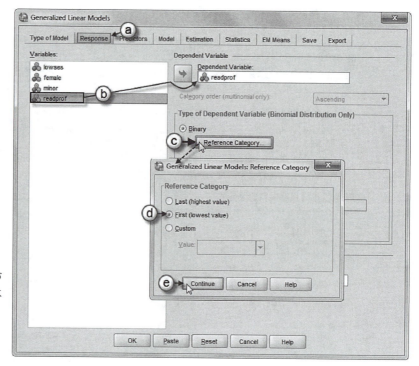

b. For this model, we will use the reading proficiency scores of students as the dependent variable. Click to select *readprof* from the *Variables* list and then click the right-arrow button to move the variable into the *Dependent Variable* box.

c. When the dependent variable has only two values, we can specify a reference category for parameter estimation. To specify a reference category, click the REFERENCE CATEGORY button, which will activate the *Generalized Linear Models Reference Category* dialog box to appear on screen.

d. We will change the reference category by clicking to select *First (lowest value)*. Interpretation of the resulting parameter estimates will be relative to the likelihood of the value "0" category.

e. Click the CONTINUE button to close the *Generalized Linear Models Reference Category* dialog box.

4a. Click the *Predictors* command tab, which enables adding predictor variables and interactions options into the model. Because this unconditional model does not have variables, we will leave the *Model Screen* as is.

b. Note that the *Include intercept in model* option is preselected. This is because the model is intercept only (default), so other model effects must be specified.

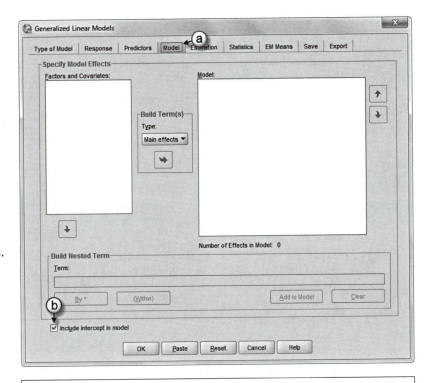

$$\eta_i = \log\left(\frac{\pi}{1-\pi}\right) = \beta_0 \quad \text{(Eq. 3.37)}$$

5a. Click the *Model* command tab (not illustrated) which enables defining different fixed effects for variables that were selected in *Predictors* command tab.

b. The default model is intercept only and no fixed effects will be specified for this model.

6a. Click the *ESTIMATION* command tab, which enables specifying estimation methods and providing initial values for the parameter estimates.

b. For this model, we will change the *Covariance Matrix* by clicking to select *Robust estimator*.

7a. Click the *Statistics* command tab, which enables specifying model effects (analysis type and confidence intervals) and print output.

b. We will retain the default *Model Effects* settings but add an additional print option to obtain exponential parameter estimates. Click to select *Include exponential parameter estimates.*

c. To generate the output shown in Tables 3.13, 3.14, and 3.17, we will at this point click to select *Iteration history* (with default *Print interval* of "1"). We will also change the *Log Likelihood Function* by clicking the pull-down menu to select *Kernal.*

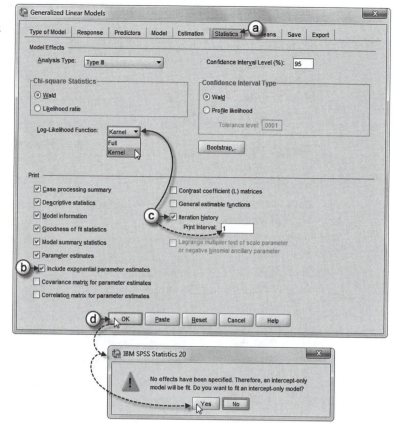

d. Click the OK button to generate the model results. A warning message is displayed to remind the user that the model has no effects. Click YES to continue.

Interpreting the Output of Model 1.1

For this first model, we provide an extended discussion of the meaning of the output. The first output of interest is the number of missing cases in the analysis. We note in Table 3.8 that there are 0 missing cases in the data set. If there were considerable missing data (perhaps 5% or more), we would want to consider imputing plausible values for the missing cases.

We can also obtain the case processing summary (Table 3.8). Of the 6,528 students in the sample, there are 2,019 students coded 0 (not proficient) and 4,509 coded 1 (proficient). The amount coded proficient represents 69.1% of the total cases (4509/6528 = 69.1%). Overall, therefore, 69.1% of the students in the sample are proficient in reading (see Table 3.9).

In Table 3.10 we provide the fixed-effect estimates. In this first example, we have only the expected value (intercept) in the sample. The intercept can be interpreted as the natural log of the odds that $Y_i = 1$ when all the other predictors in the model are equal to 0 (i.e., the point where the logistic curve crosses the y-axis). The log odds coefficient is 0.803. Because

TABLE 3.8   Selected Cases and Missing Cases: Case Processing Summary

Unweighted cases[a]		N	Percent
Selected cases	Included in analysis	6528	100.0
	Missing cases	0.0	0.0
	Total	6528	100.0
Unselected cases		0.0	0.0
Total		6528	100.0

[a] If weight is in effect, see classification table for the total number of cases.

TABLE 3.9   Model 1.1: Categorical Variable Information

			N	Percent
Dependent variable	readprof	0	2019	30.9
		1	4509	69.1
		Total	6528	100.0

log odds are a more difficult scale to interpret, they are often translated into odds ratios (e^β, where e is equal to approximately 2.71828 and β is the log odds). In this case, when we convert the log odds to an odds ratio, we see that the odds ratio is 2.233. This suggests that, in the population, students are approximately 2.3 times more likely to be proficient than not proficient.

The output also provides several indices for estimating the fit of the model. These indices can be used as a baseline against which to compare successive models, as we will demonstrate in the subsequent model building in this section.

Adding Gender to the Model

Next, we will investigate a model with a single predictor X_1 (*female*):

$$\eta_i = \log\left(\frac{\pi}{1-\pi}\right) = \beta_0 + \beta_1\,female. \tag{3.38}$$

TABLE 3.10   Model 1.1: Fixed Effects

Parameter	β	Std. error	95% Wald confidence interval		Hypothesis test			Exp(β)	95% Wald confidence interval for exp(β)	
			Lower	Upper	Wald χ^2	df	Sig.		Lower	Upper
(Intercept)	0.803	0.0268	0.751	0.856	900.283	1	0.000	2.233	2.119	2.354
(Scale)	1[a]									

Notes: Dependent variable: readprof; model: (intercept).

[a] Fixed at the displayed value.

Defining Model 1.2 With IBM SPSS Menu Commands

Note: Settings default to those used for Model 1.1.

1. Go to the tool-bar and select ANALYZE, GENERALIZED LINEAR MODELS, GENERALIZED LINEAR MODELS. This command opens the *Generalized Linear Models* (GENLIN) main dialog box.

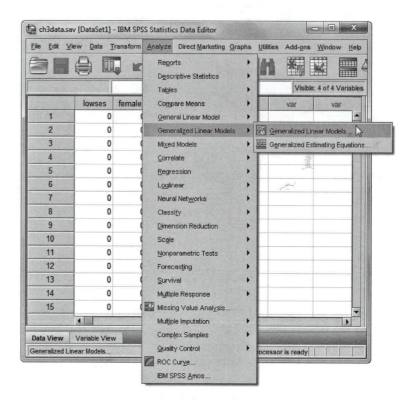

2a. The *Generalized Linear Models* display screen highlights the *Statistics* tab, which was the last command window used for Model 1.1.

b. Click the *Predictors* command tab, which enables accessing options for selecting factors and covariates in predicting the dependent variable.

c. We will use one predictor variable in this initial model. Click to select *female* and then click the right-arrow button to move the variable into the *Factors* box.

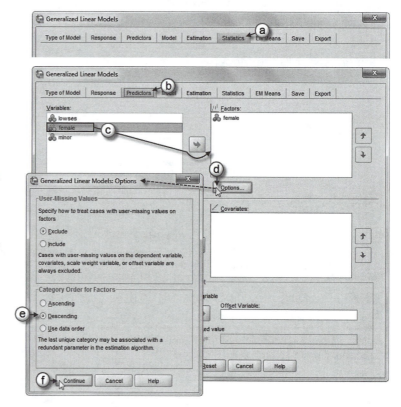

d. Because *female* has only two values, we can specify a reference category for parameter estimation. To specify a reference category, click the OPTIONS button, which will activate the *Generalized Linear Models Options* dialog box to appear on screen.

e. We will change the default category order by clicking to select *Descending*. Interpretation of the resulting parameter estimates will now be relative to the likelihood of the value "1" (female) category.

f. Click the CONTINUE button to close the *Generalized Linear Models Reference Category* dialog box.

3a. Now click the *Model* command tab, which enables specifying model effects based upon the covariate variable (*female*).

b. Note that the *Include intercept in model* option is preselected. This is because the model is intercept only (default), so other model effects must be specified.

c. We will use the default *Main effects* setting, which creates a main-effects

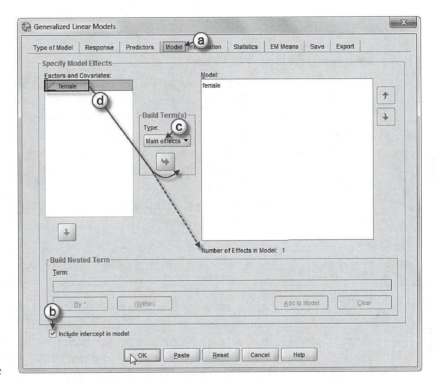

$$\eta_i = \log\left(\frac{\pi}{1-\pi}\right) = \beta_0 + \beta_1\,\textbf{\textit{female}} \qquad (\text{Eq. 3.38})$$

term for each variable selected for the model; in this case there is one: *female*.

d. Click to select *female* and then click the right-arrow button to move them into the *Model* box. Note at the bottom of the box that one effect is identified in the model (*Number of Effects in Model: 1*).

Click the OK button to generate the model results.

TABLE 3.11 Model 1.2: Fixed Effects

Parameter	β	Std. Error	95% Wald confidence interval		Hypothesis test			Exp(β)	95% Wald confidence interval for exp(β)	
			Lower	Upper	Wald χ²	df	Sig.		Lower	Upper
(Intercept)	0.614	0.0364	0.542	0.685	284.961	1	0.000	1.847	1.720	1.984
[female = 1]	0.402	0.0540	0.296	0.508	55.308	1	0.000	1.494	1.344	1.661
[female = 0]	0ᵃ	.	.	.	.	.	.	1	.	.
(Scale)	1ᵇ									

Notes: Dependent variable: readprof; model: (intercept), female.
ᵃ Set to 0 because this parameter is redundant.
ᵇ Fixed at the displayed value.

Interpreting the Output of Model 1.2

When we estimate this model, we obtain the fixed-effect estimates in Table 3.11. We note that in linear regression models, the categories for dichotomous variables are generally dummy coded (coded 0 and 1) so that the intercept in the model can be interpreted in the usual manner (i.e., the level of outcome *Y* when the *X* predictors are each 0)[2]. For categorical predictors, the default in GENLIN specifies the last category as the reference group. In this case, if the last category were used as the reference group, the intercept would be interpreted as the probability that females (coded 1) are proficient.

We can link the output back to our initial equation (Equation 3.35) and write the equation expressing the model output as follows:

$$\eta_i = 0.614 + 0.402\,(female).$$

This suggests that the intercept (β_0) log of the odds of success is 0.614. The intercept can be interpreted as the natural log of the odds that $Y_i = 1$ when all the other predictors in the model are equal to 0 (i.e., the point where the logistic curve crosses the y-axis). It is therefore important to code predictors such that 0 takes on a "meaningful" value in terms of the data being examined. In this case, the intercept represents the log odds of being proficient when female is equal to 0—that is, the probability that males are proficient in reading (0.614). We calculate this by putting the value of the variable (i.e., in this case 0) into the equation as follows:

$$\eta_i = 0.614 + (0.402)(0)\,female$$

Its exponentiated value ($e^{0.614 + .0402(0)}$, or $e^{0.614}$) can be interpreted as the predicted odds that a male student in the sample is proficient; that is, a male student is 1.847 times more likely to be proficient in reading than not proficient.

To determine the log odds of a female being proficient versus not proficient in reading, we need to add the intercept log odds ratio (0.614) to the log odds favoring female versus male (0.402), which will be 1.016. We again place the value of the variable (in this case 1) into the model to estimate the new predicted value for η_i:

$$\eta_i = 0.614 + (0.402)(1)\,female.$$

Its exponentiated value ($e^{0.614+0.402(1)}$) suggests that females are about 2.76 times more likely to be proficient than not proficient. Note that this can also be represented as follows (where we multiply the odds ratios):

$$\log\left[\frac{\pi}{1-\pi}\right] = e^{0.614}e^{0.402} = (1.847)(1.494) = 2.76$$

The slope coefficient β_1 for female in Table 3.11 represents the difference between female and male (i.e., the reference category) students in the log odds of being proficient. Its exponentiated value can be interpreted as the difference between the odds of being proficient for female versus male students (OR = 1.494); that is, the odds of being proficient are multiplied by about 1.5 times for females compared to males. We note that if we take the odds of a male being proficient versus not proficient (1.847) and multiply by the difference in the odds of being proficient favoring females over males (1.494), we obtain the odds that a female is proficient (1.847*1.494 = 2.76). One good way to think about odds ratios, then, is as the ratio of difference in the odds of success for females to the odds of success for males. In terms of the percentage change, we can say that the odds for females (1.494) being proficient are 49.4% greater than the odds for males being proficient.

A number of ways exist to test hypotheses about the ratio of the log odds estimate to its standard error. For the logistic model, the Wald test provides a test of the significance of a slope coefficient. The Wald test is similar to the *t*-test, which is often used to test individual parameters in multiple regression. For large samples, the Wald test statistic follows a χ^2 distribution with one degree of freedom, as it is a single parameter test (Azen & Walker, 2011). In this case, the Wald coefficient is 55.31 [$0.402/0.054 \sim (7.44)^2 = 55.35$]. The small discrepancy is due to rounding.

A likelihood ratio test can also be used. This approach compares a model where the slope parameter is restricted to zero (i.e., the restricted model) and a subsequent model where the parameter is freed (alternative model), and the change in model likelihoods is compared. The ratio of the two likelihoods (L_0/L_1) is referred to as the likelihood ratio. The alternative model (M1) will always fit as well as or better than (i.e., have larger log likelihood) the restricted model (M0) because the two models are nested and M1 has more free parameters. When there is no difference between models (i.e., the log likelihoods are the same), the likelihood ratio will be 1.0 and the test statistic will be 0 (Azen & Walker, 2011). The likelihood ratio statistic (G^2) can then be formulated as follows:

$$G^2 = -2\log\left(\frac{L_0}{L_1}\right) = -2[\log(L_0) - \log(L_1)] = [-2\log(L_0)] - [-2\log(L_1)]. \qquad (3.39)$$

A larger, more positive test result provides evidence against the null hypothesis that there is no difference between models.

We note that the likelihood ratio test can be used to test the effect of a single parameter or to compare two successive models (e.g., one model with three predictors versus a model with two of the predictors restricted to be 0), as long as the restricted model is nested in the unrestricted model. The degrees of freedom of the test depend on the difference in the number of parameters being tested.

For a single-parameter test, users can specify the likelihood ratio test on the Statistics screen menu in GENLIN. This test compares the current model against a baseline model that just has

TABLE 3.12   Model 1.2: Omnibus Test[a]

Likelihood ratio G[2]	df	Sig.
55.761	1	0.000

Notes: Dependent variable: readprof;
model: (intercept), female.

[a] Compares the fitted model to the intercept-only model.

the intercept. To illustrate, we will compare the current model against the intercept-only model. The difference between the two models can be tested with a likelihood ratio chi-square test. This represents the difference in $-2*\log$ likelihoods ($-2LL$) between the intercept model (L_0) and the model with female (L_1) from Equation 3.35. The test follows a χ^2 distribution with one degree of freedom because only the one parameter is restricted. As shown in Table 3.12, the estimate is 55.761 ($p < .001$).

We can obtain information from the output that shows how the likelihood ratio test is conducted. Although we can not obtain the $-2LL$ directly in GENLIN, we can obtain the kernel log likelihood and multiply it by -2. The final value of the kernel log likelihood in Table 3.13 is -4009.856. We multiply it by -2 and get the $-2LL$ estimate of 8,019.712.

Next, in Table 3.14, we provide the final kernel log likelihood for the intercept-only model. This can also be multiplied by -2 ($-4037.7365*-2 = 8,075.473$).

TABLE 3.13   Model 1.2: Iteration History

Iteration	Update type	Number of step-halvings	Log likelihood[b]	Parameter (Intercept)	[female = 1]	(Scale)
0	Initial	0	−4019.033	0.723724	0.414924	1
1	Scoring	0	−4009.862	0.611724	0.399948	1
2	Newton	0	−4009.856	0.613747	0.401786	1
3	Newton	0	−4009.856	0.613747	0.401789	1
4	Newton[a]	0	−4009.856	0.613747	0.401789	1

Notes: Redundant parameters are not displayed. Their values are always 0 in all iterations. Dependent variable: readprof; model: (intercept), female.

[a] All convergence criteria are satisfied.

[b] The kernel of the log likelihood function is displayed.

TABLE 3.14   Model 1.1: Iteration History

Iteration	Update type	Number of step-halvings	Log likelihood[b]	Parameter (Intercept)	(Scale)
0	Initial	0	−4048.311	0.927626	1
1	Scoring	0	−4037.744	0.800210	1
2	Newton	0	−4037.737	0.803471	1
3	Newton	0	−4037.737	0.803473	1
4	Newton[a]	2	−4037.737	0.803473	1

Notes: Redundant parameters are not displayed. Their values are always 0 in all iterations. Dependent variable: readprof; model: (intercept).

[a] All convergence criteria are satisfied.

[b] The kernel of the log likelihood function is displayed.

We can now conduct a likelihood ratio test. Using Equation 3.36, the likelihood ratio statistic is calculated as follows:

$$G^2 = 8,075.473 - 8,019.712 = 55.761$$

For one degree of freedom, the required coefficient is 3.84 at $p = .05$. The null hypothesis is that there is no difference between the two models. As shown in Table 3.12, the likelihood ratio chi-square is the same as the estimate calculated from the GENLIN output. We can conclude that there is a significant difference between gender and the probability of being proficient in reading.

Obtaining Predicted Probabilities for Males and Females

It is important to note that the log odds coefficients in Table 3.11 are described in terms of the underlying continuous predictor (η_i) of Y. We can translate the odds ratio back to the predicted probabilities of $Y = 1$. For this model, the predicted probability will be equal to $1/[1 + \exp(-\eta)]$, which for our specified model, $\eta_i = \beta_0 + \beta_1 female$, can also be written as the following:

$$\pi = \frac{1}{1 + e^{-(\beta_0 + \beta_1 x_1)}}. \tag{3.40}$$

Because for gender (X_1), males are coded 0, the predicted probability will simply make use of the intercept log odds. More specifically, the coefficient $\beta_1 x_1$ will be 0 for males, so just the intercept is needed to calculate the probability of males being proficient. The equation thus reduces to $1/(1 + e^{-(\beta_0)})$. The predicted probability is therefore 0.649 [calculated as $1/1 + e^{-(0.614)})$, or $1/1.541$]. This result represents the predicted probability that $y = 1$ when $x = 0$ (or male). For females, the predicted probability when $x = 1$ will be $1/(1 + e^{-(\beta_0 + \beta_1 x_1)})$. For females (coded 1), the combined log odds are $0.614 + 0.402$, or 1.016. The predicted probability will then be 0.734 [calculated as $1/(1 + e^{-(1.016)})$, or $1/1.362$]. This represents that proportion of females who are proficient.

We can confirm the estimates in this simple analysis by looking at a cross tabulation of reading proficiency by gender in Table 3.15. As the table suggests, the percentage of males (coded 0) who are proficient (coded 1) is 64.9% (0.649). The percentage of females who are proficient is 73.4% (0.734).

Adding Additional Background Predictors

Next we will add two other background variables to the analysis (low SES coded 1, Else = 0; minority by race/ethnicity coded 1, Else = 0). The model will look like the following:

$$\eta_i = \log\left(\frac{\pi}{1 - \pi}\right) = \beta_0 + \beta_1 female + \beta_2 lowses + \beta_3 minor. \tag{3.41}$$

TABLE 3.15 Cross Tabulation of Reading Proficiency by Gender

			Female		
			0	1	Total
readprof	0	Count	1166	853	2019
		Within readprof	57.8%	42.2%	100.0%
		Within female	35.1%	26.6%	30.9%
	1	Count	2154	2355	4509
		Within readprof	47.8%	52.2%	100.0%
		Within female	64.9%	73.4%	69.1%
Total		Count	3320	3208	6528
		Within readprof	50.9%	49.1%	100.0%
		Within female	100.0%	100.0%	100.0%

Defining Model 1.3 With IBM SPSS Menu Commands

Note: Settings default to those used for Model 1.2.

1. Go to the tool-bar and select ANALYZE, GENERALIZED LINEAR MODELS, GENERALIZED LINEAR MODELS. This command opens the *Generalized Linear Models* (GENLIN) main dialog box.

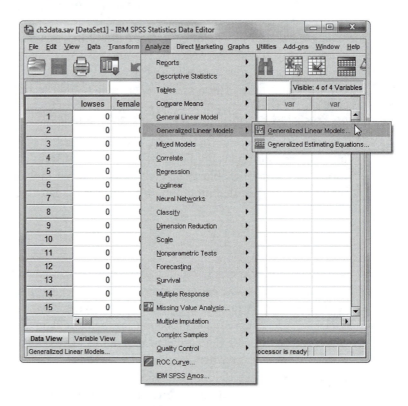

2a. The *Generalized Linear Models* display screen highlights the *Model* tab, which was the last command window used for Model 1.2.

b. Click the *Predictors* command tab, which enables selecting factors and covariates to use in predicting the dependent variable.

c. We will add two predictor variables (*lowses, minor*) to the model. Click to select *lowses* and *minor* and then click the right-arrow button to move the variables into the *Factors* box.

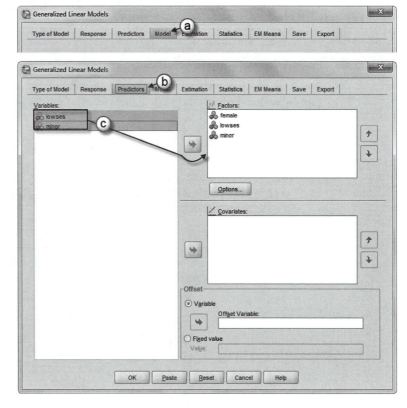

3a. Now we will adjust the model's fixed effects. First, click the *Model* command tab, which enables specifying model effects based upon the covariate variable (female).

b. Click to select *lowses* and *minor* and then click the right-arrow button to move them into the *Model* box. Note that the three effects are now identified in the model (*Number of Effects in Model: 3*).

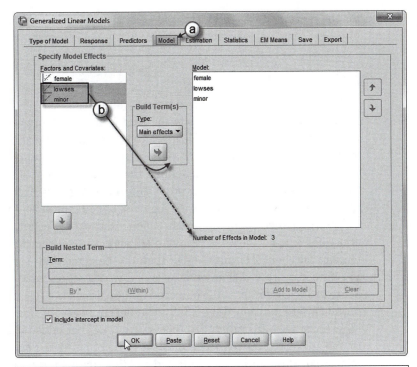

$$\eta_i = \log\left(\frac{\pi}{1-\pi}\right) = \beta_0 + \beta_1 female + \pmb{\beta_2 lowses} + \pmb{\beta_3 minor} \quad \text{(Eq. 3.41)}$$

Click the OK button to generate the model results.

Interpreting the Output of Model 1.3

The results are presented in Table 3.16. It is useful to consider the interpretation of the intercept for individuals who are 0 on all the predictors in the model. In this case, the intercept predicts the log odds of being proficient for males (0) of average or high socioeconomic status (0) and who are not a minority by race/ethnicity (0). The log odds coefficient of being proficient in reading for those individuals is 1.545. We can once again calculate the probability of being proficient

TABLE 3.16 Background Variables Explaining Probability of Being Proficient

Parameter	β	Std. error	95% Wald confidence interval Lower	Upper	Hypothesis test Wald χ²	df	Sig.	Exp(β)	95% Wald confidence interval for exp(β) Lower	Upper
(Intercept)	1.545	0.0559	1.436	1.655	764.787	1	0.000	4.690	4.203	5.232
[female = 1]	0.465	0.0572	0.353	0.577	66.136	1	0.000	1.592	1.423	1.780
[female = 0]	0ᵃ							1		
[lowses = 1]	−1.053	0.0593	−1.170	−0.937	315.439	1	0.000	0.349	0.310	0.392
[lowses = 0]	0ᵃ							1		
[minor = 1]	−0.770	0.0595	−0.886	−0.653	167.422	1	0.000	0.463	0.412	0.520
[minor = 0]	0ᵃ							1		
(Scale)	1ᵇ									

Notes: Dependent variable: readprof; model: (intercept), female, lowses, minor.
ᵃ Set to 0 because this parameter is redundant.
ᵇ Fixed at the displayed value.

for the person described by the intercept. Because all of the predictors are equal to 0, only the intercept contributes to the probability that $\pi = 1$. This estimate will be

$$\pi = 1 / (1 + e^{-(1.545)}),$$

or 1/1.213, which equals approximately 0.824.

As we have noted, this refers to the proportion of individuals who are proficient in reading and who are coded = 0 on each predictor. Holding the other variables constant, the log odds coefficient for females suggests a significant increase in logit units of being proficient compared with males (β = 0.465, $p < .001$). In contrast, holding the other variables constant, low SES is negatively related to the log of the odds of being proficient in reading ($\beta = -1.053$, $p < .001$). Being a minority by race or ethnicity also decreases the log odds of being proficient in reading ($\beta = -0.770$, $p < .001$).

We emphasize again that each exponentiated coefficient, or odds ratio, is the ratio of two odds. For female, the odds ratio (1.592) suggests that the expected odds of being proficient for females are about 1.6 times higher than the expected odds for males (or about a 59.2% increase). We note in passing that an odds ratio of 2 would represent a 100% increase in the probability of being proficient for females compared with males. The odds ratio suggests that the odds of being proficient are multiplied by 0.349 (or reduced by 65.1%) for low SES students compared with their more average or high SES peers. We note that odds ratios below 1.0 are sometimes easier to explain by converting them into the odds of being *not proficient*. Because the relationship is reciprocal, this can be accomplished by dividing 1 by the obtained odds ratio (1/0.349). This will change the odds ratio to 2.865, suggesting that the odds of being *not proficient* are almost 2.9 times higher for low SES students compared with their peers of average or higher SES background. Finally, for minority by race/ethnicity, the odds of being proficient are multiplied by 0.463 (or reduced by 53.7%) compared with students who are not minority by race/ethnicity, holding other variables constant.

In addition to examining the statistical significance of the individual variables in the model, we can also obtain other information to evaluate how well the model fits the data, such as the ability to classify individuals correctly into likelihood to be proficient or not, or the model's –2*log likelihood (–2LL), which can be used to evaluate a series of model tests. In this case, we know that the –2LL for Model 1 (with gender) is 8,019.712 with two degrees of freedom. We can again obtain the kernel log likelihood for Model 2 (with three predictors). This value is –3,668.478 from the iteration history of the parameter estimation summarized in Table 3.17. If we multiply the value by –2, we obtain the –2LL as 7,336.956.

From Equation 3.36, we can obtain a likelihood ratio test comparing the previous model with one predictor against the current model:

$$8,019.712 - 7,336.956 = 682.756,$$

TABLE 3.17 Model 1.3: Iteration History

| Iteration | Update type | Log likelihood[b] | Parameter | | | | |
			(Intercept)	[female = 1]	[lowses =1]	[minor = 1]	(Scale)
0	Initial	–3669.348	1.551077	0.426674	–1.034508	–0.728849	1
1	Scoring	–3668.478	1.545166	0.464706	–1.053093	–0.769502	1
2	Newton	–3668.478	1.545353	0.464819	–1.053351	–0.769556	1
3	Newton[a]	–3668.478	1.545353	0.464819	–1.053351	–0.769556	1

Notes: Redundant parameters are not displayed. Their values are always 0 in all iterations. Dependent variable: readprof; model: (intercept), female, lowses, minor.
[a] All convergence criteria are satisfied.
[b] The kernel of the log likelihood function is displayed.

TABLE 3.18 **Classification Percentage Classification Table**[a]

Observed			Predicted readprof 0	Predicted readprof 1	% Correct
Step 1	readprof	0	529	1490	26.2
		1	437	4072	90.3
	Overall percentage				70.5

[a] The cut value is .500.

which, because Model 1.2 has two degrees of freedom and Model 1.3 has four degrees of freedom, results in a likelihood ratio coefficient of 682.756.

This coefficient is easily beyond the required chi-square coefficient of 5.99 (for two degrees of freedom at $p = .05$). We can therefore reject the null hypothesis of no difference between models and conclude that the model with three predictors is a significant improvement in explaining probability of being proficient compared with Model 1.2 with only gender as a predictor.

We can compare the initial categories (69.1% proficient, 30.9% not proficient) against the number of individuals classified correctly using the set of background predictors. Using the overall percentage of individuals correctly classified in Table 3.18, we note the model correctly classified 70.5% of the participants. The accuracy was 90.3% for classifying proficient students.

In contrast, however, the model only correctly classified 529 of 2,019 students who were not proficient (26.2%). This suggests that the overall number of individuals correctly classified by this model (70.5%) does not substantively increase the percentage of students who would be classified as proficient due to chance alone (because 69.1% are proficient). The ability to classify individuals correctly is important to keep in mind because, with almost 70% proficient to begin with, if we guessed each successive individual's likelihood to be proficient as "yes," we would be right about 70% of the time with no other information about each one.

This is important to keep in mind as proportions comprising a dichotomous outcome become more extreme (e.g., beyond 0.90 to 0.10) because we may find that we cannot correctly classify any individual in the extremely small group using the proposed model. Smaller sample sizes can add to the inaccurate estimation of relatively rare events. This would call into question the model's usefulness in accounting for the events comprising the outcome. In these cases, procedures such as logistic regression can inaccurately estimate the probability that rare events (e.g., events occurring less than 5% of the time in the data set) will occur. This can necessitate more specialized adjustments to estimate extremely rare events.

Testing an Interaction

Most often, the variables in our models function as covariates; that is, they are expected to account for variance in the outcome, but they are generally not expected to be related to the other X variables under consideration in a model. It is often the case, however, that a second predictor added to the model may influence the relationship between a primary variable of interest and the outcome. In our model, we may suspect that the effect of gender on proficiency might also be related to levels of socioeconomic status. In a multilevel setting, the moderator might also be a variable at a higher level of the data hierarchy that enhances or diminishes the strength of a relationship between X and Y at a lower level. In our present model, therefore, we can also test for particular interactions (e.g., *female*lowses*, *minority*lowses*). A significant interaction suggests that the effect of a predictor variable on the outcome is contingent on the level of another predictor. These are often referred to as moderators because they may be either a categorical or continuous variable that changes the direction or strength of the relationship of X to Y.

In such cases, where interaction effects may be of interest, analysts will often estimate a model with main effects and interactions and then examine whether a more simplified model (with nonsignificant interactions removed) appears to fit the data better. This can be accomplished by examining a Wald test for the interaction parameter in the full model or using a likelihood ratio test to compare the simplified model without an interaction between two variables against the model with the interaction present.

We show this briefly, comparing the current model against a model that includes an interaction for *lowses*minor*. We do not focus on specifying the model but rather on testing for the model with and without the interaction. We specified an interaction for minority as follows (model syntax is provided in Appendix A):

$$\eta_i = \log\left(\frac{\pi}{1-\pi}\right) = \beta_0 + \beta_1 \textit{female} + \beta_2 \textit{lowses} + \beta_3 \textit{minor} + \beta_4 \textit{lowses} * \textit{minor}. \qquad (3.42)$$

We provide the results of comparing Model 1.3 (no interaction) and Model 1.4 (with the interaction) (Table 3.19). We can use the Wald test or a likelihood ratio test to test the null hypothesis that there is no difference between the restricted and alternative models. Based on our parameter tests, we can reject the null hypothesis and conclude that the model with interaction provides an improved fit to the data compared with the main-effects model. We will discuss the interpretation of a significant interaction effect in the next chapter.

Limitations of Single-Level Analysis

Single-level analyses are appropriate when there is little interest in considering group-level influences on the outcomes. The key point about a single-level model is that the estimates of the intercept and slope are each fixed to one value that describes the average for the sample. More specifically, the log odds expressing the relationship between SES and reading proficiency ($\beta = -1.053$ in Table 3.16) will be the same across all cases. As we have noted, however, there would likely be problems in the accuracy of this analysis if the students are nested within schools.

The assumptions necessary for single-level models to yield optimal estimates are most realistic when the data have been collected through simple random sampling. Random sampling assumes that subjects in a study are independent of each other. As groups are added as a feature of a study, however, this assumption becomes more tenuous. In large-scale educational research, simple random sampling would be rarely used. Instead, various types of complex sampling strategies are employed to select schools, classrooms, and students. These can include multistage sampling strategies, where individuals may be sampled within various groups with likely oversampling of some groups. Clustered data, therefore, result from the strategies used in large-scale databases, as well as the natural groupings of students in classrooms and schools.

The single-level logistic regression model does not take into consideration that the students may be clustered within a number of schools with other students having similar backgrounds and math scores. In our example, the analysis does not take into consideration that students are nested within the elementary schools in the sample (or that the estimated SES-proficiency slope

TABLE 3.19 Comparison Between Main-Effect Model and Interaction Model

Random and Residual Effects	Estimate	Std. Error	Wald	−2LL
Model 1.3				
Main effects only				7336.956
Model 1.4				
lowses*minor	0.342	0.118	8.347	7328.606

Notes: Kernel log likelihood = −2LL/−2. G^2 (1 df) = 8.35, $p < .01$.

might be different between these schools). In samples where individuals are clustered in higher order social groupings (a department, a school, or some other type of organization), simple random sampling does not hold because people clustered in groups will tend to be similar in various ways. For example, they attend schools with particular student compositions, expectations for student academic achievement, and curricular organization and instructional strategies. If the clustering of students is ignored, it is likely that bias will be introduced in estimating model parameters. In other words, when there are clustered data, it is likely that there is a distribution of both intercepts and slopes around their respective average fixed effects. In this situation, we might wish to investigate the *random* variability in intercepts and slopes across the sample of higher level units in the study.

Summary

This chapter summarized the different types of categorical outcomes covered by the GENLIN program. We build a series of categorical single-level and multilevel models in subsequent chapters. We have found that in many instances multilevel investigations unfold as a series of analytic steps. Of course, there may be times when the analyst might change the specific steps, but, in general, this overall model development strategy works pretty well. We illustrate this strategy to explore a random intercept and a random slope in the next chapter.

Note

1. We note that the natural logarithm is referred to in various ways as log, ln, or $\log_e$. This has caused considerable confusion with the common logarithm ($\log_{10}$, which is log base 10). We will use log to refer to the natural logarithm because this is how it is referred to in IBM SPSS. The base (e) for the natural logarithm is approximately 2.71828.
2. Some fields may prefer effect coding, where the 0 category is coded −1 instead. Effect coding tests whether the mean of the category deviates from the "grand mean," rather than testing the difference in means between the two categories (as in dummy coding).

CHAPTER 4

Multilevel Models With Dichotomous Outcomes

Introduction

In this chapter, we develop a two-level model to examine the event probability for a response variable with two possible outcomes. We then extend the basic two-level model to a three-level formulation. The general concepts we present in this chapter become more familiar as one reads more research studies that use multilevel techniques and has more opportunities to work with multilevel data. The extension of GLM techniques to multilevel data structures is referred to as generalized linear mixed modeling (GLMM). Introductions to these models are described elsewhere in the literature (e.g., Goldstein, 1991; Heck & Thomas, 2009; Hedeker, 2005; Hox, 2010; Longford, 1983; Mislevy & Bock, 1989; Muthén & Muthén, 1998–2006; Raudenbush & Bryk, 2002; Wong & Mason, 1989).

GLMM describes models for categorical data where the subjects are nested within groups or where repeated measures are nested within individuals (and perhaps within group structures). In these types of analyses, the focus is often on the variability in the effects of interest (i.e., random intercepts, regression coefficients) at the group level. The result is a mixed-effect model, which includes both the usual fixed effects for the predictors and the variability in the specified random effects—variability that occurs across higher level units or across time periods. In this chapter, our intent is to develop the rationale behind the specification of this general class of models in a relatively nontechnical manner and to illustrate its use in an applied research setting. We believe that this chapter is especially important because the multilevel methods presented here will provide a basis for the application of these techniques to a wider set of research problems with other categorical outcomes in the chapters that follow.

Components for Generalized Linear Mixed Models

The GLMM procedure has the same three GLM components, which include:

- The *random* component that specifies the variance in terms of the mean (μ_{ij}), or expected value, of Y_{ij} for individual i in group j;
- The *structural* component that relates the transformed predictor η_{ij} of Y to a set of predictors;
- The *link function* $g(\cdot)$, which connects the expected value (μ_{ij}) of Y to the transformed predicted values of η_{ij}, therefore providing the relationship between the linear predictor and the mean of the distribution function.

Specifying a Two-Level Model

We start here by introducing the basic specification for a two-level model. The Level-1 model for individual i nested in group j is of the general form:

$$\eta_{ij} = x'_{ij}\beta, \tag{4.1}$$

where x'_{ij} is a $(p + 1)*1$ vector of predictors for the linear predictor η_{ij} of Y_{ij} and β is a vector of corresponding regression coefficients. An appropriate link function is then used to link the expected value of Y_{ij} to η_{ij}. In this case, we will use the logistic link function:

$$\eta_{ij} = \log\left[\frac{\pi_{ij}}{1 - \pi_{ij}}\right] = \beta_{0j} + \beta_{1j}X_{1ij} + \beta_{2j}X_{2ij} + \cdots + \beta_{qj}X_{qij}. \tag{4.2}$$

There is again no residual variance term included at Level 1 because the underlying probability distribution associated with Y_{ij} is not normally distributed.

At Level 2, the Level-1 coefficients β_{qj} can become outcome variables. Following Raudenbush, Bryk, Cheong, and Congdon (2004), a generic structural model can be denoted as follows:

$$\beta_{qj} = \gamma_{q0} + \gamma_{q1}W_{1j} + \gamma_{q2}W_{2j} + \cdots + \gamma_{qS_q}W_{S_qj} + u_{qj}, \tag{4.3}$$

where γ_{qS} ($q = 0, 1,\ldots, S_q$) are the Level-2 coefficients, W_{S_qj} are Level-2 predictors, and u_{qj} are Level-2 random effects.

Unlike a single-level model, in a multilevel model we also have to think about the specification of the Level-2 covariance matrix. The specification depends on the number of randomly varying effects at Level 2. The maximum dimension is $(Q + 1)\times(Q + 1)$. For example, for a random intercept and single random slope, we could specify an unstructured covariance matrix, which has an intercept variance (σ_I^2), a slope variance (σ_S^2), and a covariance between them (σ_{IS}):

$$\begin{bmatrix} \sigma_I^2 & \sigma_{IS} \\ \sigma_{IS} & \sigma_S^2 \end{bmatrix}. \tag{4.4}$$

Notice that a covariance matrix is a square matrix, which means that the variances are specified as diagonal elements of the matrix and the off-diagonal elements are the same above and below the diagonals. We also specify a diagonal matrix, which assumes that the covariance between the random intercept and slope is 0:

$$\begin{bmatrix} \sigma_I^2 & 0 \\ 0 & \sigma_S^2 \end{bmatrix}. \tag{4.5}$$

Specifying a Three-Level Model

In the case where there is a third level or more levels, the notation can get cumbersome (Hox, 2010). Hox notes that two different types of notational systems are typically used. One is to use different Greek letters for the regression coefficients and variance terms at each level. It is often easier, however, to use the same coefficients as in Equation 4.3 (e.g., γ for the intercepts and regression slopes and u for the variance terms) and let the numbering of the coefficients identify the level to which each belongs (e.g., γ_{00} for Level 2 and γ_{000} for Level 3).

Model Estimation

Multilevel models with categorical outcomes can be more challenging to estimate than models with continuous outcomes. Combining multilevel data structures and generalized linear models can lead to much greater complexity in estimating model parameters and considerably more computer time in providing model estimation. Because the link functions estimated are nonlinear, the prevailing approach in a number of software programs (e.g., HLM, MlwiN) is to approximate the nonlinear link by a nearly linear function and embed the multilevel estimation for that function in the generalized linear model (Hox, 2010). This approach is a type of quasilikelihood approach relying on Taylor series expansion. Other software packages (e.g., Mplus) use numerical integration (i.e., routines to approximate the solution of functions that cannot be found analytically) for computing the likelihood because the link function within groups is nonlinear.

The default estimation method for categorical outcomes used in IBM SPSS's GENLIN MIXED is referred to as active set method (ASM) with Newton-Raphson estimation. ASM is a well-known theoretical approach for solving a constrained optimization problem. It does this by moving constraints in and out of the active constraint set to make the solution in the new iteration better along a feasible search direction. ASM proceeds by finding a feasible starting point and repeating until it reaches a solution that is "optimal enough" according to a set of criteria. More specifically, at each iteration, the approach proceeds by partitioning inequality constraints into an active (or sufficiently close to be deemed active for this iteration) or inactive set (Nocedal & Wright, 2006).

The inactive constraints are ignored for the iteration. The active set is particularly important in optimization theory, as it determines which constraints will influence the final result of optimization. The active set at each iteration is sometimes referred to as the *working set*. Then the new point is selected by moving on the surface defined by the working set. This type of iteration continues until a solution is reached. We note in passing the reported occasional instability of ASM solutions that can occur mainly due to numerical error during estimation (Lau, Yue, Ling, & Maciejowski, 2009). This type of problem can be reflected in an output statement that "a numerical error occurred." This is actually a processor limitation that results in something akin to rounding error. This error propagates through the successive iterations in the convergence algorithms until the emergent solution becomes so unstable that calculations cannot continue.

Such multilevel estimation approaches with categorical outcomes can take considerable time to converge on a solution. The solutions are more challenging to approximate when there are larger samples and more random components in the model. In some instances, these techniques bring us to the frontier of the computational capacities of today's computing hardware and programming. The consequences of this reality become clearer as models become more complex and computationally intensive. An immediate symptom of this is found in relatively long processing times and the occasional error in estimation that can bring the analysis to an abrupt halt.

Building Multilevel Models With GENLIN MIXED

We next provide a general overview of the GENLIN MIXED program and then build an example of two- and three-level models to examine students' proficiency in reading. In samples where individuals are clustered in higher order social groupings (e.g., a department, a school, or some other type of organization), simple random sampling does not hold because people clustered in groups will tend to be similar in various ways. For example, they attend schools with particular student compositions, expectations for student academic achievement, and curricular organization and instructional strategies. If the clustering of students is ignored, it is likely that bias will be introduced in estimating model parameters. In other words, where there are

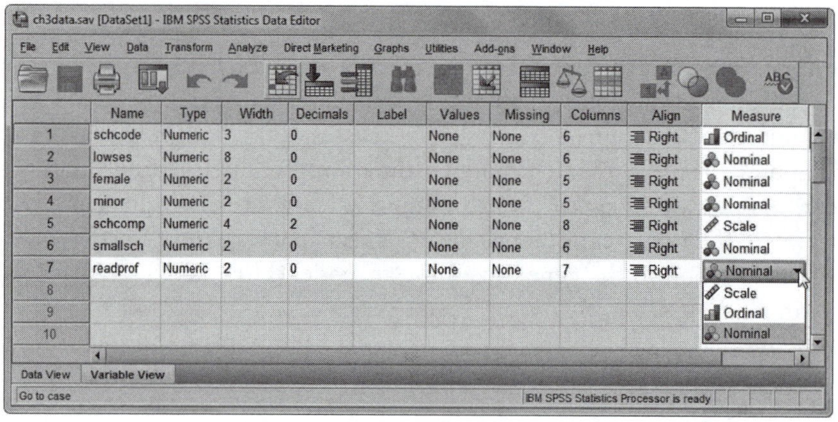

FIGURE 4.1 Defining the measurement setting as nominal for selected variable (*readprof*).

clustered data, it is likely that there is a distribution of both intercepts and slopes around their respective average fixed effects.

In this situation, we might wish to investigate the *random* variability in intercepts and slopes across the sample of higher level units in the study. Once we determine that variation exists in the parameters of interest across groups, we can build Level 2 models to explain this variation. In some cases, we may have a specific theoretical model in mind that we wish to test. In other cases, however, we might be trying to explore possible new mechanisms that explain this observed variation in parameters across groups.

As a first step, it is important to make sure that the variables in the data set are explicitly defined in terms of their scale of measurement. Each of the variables must be designated as continuous (scale), ordinal, or nominal as shown in Figure 4.1. There are a few considerations to keep in mind. First, if dependent variables are not properly defined, the program may not be able to run the proposed model. For example, defining an outcome as nominal (unordered categories) or ordinal (ordered categories) would make a difference in the type of link function required to estimate the model appropriately. The coding of the dependent variable will also determine which category serves as the default reference group for the dependent variable (although this can be changed within the program).

Second, proper designation of the scale of measurement and coding of the categorical variables are important in terms of understanding the meaning of the output produced for predictors. For independent variables, predictors referred to as nominal or ordinal will produce estimates for one less than the total number of categories ($C-1$). Similarly to defining a multiple regression model, a dichotomous predictor such as gender can be defined as either scale (continuous) or nominal. If it were dummy coded (0, 1) and defined as scale, the reference group will be the category coded 0, and the estimate produced will be for the category coded 1. If it is defined as nominal, it will be considered as a factor, so the default reference group will be the category coded 1, and the estimate produced will be for the category coded 0.

Regarding a predictor like gender, for example, users should consider how they want the coefficients to look in the fixed-effect tables in terms of their theoretical arguments being advanced. To illustrate, a negative coefficient for females in the output might be considered as a *disadvantage* in terms of a line of reasoning about the probability of being promoted in the workplace. In contrast, a positive coefficient for males might be considered as an advantage favoring males in the promotion process.

Third, as we subsequently demonstrate in this example, how the variables are formally designated within SPSS makes a difference in how the program calculates the total number of

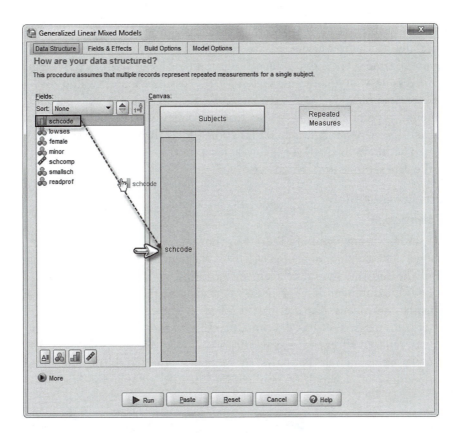

FIGURE 4.2 The GENLIN MIXED *Data Structure* command screen.

parameters estimated in the model. The user should therefore make sure the program is referring to the categories of the independent variables and the dependent variable in the manner that is intended in defining them.

Data Structure Command Tab

After defining variables, the GLMM option in MIXED first requires the user to specify the data structure (see Figure 4.2). To access the GENLIN MIXED feature, go to the IBM SPSS toolbar and select ANALYZE > MIXED MODELS > GENERALIZED LINEAR [MIXED MODELS]. The command will cause the *Data Structure* screen to appear (Figure 3.10, in Chapter 3), which is the default screen when creating a model for the first time.

The basics of multilevel modeling involve the investigation of randomly varying outcomes (i.e., intercepts and slopes) across groups. The *Subjects* specification allows the user to define an observational unit that can be considered independent of other subjects. In multilevel modeling, these are typically group structures that contain clustered observations (e.g., students nested in schools). In this case, the subject variable is a school identifier (*schcode*). In some cases, there could also be a teacher identifier to cover the situation where individuals are nested in classrooms and schools. In a typical single-level study, the analyst does not have to be concerned about the data structure because all observations are considered to be independent of each other.

One attractive feature of the MIXED routine in SPSS is that it is not limited in terms of the number of levels in a nested or cross-classified data structure that can be analyzed simultaneously. Readers who are not familiar with cross-classified data structures can consult our first workbook for further discussion and examples (Heck, Thomas, & Tabata, 2010). The ability to specify multiple *Random* commands (i.e., randomly varying intercept and slopes) at successive levels of a data hierarchy facilitates the investigation of a variety of multilevel models that are difficult (or not currently possible) to estimate optimally in other software packages. As with most current software programs, however, adding levels to examine in a data hierarchy, randomly varying parameters, and cross-level interactions can greatly increase the amount of time it takes to produce a solution (i.e., from seconds to several hours) and will require increasingly large amounts of RAM, processing capability, and disk space as models become more complex.

A different type of clustering results from repeated observations (which are nested within individuals). The *Repeated Measures* designation allows the analyst to specify repeated observations nested within individuals at the lowest level. In this case, each subject will occupy multiple lines in the data set (one for each observation).

Fields and Effects Command Tab

Target Main Screen

The next screen, summarized in Figure 4.3, shows how to define the outcome variable, as well as the model's fixed and random effects. This is the default screen when creating a model for the first time. The outcome should be a single categorical variable (referred to as a "target"), which can have any measurement level or an events/trials specification—in which case the events and trials must be continuous. The user can also specify the corresponding probability distribution (e.g., binomial, Poisson, negative binomial) and link function (e.g., identity, log), as well as the fixed effects, random effects, offset variable, or analysis weight (IBM SPSS, 2010).

We emphasize again that the analyst should make sure that the dependent variable is defined as scale (continuous), ordinal, or nominal as intended so that the analysis will be properly

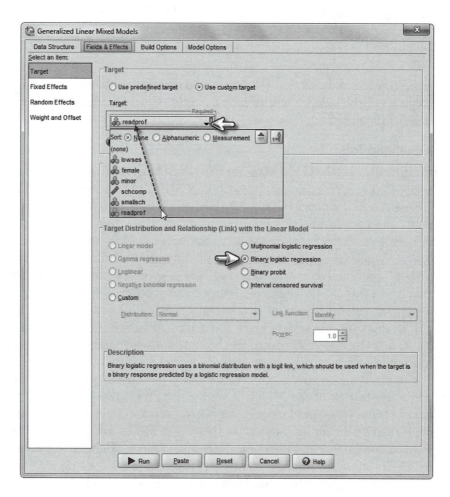

FIGURE 4.3 Defining the target variable (*readprof*) and target distribution (*Binary logistic regression*) in GENLIN MIXED.

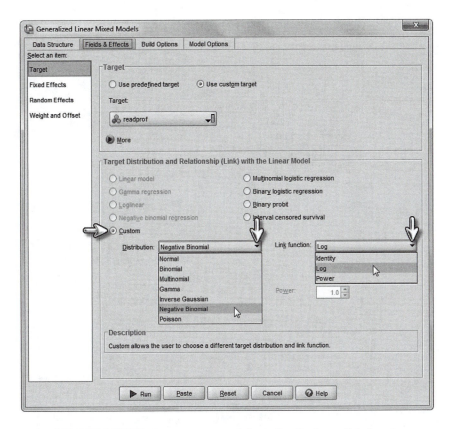

FIGURE 4.4 Using the GENLIN MIXED *Custom* setting to define a distribution to link function.

conducted. One can manually assign the measurement levels in *Variable View* within the data editor (as illustrated in Figure 4.1). The program can also scan the data in the active data set and assign a default measurement level to variables whose measurement levels are unknown. The target variable can have any measurement level. The measurement level of the dependent variable, however, restricts the set of distributions and link functions that are appropriate for the given analysis. GENLIN MIXED offers several common distribution and link function models that can be selected. For example, with a dichotomous outcome (reading proficiency), the selection *Binary logistic regression* automatically couples a binomial probability distribution with the logit link function (which is the canonical link function for the binomial distribution).

If the user selects *Custom*, as shown in Figure 4.4, the program will allow selecting other possible distribution–link function combinations that may be appropriate for a particular modeling situation. The user should select a combination, however, that can be justified by theoretical, methodological, and empirical considerations.

Fixed Effects Main Screen

Multilevel modeling contributes to our understanding of hierarchical data structures by allowing the analyst to estimate structural and variance/covariance parameters (e.g., residual variance in intercepts or slopes at Level 2, covariance between intercepts and slopes at Level 2) more efficiently and accurately. In a multilevel model, we typically focus on output concerning two types of model parameters. Structural parameters are referred to as the model's *fixed effects*. These include intercept coefficients (e.g., a group mean) or slope coefficients (e.g., the relationship between gender and achievement). The complete set of variances and covariances in model parameters is referred to as its *covariance components* (Singer & Willett, 2003). Specific parameters can be designated as *randomly varying*, which means that the sizes of the estimates are

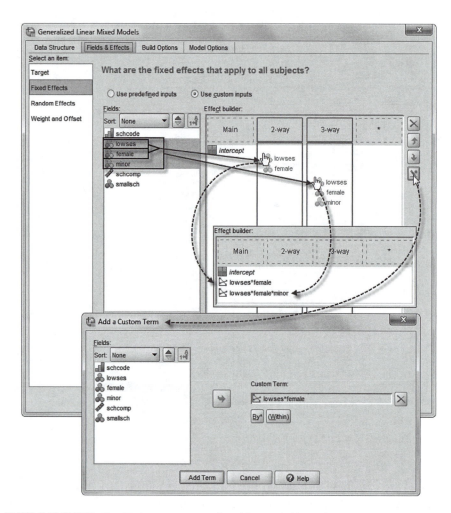

FIGURE 4.5 GENLIN MIXED *Fixed Effects* main screen for adding variables and cross-level interactions into a model (i.e., drag and drop *Add a Custom Term*).

allowed to vary across groups. Investigating these randomly varying intercepts and slopes is at the center of our general multilevel modeling strategy in subsequent chapters.

The fixed effects (see Figure 4.5) are the estimates of the influence of the independent variables (and intercept) in explaining the outcome. Categorical variables (e.g., nominal or ordinal measurement) are represented in the program as *Factors*, and continuous variables (scale measurement) are referred to as *Covariates*. A number of different kinds of fixed effects can be defined. *Main effects* refer to the effect of the variable alone in terms of the outcome. *Two-way* interactions refer to possible pairs of variable effects (e.g., *lowses*female*), and *three-way* interactions refer to possible triplets of variables (e.g., *lowses*female*minor*). Custom terms can also be added to the model by either dragging and dropping variable combinations or building them separately using the *Add a Custom Term* dialog window and then adding them to the model. We illustrate both approaches here, but for consistency, we will use the "drag and drop" method in subsequent chapters of the workbook.

Nested terms refer to situations where the effect of a factor or covariate is not assumed to interact with the levels of another factor. This is shown as A(B). There are some limitations. First, factors within nested effects must be unique, so if A is a factor, specifying A(A) is invalid. Second, no effect can be nested within a covariate. Therefore, if A is a factor and X is a covariate, specifying A(X) is invalid. Interactions may also be specified, such as higher order polynomial terms for the same covariate (e.g., *age*age*). It should be noted that all factors within an interaction must be unique such that if A is a factor, A*A would be invalid.

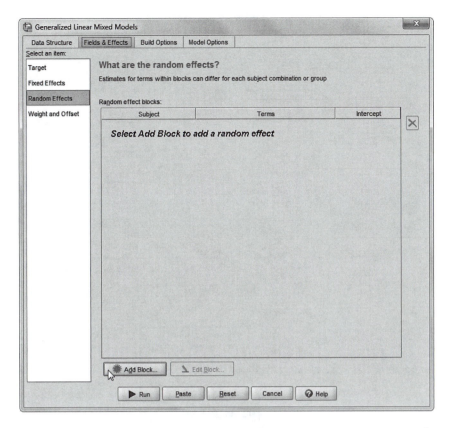

FIGURE 4.6 Preliminary *Random Effects* screen (full screen view; subsequent illustrations are truncated).

Random Effects Main Screen

The random effects specification (Figure 4.6) is used to determine whether the intercept and possible Level-1 slopes are allowed to vary randomly across groups at higher levels of the data hierarchy.

Random effects. We can click on the *Add Block* button to open the *Random Effect Block* screen to add a random intercept or a random Level-1 slope. Random effects can be created by selecting the effect (e.g., *female*) and dragging it into the Random box under *Main Effects*. This is shown in Figure 4.7. The type of effect depends upon the specific column (i.e., main, two-way, three-way) in which the variables are placed. Variables placed in the *Main* section appear as separate main effects. Pairs of variables dropped in the *2-way* area appear as two-way interactions, and triplet variables placed in the *3-way* section appear as three-way interactions. We note that it is also possible to select an interaction as randomly varying across Level-2 units.

Covariance structure. After specifying the random variables in the model, the user can select the *Subject Combination* (i.e., at what level the random effect will vary). In this case, the intercept is checked as a random effect that will vary across schools, which is noted by selecting the school identifier (*schcode*) as the subject combination variable. The user can also select an appropriate covariance structure for the random coefficients at successive levels of the data hierarchy. The default setting is *Variance component,* which provides a variance estimate for each random effect specified. For one random effect, this is the same as specifying an identity covariance matrix. For two or more random effects, it is the same as specifying a diagonal covariance matrix. Other types include *Identity, Diagonal,* and *Unstructured.* If the outcome consists of repeated measures data, some commonly used covariance structures might be *First-order autoregressive* (AR1), *Autoregressive moving average* (1,1) (ARMA11), *Unstructured,* or *Compound symmetry.*

Define covariance groups by. Using the *Define covariance groups by* option in Figure 4.7 allows the user to define independent sets of repeated effects covariance parameters—one for each

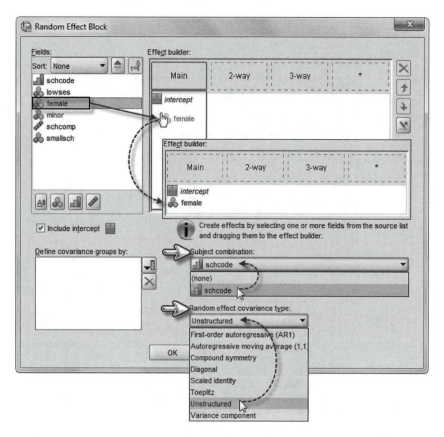

FIGURE 4.7 Selecting *Random Effects, Subject,* and *Covariance Structure* settings.

category defined by the cross-classification of the grouping effects. More specifically, all subjects will have the same covariance type; subjects within the same covariance grouping will have the same values for these parameters.

Weight and Offset Main Screen

Figure 4.8 displays the *Weight and Offset* main screen. When the response is a number of events occurring in a set of trials, the dependent variable contains the number of events. The analyst can select an additional variable to represent the number of trials. Alternatively, if the number of trials is the same across subjects, then trials can be specified using a fixed value (i.e., the number of trials should be greater than or equal to the number of events for each case). Events should be non-negative integers and trials should be positive integers.

Analysis weight. The *Analysis Weight* option here is not to be confused with the sample weight (frequency or probability) used in the *Weight By* command outside the GLMM routine. The *Analysis Weight* is also known as a scale parameter—an estimated parameter related to the variance of the response. Cases with scale weights that are less than or equal to 0 or missing are not used in the analysis.

Offset. The offset term is a structural parameter. Its coefficient is not estimated, but it is assumed to have a value of 1 (i.e., the values of the offset are simply added to the linear predictor of the dependent variable). The offset can be useful in Poisson regression models where each case may have different levels of exposure to the event of interest. For example, when modeling course failures for individual students, there would be an important difference between a student who had one course failure in the first few units attempted in college (a first-time freshman) versus a student who had one course failure in 120 units attempted (a graduating senior).

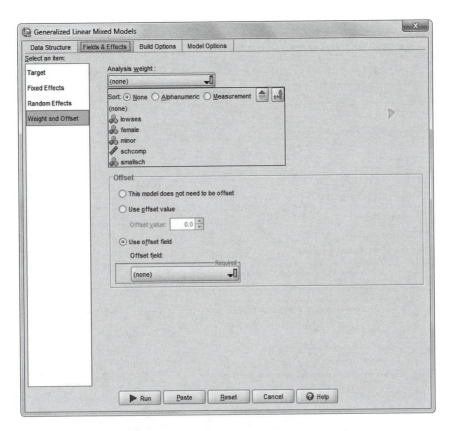

FIGURE 4.8 Selecting an analysis weight and offset setting within the *Weight and Offset* main screen.

When the target outcome variable is a count—that is, it represents a number of events occurring in a set of trials—the target field contains the number of events and the user can select an additional field that specifies the number of trials. For example, the number of course failures can be modeled as a Poisson response variable if the number of units the student has attempted is included as an offset term. Alternatively, if the number of trials is the same across all subjects, then trials can be specified to be fixed at some value. It is important to note that the number of trials should be greater than or equal to the number of events for each record. Events should be non-negative integers (i.e., including 0) and trials should be positive integers $(1, 2,\ldots, N)$.

Build Options Command Tab

Selecting the Sort Order

GENLIN MIXED allows the analyst to change the default reference category of the outcome and any categorical predictors. In Figure 4.9, the *Sorting Order* refers to designating how the categories of the categorical outcome and any categorical predictors will be defined. In a categorical variable, the first value encountered in the data defines the first category and the last value encountered defines the last category. It should be noted that with a binary or multinomial outcome, the analyst can select either the first or last category to be the reference category. In models with binary outcomes, the first group (coded 0) is typically selected as the reference group. The results will then be reported in terms of the likelihood of the event coded 1.

To illustrate, if reading proficiency is coded 1 = proficient and 0 = not proficient, then the model parameters would be in terms of the probability of an individual being proficient (i.e., the probability that $Y = 1$) for a unit change in each predictor. If the predictor is dichotomous, such as female, it would provide the likelihood of the individual being proficient (coded 1) if she is

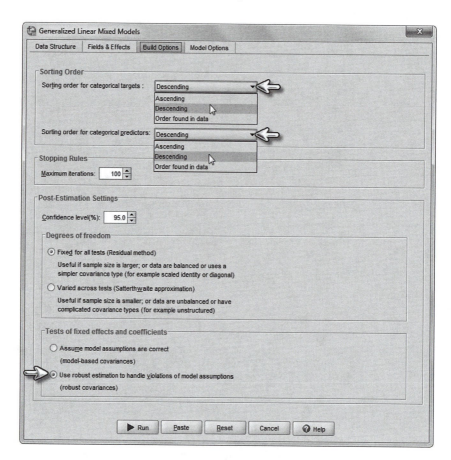

FIGURE 4.9 Available settings in the *Build Options* screen for model construction.

female (coded 1). It is important to note that this is the default for the single-level dichotomous logistic regression program in IBM SPSS but not for the GENLIN MIXED program.

GENLIN MIXED reports outcome results in terms of the *first* category of the outcome and categorical predictors, instead of the second (or last) category, unless *descending* is used in referring to the outcome categories. If the analyst wishes the first category (not proficient) to be the reference group, then the term *descending* should be selected, as shown in Figure 4.9. The resulting model estimates will then be in terms of the likelihood of being proficient for the last category of a dichotomous predictor, such as female, or the other categories except the reference category, where there are three or more categories.

It is important to note that this language is used separately for the set of categorical predictors and the categorical outcome variable; that is, one could report the model effects in terms of the last category of the outcome and the first category of the predictors. Our advice is to be consistent in selecting either descending or ascending categories to facilitate the reporting of the output. We reiterate that dummy coded categorical variables (i.e., 0, 1) can be entered as covariates and considered as *scale*, as is common in multiple regression models. In this case, the estimate reported will be in terms of the predictor category coded 1, producing the same estimate as when the variable is entered as *nominal* (with *descending* categories).

Similarly, on a multinomial outcome, the analyst can decide whether to make the first category or the last category the reference group. Often with multinomial outcomes, analysts will select the last group as the reference. This can be accomplished by using the default *ascending* (last category as reference) categories. Our preference, however, is to use the first category as the reference group because we typically define the outcome such that the last category is the favored result. For example, if we were looking at students' postsecondary plans, we would likely code "enter work force" as the reference (0), "2-year or technical school" (1), and "4-year college" (2). Our multinomial analysis would then compare two-year or technical schools against the reference category and four-year schools against the reference category. This can be accomplished by selecting *descending* as the sorting order (i.e., the first category will then be the reference group).

Stopping Rules

This refers to the maximum number of iterations before the model is stopped. Typically, if a model does not converge on a solution by 100 iterations, it likely is not specified correctly.

Confidence Intervals

This allows the user to set a confidence interval band around the estimates produced. Commonly reported confidence intervals are 95% (the default as shown) or 99%.

Degrees of Freedom

This refers to the degrees of freedom in the denominator used to calculate significance tests for parameters. The default is for common degrees of freedom (referred to as *Residual*). This is useful in situations where there are ample numbers of Level-2 units (say, 100 or more) and balanced data in terms of the number of individuals in each unit, and where more simplified covariance matrices of relationships are implied (identity, diagonal). The program also supports the application of the Satterthwaite correction to denominator degrees of freedom in situations where there are few Level-2 units, unbalanced data, or more complex covariance matrices of effects (e.g., unstructured).

Tests of Fixed Effects

Finally, on this screen the user can specify the IBM SPSS default model-based standard errors setting (where there are large samples and data are normal) or robust standard errors, where there may be departures from normality. This latter approach is a more conservative approach to the calculation of standard errors. Standard errors are important to consider because they indicate the relative variability (or imprecision) in estimating regression coefficients. The calculation of the standard errors directly affects the efficacy of hypothesis tests concerning the predictors in the model because hypothesis tests are formulated as the ratio of the slope estimate to its standard error. Because statistical tests of model parameters are based on these ratios, the underestimation of standard errors will often lead to more findings of significance than would be observed if clustering were considered. We generally recommend using robust standard errors to be more conservative with respect to possible departures from normality. When there are only a small number of Level-2 units, however, we suggest using caution, as this approach may result in inaccurate estimates.

We note that tests of the significance of model parameters are typically constructed using the standard normal distribution. In some situations (e.g., small sample sizes, probability distributions other than normal), however, the test statistics may not approximate a normal distribution very accurately. GENLIN MIXED uses a *t*-test to construct significance tests involving the ratio of a fixed-effect parameter to its standard error. For example, if the estimate of the effect of being female on a reading score is 0.4 and the standard error is estimated as 0.2, this would result in a *t*-ratio of 2.0 (i.e., the ratio of the estimate to its standard error). At a commonly adopted significance level of $p = .05$ and a sample size of 65 individuals, the required *t*-ratio would be 2.0. So for small sample sizes, we note that *t*-values above the typical 1.96 in a normal distribution with large samples will be required.

The program also provides a correction to the degrees of freedom used in constructing statistical tests in multilevel models. This is known as the Satterthwaite approximation (Satterthwaite, 1946); it offers a correction for calculating degrees of freedom in constructing hypothesis tests that provides generally a more conservative estimate of the standard errors. This can be useful with Level-2 units that vary in size considerably or when the covariance matrix chosen is more complex (such as unstructured). With large sample sizes, it typically does not make much difference in correcting the estimates.

Readers should also keep in mind that, for multilevel models with categorical outcomes, particular estimation methods (e.g., those featuring quasilikelihood approximations) may make the use of model testing techniques that depend on the real likelihood function suspect. In particular, this affects direct comparisons of models (or single parameters) using the deviance statistic (–2LL) and the likelihood ratio test. In general, methods relying on full information ML and numerical integration will produce results that support direct comparisons of model fit using the difference in model log likelihoods. Because multilevel modeling with categorical outcomes is relatively new, it is likely that continued methodological progress will be made in providing expanded estimation options for these models. This should result in more accurate comparisons of model fit.

Tests of Variance Components

At Level 2 (and successive levels), GENLIN MIXED provides a z-statistic to test hypotheses about the variance of randomly varying slopes and intercepts across groups. As we have noted, this test is based on the assumption of an underlying normal distribution. However, the sampling variance of Level-2 variances is often skewed, especially when the variance is small and when there may be a small number of Level-2 groups (Hox, 2010; Raudenbush & Bryk, 2002). In these situations, significance tests typically have very low power to detect an effect.

When the variance of a random Level-2 parameter is near 0, model convergence problems may also occur. We also note that chi-square difference tests of variance parameters are more complicated in multilevel models with categorical outcomes because they depend on relatively large sample sizes and normal distributions, and because negative variances are not permissible in examining variance components, the resulting p-values are too high (Berkhof & Snijders, 2001; Hox, 2010). Currently, the program does not support chi-square estimates of variance and covariance parameters with multilevel categorical outcomes. We also note that because variance estimates cannot be negative, some software programs will correct them to 0. Because 50% of the distribution is removed in this instance, one possible remedy is to adjust the significance level at which the test is conducted (Agresti, 2002; Hox, 2010), especially in small level-2 sample sizes. As Hox (2010) notes, when the Wald Z is used to test a variance component, it is appropriate to use a one-tailed significance test (or divide the p-value by two).

Model Options Command Tab

Estimating Means and Contrasts

This screen offers a number of options for investigating means and performing various contrast tests. These are summarized in Figure 4.10. Charts can be displayed for up to 10 fixed all-factor effects, beginning with three-way interactions, two-way interactions, and main effects (IBM SPSS, 2010). The charts can display model-estimated values of the dependent variable on the vertical axis for each value of the main effect (or first listed effect in an interaction) on the horizontal axis. A separate line is produced for each value of the second listed effect in an interaction, with a separate chart produced for the third effect in an interaction. This can be helpful in considering the effect of each predictor on the outcome. If no effect is significant, no estimated means are produced.

Custom effects can also be generated. If contrasts are requested, charts can be displayed to compare levels of the contrast fields. Contrasts include pairwise, deviation, and simple. Alternatively, for contrasts, tables can be generated with the model-estimated value of the dependent variable, its standard error, and confidence interval for each level combination of the fields in the effect (with other predictors held constant). A table with the overall test results can be displayed, as well as, for interactions, a separate overall test for each level combination of the effects other than the contrast field. Confidence intervals can also be displayed for the marginal

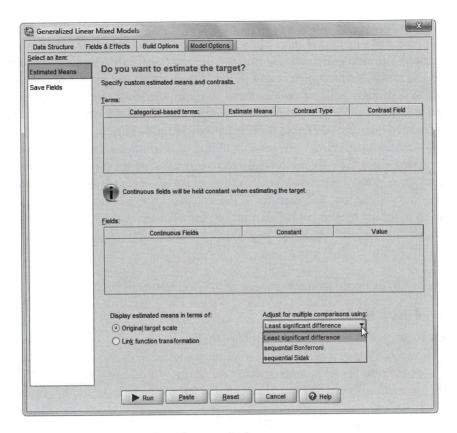

FIGURE 4.10 The *Estimating Means and Defining Contrasts* display screen.

means (using the confidence level specified in the model-building options). Readers can find further information about these display options in the SPSS discussion of GLMM in the user's guide of the Advanced Module (IBM SPSS, 2010).

Save Fields

Finally, the program also allows users to save a variety of predicted values, confidence intervals, and Pearson residuals (which should be 0 if the model fits the data well) for use in further analyses.

Examining Variables That Explain Student Proficiency in Reading

We return to our analysis from Chapter 3, where the researcher wishes to determine whether there is a relationship between elementary-aged students' background variables (e.g., gender, socioeconomic status) and whether school variables might explain their likelihood to be proficient in reading.

Research Questions

Our presentation is built around a series of steps to illustrate multilevel analyses using GENLIN MIXED. Our first research question might simply focus on whether student proficiency in reading varies across a set of schools. We can ask: Does student proficiency in reading vary across schools? To answer this type of question, we would have to employ a mixed (or multilevel) modeling framework. Second, we might ask: Does background affect students' probability of being proficient? In this case, we add variables such as socioeconomic status and gender. Of course, we can answer this question with a single-level analysis, but we would have to make the assumption that students are randomly sampled and that there are not higher level groupings of students present. Third, extending this analysis of randomly varying proficiency intercepts, we might ask: Do school-level context and process variables influence student proficiency in reading? Finally, we might investigate whether school context variables (i.e., student composition) and academic processes affect the relationship between a student background variable (e.g., SES, race/ethnicity) and probability of attaining proficiency. More specifically, we ask: Do features of schools' contexts and academic environments moderate the relationship between individual student race/ethnicity and likelihood to be proficient in reading?

Our research questions therefore provide an illustration of building a two-level model to investigate (1) a randomly varying intercept (likelihood to be proficient) and, subsequently, (2) a randomly varying slope (i.e., the individual race/ethnicity–reading proficiency relationship) across school settings. We provide considerable detail of these previously mentioned steps in building modeling in this chapter, but then leave it to readers to duplicate (and extend) these specific types of steps for the models we build in subsequent chapters. Often, as readers will note in their research, they may adapt these basic procedures in various ways as their analyses proceed.

The Data

The data in this extended example consist of 6,528 students in 122 schools (Table 4.1).

The Unconditional (Null) Model

We next present an overall modeling strategy that we have used to investigate multilevel models. Our first research question focused on whether students' likelihood to be proficient in reading varies across schools (Does student reading proficiency vary across schools?). We can therefore begin by estimating an unconditional (no predictors) model to examine the extent of variability of the dichotomous outcome across Level-2 units.

The expected probability that individual i in school j is proficient in reading can be expressed as

$$E(Y_{ij}) = \pi_{ij}, \tag{4.6}$$

where π_{ij} is the probability that $Y_{ij} = 1$, and $1 - \pi_{ij}$ will be the probability of not being proficient ($Y_{ij} = 0$).

TABLE 4.1 Data Definition of ch4data1.sav ($N = 6,528$)

Variable	Level[a]	Description	Values	Measurement
schcode	School	School identifier (122 schools)	Integer	Ordinal
lowses	Individual	Predictor dichotomous variable measuring student's social economic status (SES)	0 = not low socioeconomic status 1 = low socioeconomic status	Nominal
female	Individual	Demographic predictor variable representing student's gender	0 = male 1 = female	Nominal
minor	Individual	Demographic predictor variable representing student's ethnicity	0 = nonminority 1 = minority	Nominal
schcomp	School	Weighted composite representing the percentage of students receiving additional support services related to economic status, language background, or other special learning needs	–2.80 to 2.59	Scale
smallsch	School	Predictor dichotomous variable representing school size based on the number of students	0 = not small school (>599) 1 = small school (<600)	Nominal
readprof	Individual	Two-category dependent variable referring to student's reading proficiency	0 = not proficient 1 = proficient	Nominal

[a] Individual = Level 1; school = Level 2.

For multilevel models with categorical outcomes, it is important to note that only the Level-1 model (i.e., the model for individuals) differs from the more familiar multilevel model with continuous outcomes (Raudenbush et al., 2004). The higher levels are specified in the same way as typical multilevel models with continuous outcomes.

At Level 1, the unconditional model to relate the transformed predicted values to an intercept parameter is the following:

$$\eta_{ij} = \log\left(\frac{\pi_{ij}}{1 - \pi_{ij}}\right) = \beta_{0j}. \tag{4.7}$$

We remind readers that multilevel models with categorical or count outcomes do not assume normal distributions at Level 1. Because of this, with a dichotomous outcome, there is no Level-1 residual in the Level-1 model because, in a binomial distribution, the variance of the observed proportion at Level 1 is determined by the estimated value of the population proportion (π_{ij}). The initial output confirms that we have defined a model with binomial probability distribution and logit link function.

At Level 2, the addition of a variance parameter (u_{0j}) suggests the intercept (γ_{00}) varies between schools:

$$\beta_{0j} = \gamma_{00} + u_{0j}. \tag{4.8}$$

Through substitution of Equation 4.8 into Equation 4.7, the combined single equation is the following:

$$\eta_{ij} = \gamma_{00} + u_{0j}. \tag{4.9}$$

This suggests that there are two parameters to be estimated in the unconditional model: one fixed effect for the Level-2 intercept and the Level-2 variance for the intercept.

For this first model, we will describe how to open the GENLIN MIXED program, select the Level-2 group identifier, and select the appropriate outcome (*readprof*), distribution (binomial), and link function (log). In subsequent models, we will expect readers to access the program and begin at the *Field and Effects* screen or another specified screen. Readers may wish to "paste" a copy of the syntax and then save it before running each model so that a record of each subsequent model examined in the series of analyses can be kept.

Defining Model 1.1 With IBM SPSS Menu Commands
Note: We will use the *ch4data1.sav* data file.

1. Go to the toolbar and select ANALYZE, MIXED MODELS, GENERALIZED LINEAR. This command opens the *Generalized Linear Mixed Models* (GENLINMIXED) main dialog box.

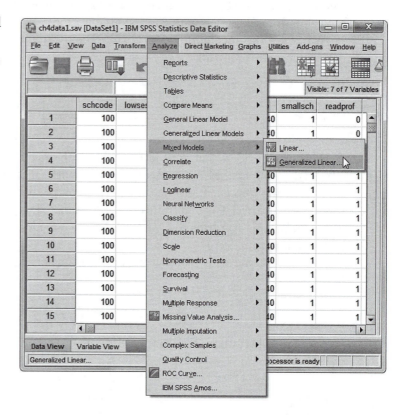

2. The *Generalized Linear Mixed Models* display screen shows four command tabs: *Data Structure, Fields & Effects, Build Options,* and *Model Options*.

a. The *Data Structure* command tab is the default screen display when creating a model for the first time. The options enable specifying structural relationships between data set records when observations are correlated.

b. To build our model, begin by defining the subject variable. A "subject" is an observational unit that can

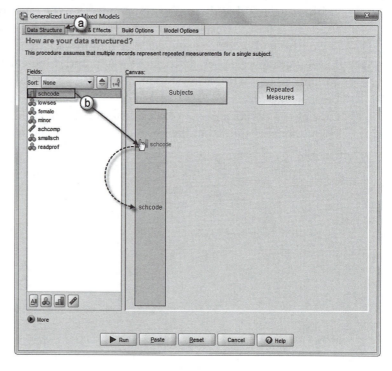

be considered independent of other subjects. Click to select the *schcode* variable from the left column and then drag and drop *schcode* onto the canvas area beneath the *Subjects* box.

3a. Click the *Fields & Effects* command tab, which displays the default *Target* display screen. The settings allow specifying a model's target (dependent variable), distribution, and relationship to model predictors through the link functions.

b. For this model we will use student reading proficiency (*readprof*) as the dependent variable. Click the *Target* pull-down menu and select *readprof*. *Note:* Selecting the

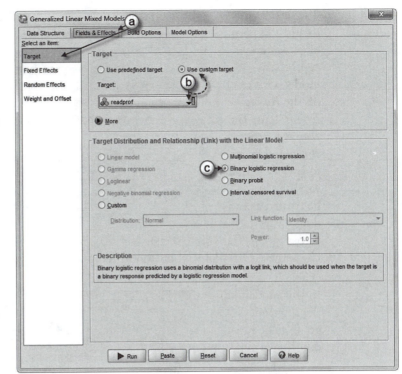

target variable will automatically change the setting to *Use custom target*.

c. For this first model we will designate a binary logistic analysis due to the dichotomous aspect of the dependent variable (*readprof*) to be used for this model. Click to select *Binary logistic regression*.

4a. Now click the *Fixed Effects* item to access the main dialog box.

b. Note that the *Include intercept in model* option is preselected and displayed in the *Effect builder* as *intercept*. This is because a Level-1 model is intercept only (default), so other model effects must be specified. Because this is the null model, no predictor variables will be used at this time.

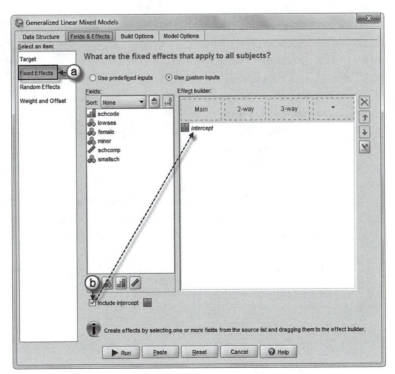

$$\eta_{ij} = \log\left(\frac{\pi_{ij}}{1 - \pi_{ij}}\right) = \beta_{0j} \quad \text{(Eq. 4.7)}$$

5a. The next step is to add a random effect (intercept) to the model. Click on the *Random Effects* item. The *Random Effects* command is used to specify which variables are to be treated as randomly varying across groups (random effects).

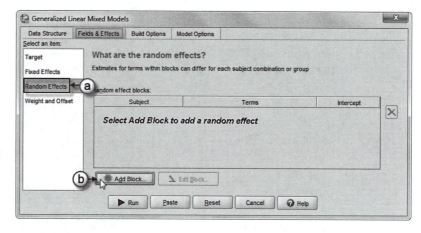

b. Click the ADD BLOCK button to access the *Random Effect Block* main dialog box.

c. In this model only the intercept is going to be randomly varying, so click to select the *include intercept* option which changes the setting within the *Effect builder* from *no intercept*. Because there is only one random effect (the intercept), we can use the default variance component (VC) setting. This specifies a diagonal covariance matrix for the random effects;

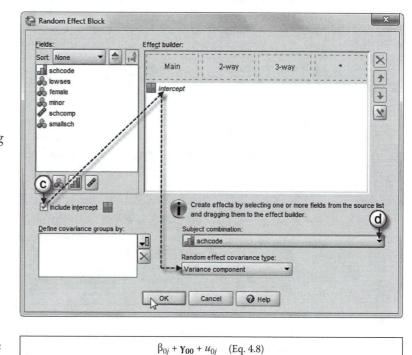

$$\beta_{0j} + \gamma_{00} + u_{0j} \quad \text{(Eq. 4.8)}$$

that is, it provides a separate variance estimate for each random effect, but not covariance between random effects. With one random effect, it becomes an identity covariance matrix.

d. Select the *Subject combination* by clicking the pull-down menu and selecting *schcode*.

Click the OK button to close the *Random Effect Block* dialog box and return to the *Random Effects* main screen.

e. In the main screen notice that the *Random effect blocks* displays a summary of the block just created: *schcode* (Subject) and *intercept* (Yes).

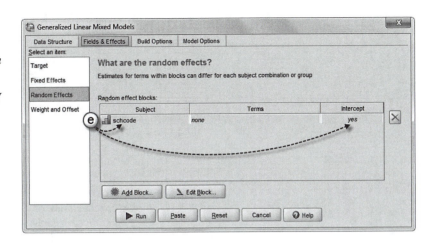

6a. Finally, click the *Build Options* command tab located at the top of the screen to access advanced criteria for building the model.

b. Change the sorting order for both categorical targets and categorical predictors. Click the pull-down menu for each option and select *Descending*. We will retain the default settings for *Stopping Rules, Post Estimation,* and *Degrees of Freedom.*

c. Change the *Tests of fixed effects and coefficients.* Click to select *Use robust estimation to handle violations of model assumptions.*

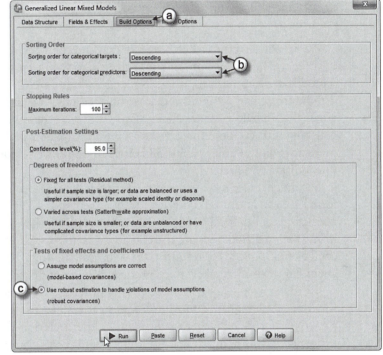

Click the RUN button to generate the model results.

Interpreting the Output of Model 1.1

Table 4.2 presents information about the fixed and random effects in the proposed model, as well as the number of Level-2 units. It is important to confirm that what was estimated was what was intended. The table indicates one fixed effect (the school mean probability of being proficient) and one random effect (the randomly varying intercept) to be estimated. The number of Level-2 units (i.e., *Common Subjects*) is also summarized ($N = 122$). The information at the bottom of the table explains how the subject units are specified and the number of columns in the data matrix per subject unit. Because this information does not change, we will not include it in subsequent tables describing the model's parameters in this chapter.

TABLE 4.2   Model 1.1: Covariance Parameters

Covariance parameters	
Residual effect	0
Random effects	1
Design matrix columns	
Fixed effects	1
Random effects	1[a]
Common subjects	122

Note: Common subjects are based on the subject specifications for the residual and random effects and are used to chunk the data for better performance.

[a] This is the number of columns per common subject.

As we noted in the last chapter, we may interpret the results from a logistic regression model in three ways. First, we can interpret the equation in terms of log odds. In Table 4.3 the intercept log odds are summarized. The estimated log odds are 0.904 (SE = 0.079, $t = 11.377$, $p < .001$). The intercept can be interpreted as the predicted log of the odds that a student is proficient when that individual has a standing of 0 on the other predictors and holding the *random effect* constant at 0. We note that this assumption is made when considering the effects of individual predictors on the unit-specific outcomes in multilevel models. Of course, there are none in the unconditional model.

Second, because most people do not find it natural to think in terms of log odds, we can transform the log odds into an odds ratio by exponentiation. With respect to odds, the influence of each predictor is multiplicative, rather than additive, as for the log odds. Thus, for each one unit increase in X_{ij}, the predicted odds are increased by a factor of e^β. In this case, exponentiating the log odds coefficient ($2.71828^{0.904}$) results in the odds ratio (2.469). The odds ratio suggests that students are about 2.5 times more likely to be proficient than not proficient within the average school.

Third, we can also express the results in terms of probabilities by use of the inverse logistic function. Although this form may seem easiest because it results in predicted probabilities for $Y_{ij} = 1$, unfortunately there is no simple interpretation for the logistic regression coefficients expressed in this manner. Because the model is nonlinear in expressing the probabilities, in this format we must choose a reference point for calculating the effects of unit changes in X_{ij}. The most common choice of a reference point is a person who has the sample mean values on all X_{ij} variables. For a single predictor, however, we can generate a plot with X_{ij} on the horizontal axis and the probability $Y_{ij} = 1$ represented on the vertical axis. The resulting curve is S-shaped (e.g., see Figure 3.8 in Chapter 3).

TABLE 4.3   Model 1.1: Fixed Effects

Model term	Coefficients	Std. error	T	Sig.	Exp(coefficient)	95% Confidence interval for exp(coefficient)	
						Lower	Upper
Intercept	0.904	0.079	11.377	0.000	2.469	2.113	2.885

Note: Probability distribution: binomial; link function: logit.

TABLE 4.4 Model 1.1: Variance Components

Random effect	Estimate	Std. error	z-Test	Sig.	95% Confidence interval	
					Lower	Upper
Var(Intercept)	0.636	0.106	6.025	0.000	0.459	0.881

Note: Covariance structure: variance components; subject specification: schcode.

In this case, when we convert the predicted log odds back to the average school-level probability that students are proficient in reading [$1/(1 + e^{-(0.904)})$, we obtain the estimated probability of 0.712 (or 71.2% proficient). We note in passing that this coefficient is slightly larger than the probability of being proficient obtained in the single-level analysis in Chapter 3 when no predictors were included in the model (i.e., 0.691 or 69.1% in Table 3.8). This is because it represents the average unit-level outcome as opposed to the average population outcome.

The variance components are summarized in Table 4.4. The z-test ($z = 6.025$, $p < .001$) suggests that the intercept variance varies between Level-2 units (σ_B^2). Therefore, in this case, we can assume statistically significant variability in intercepts across schools, which justifies developing a multilevel model.

We note that the residual variance at Level 1 is scaled to 1.0, so it cannot be tested for statistical significance. It is merely a means of providing a metric for the underlying latent predictor (η_{ij}). It turns out, however, that the scale factor can be used to calculate an intraclass correlation (ICC). The ICC describes the proportion of variance that lies between units ($\sigma_{Between}^2$) relative to the total variance (i.e., $\sigma_{Between}^2 + \sigma_{Within}^2$). The variance of a logistic distribution with scale factor 1.0 is $\pi^2/3$, or approximately 3.29 (Hox, 2002; Hedeker, 2007), so we can estimate the ICC (ρ) as follows:

$$\rho = \sigma_{Between}^2/(\sigma_{Between}^2 + 3.29_{Within}) \tag{4.10}$$

For these data, the ICC can then be calculated as 0.162 [$0.636/(0.636 + 3.29)$], suggesting that about 16.2% of the variability in reading proficiency lies between schools.

Defining the Within-School Variables

Given that there is considerable variability between schools in likelihood for students to be proficient in reading, we can develop a multilevel model to explain this variability. The second step is generally to estimate a model with only the Level-1 predictors. In this case, the model consists of the three background variables already used in the single-level analysis. The Level-1 model can be written as follows:

$$\eta_{ij} = \log\left[\frac{\pi_{ij}}{1 - \pi_{ij}}\right] \beta_{0j} + \beta_{1j}lowSES_{ij} + \beta_{2j}female_{ij} + \beta_{3j}minor_{ij}. \tag{4.11}$$

In this case, we will define the three predictors as nominal, with the reference categories coded as 0 and the categories named in Equation 4.10 coded 1. Note that we already set the SPSS defaults to estimate the output terms of the categories coded 1 by selecting "descending" categories for the target response variable and the categorical predictors.

Once again, the Level-2 model has only a random intercept, as in Equation 4.8. The Level-1 slopes for the background variables ($\beta_{1j} - \beta_{3j}$) are typically defined as fixed initially; that is, no corresponding school-level error terms ($u_1 - u_3$) are defined for them:

$$\beta_{1j} = \gamma_{10}$$
$$\beta_{2j} = \gamma_{20} \quad\quad (4.12)$$
$$\beta_{3j} = \gamma_{30}.$$

Substituting Equation 4.12 and Equation 4.8 into Equation 4.11 therefore results in the combined school-level model with four fixed effects and one random effect (u_{0j}):

$$\eta_{ij} = \gamma_{00} + \gamma_{10}lowses_{ij} + \gamma_{20}\,female_{ij} + \gamma_{30}minor_{ij} + u_{0j}. \quad\quad (4.13)$$

Defining Model 1.2 With IBM SPSS Menu Commands

Note: IBM SPSS settings default to those used in Model 1.1.

1. Go to the toolbar and select ANALYZE, MIXED MODELS, GENERALIZED LINEAR. This command opens the *Generalized Linear Mixed Models* (GENLIN MIXED) main dialog box.

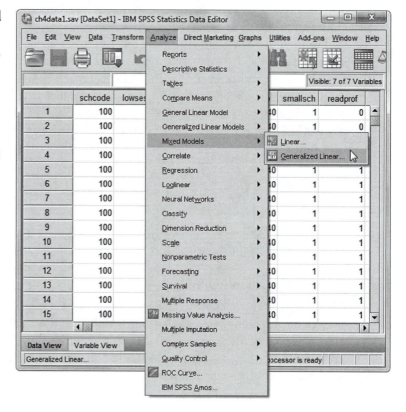

2a. The *Generalized Linear Mixed Models* dialog box displays the *Build Options* command tab, which was the last setting used in the prior model.

b. We will begin to build the model by adding three predictor variables. First, click to select the *Fields & Effects* command tab, which displays the *Random Effects* main screen (not illustrated).

Note: The *Random Effects* item is preselected (represented in the illustration by the dashed-line box but would appear as grayed-out on the computer screen) as it was the last setting used from the prior model.

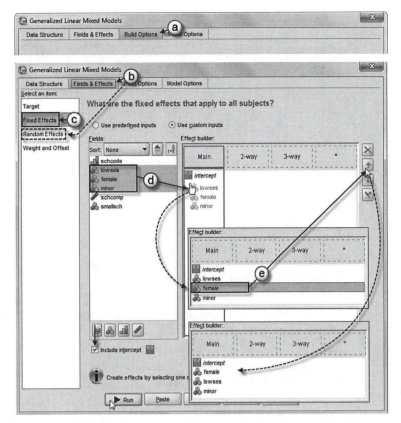

$$\eta_{ij} = \log\left[\frac{\pi_{ij}}{1-\pi_{ij}}\right]\beta_{0j} + \beta_{1j}lowSES_{ij} + \beta_{2j}female_{ij} + \beta_{3j}minor_{ij} \quad \text{(Eq. 4.11)}$$

c. Click to select the *Fixed Effects* item to display the main screen, where we may add variables to the model.

d. Click to select *lowses, female,* and *minor* and then drag and drop them into the *Main* (effect) column of the *Effect builder*.

e. To facilitate reading of the output tables, we will rearrange the variable order. Click to select *female* and then click the up-arrow button on the right to place *female* above *lowses*. The variable order is now *lowses, female, minor.*

Note: The order may also be controlled by entering the variables into the *Effect builder*, beginning with *female* and followed by the remainder (*lowses, minor*).

Click the RUN button to generate the model results.

Interpreting the Output of Model 1.2

The output provides a dimensions table (Table 4.5) that can be checked to make sure the model estimated is the one intended. For Model 2.2, as we noted from Equation 4.12, four fixed effects and one random effect are estimated. However, because we specified the predictors as dichotomous factors, the design matrix suggests that there are seven fixed effects. These consist of the intercept (which is continuous) and the three predictors (i.e., *lowses, female, minor*), which consist of two categories each. There is also one random effect (i.e., the randomly varying intercept) in the covariance parameters portion of the table. Because one category is redundant for each dichotomous predictor, there are only five total parameters to be estimated (i.e., four fixed effects

TABLE 4.5 Model 1.2: Covariance Parameters

Covariance parameters	
Residual effect	0
Random effects	1
Design matrix columns	
Fixed effects	7
Random effects	1
Common subjects	122

and one school-level random effect), as summarized in Equation 4.9. Once again, there is no Level-1 residual listed in the table (because it is fixed to 1.0).

Table 4.6 provides the fixed-effect output for this model. The estimates for the three predictors in the model are given in terms of the group coded 1 (i.e., females, low SES students, and students who are minority by race/ethnicity). The reference parameters are 0 because they are redundant. We can say that being female increases the log odds that a student will be proficient by 0.474 units ($\beta = 0.474$), holding the other effects constant. Exponentiating the log odds results in an odds ratio of 1.606, suggesting that, for females, the odds of being proficient are multiplied by 1.606 (or the odds are increased by about 61%) compared with males. Another way of expressing this is to say that the odds of being proficient are about 1.6 times higher for females than for males, holding race/ethnicity and SES constant.

Regarding SES, we can note that the log odds of being proficient for low SES students is decreased by 0.863 units ($\beta = -0.863$) compared with their more average or high SES peers, holding the other effects constant. The odds ratio suggests that the odds that a low SES individual is proficient in reading (adjusted for race/ethnicity and gender) are multiplied by 0.422 (or reduced by about 57.8%) compared with their higher SES peers. Finally, for students who are underrepresented by race/ethnicity, the log odds of being proficient are decreased by 0.627 units compared to their majority peers ($\beta = -0.627$), adjusting for SES and gender. For students who are minority by race/ethnicity, the adjusted odds of being proficient are multiplied by 0.534 (or reduced by about 46.6%), which suggests that they are only about half as likely to be proficient as their majority peers.

We note that an odds ratio of 0.5 to be proficient would be the same as an odds ratio of 2.0 favoring being not proficient (1/0.5 = 2.0). If desired, we can convert the odds ratio to express the likelihood of students who are minority by race/ethnicity not to be proficient in reading

TABLE 4.6 Model 1.2: Fixed Effects

Model term	Coefficient	Std. error	T	Sig.	Exp(coefficient)	95% Confidence interval for exp(coefficient)	
						Lower	Upper
Intercept	1.419	0.093	15.258	0.000	4.132	3.443	4.958
female = 1	0.474	0.058	8.178	0.000	1.606	1.434	1.800
female = 0	0ᵃ						
lowses = 1	−0.863	0.064	−13.442	0.000	0.422	0.372	0.479
lowses = 0	0ᵃ						
minor = 1	−0.627	0.080	−7.831	0.000	0.534	0.457	0.625
minor = 0	0ᵃ						

Note: Probability distribution: binomial; link function: logit.
ᵃ This coefficient is set to 0 because it is redundant.

(1/.534). The result indicates that they are about 1.87 times more likely to be *not proficient* than their majority peers, again adjusting for SES and gender.

We can also use the log odds to predict the probability of a student being proficient for different levels or categories of the predictors. We can rearrange the log odds equation to predict the probability $Y = 1$ (the student is proficient) for given levels of a predictor. We use the following equation where π is the probability $Y = 1$:

$$\pi = \frac{\exp(\beta_0 + \beta X)}{1 + \exp(\beta_0 + \beta X)} \tag{4.14}$$

For example, if we want to know the probably of a student of average SES (0) being proficient, we only need the intercept log odds (β_0), since $\beta X = 0$ at the intercept (and the other variables are also held constant at 0). In Table 4.6, the log odds is 1.419. Using Equation 4.14, we can write this as ($e^{1.419}/(1 + e^{1.419})$). This would be 4.132/5.132, which is 0.805. This suggests the probability that $\pi_{ij} = 1$ (i.e., proficient in reading) is about 0.81, when the other predictors are held constant at 0. For a student who is low SES (coded 1), we can add the two log odds [1.419 + (−0.863) = 0.556]. If we exponentiate this ($e^{.556}$), we obtain 1.744. Once again, from Equation 4.14, the probability of a person who is low SES being proficient is 1.744/2.744 or 0.636. Another way of looking this is to multiply the intercept odds ratio by the low SES odds ratio (4.132 × 0.422 = 1.744). Using the previous equation, we again have the value of the numerator as 1.744 and the denominator as 2.744, which also yields the probability of $Y = 1$ as 0.636.

The Level-2 variance component for the intercept is summarized in Table 4.7. Once again, the z-test is significant, suggesting significant variance in the probability of being proficient across groups. Readers should keep in mind, however, that the variance at Level 1 is rescaled each time variables are added to the model, which also affects the variance estimate at Level 2 (Hox, 2010). Because of this, the notion of examining the reduction of variance as successive sets of predictors are added to the model can be misleading when estimating models with categorical outcomes (Hox, 2010).

As Hox (2010) notes, if the analyst starts with an intercept-only model and then adds a set of predictors, we expect the available variance to diminish in models with continuous outcomes. In logistic and probit (or probability unit) models (and other types of categorical models that depend on an underlying continuous variable), the underlying variable is rescaled to 1.0, so the lowest level variance is again 3.29 in the logistic distribution. Models using other link functions have different variances from the logistic distribution (Evans, Hastings, & Peacock, 2000). For example, for the probit model, the variance is scaled to 1.0. Because the Level-1 variance is rescaled, this has the effect of rescaling the higher level variances—in addition to any actual changes in variance that take place due to adding the predictors to the model (Hox, 2010). These scale changes make it impossible to compare regression coefficients across models or examine reduction in variance in the same way as models that have continuous outcomes (Hox, 2010; Snijders & Bosker, 1999).

For single-level models with categorical outcomes, a number of pseudo *r*-square formulas have been developed; these are based on the model log likelihood. They can aid in interpreting

TABLE 4.7 Model 1.2: Variance Components

Random effect	Estimate	Std. error	z-Test	Sig.	95% Confidence interval	
					Lower	Upper
Var(Intercept)	0.275	0.059	4.678	0.000	0.181	0.418

Note: Covariance structure: variance components; subject specification: schcode.

the worth of particular models. These pseudo *r*-square indices, however, are not the same as explained variance in a multiple regression model (Tabachnick & Fidell, 2007). As Hox (2010) notes, the formulas can be applied to multilevel logistic and probit regression models to provide a constant scale of comparison across several models, if a reliable estimate of the log likelihood is available.

This latter point is important because when pseudolikelihood estimation is used, calculation of the reduction in variance may not be very accurate. Hox (2010) notes that a scale correction factor can be calculated that places subsequent models on the same scale as the null model (p. 136). Our own view is that analysts should always rely most heavily on the substantive sensibility of particular models (see Hox, 2010, for further discussion of pseudo *r*-square measures for categorical multilevel models).

As we have suggested previously, we could also define the three background predictors as *scale* variables in the data set (even though they are dichotomous) rather than nominal. The fixed-effect and random coefficient output is the same, but the change in defining the predictors as scale instead of categorical will affect the model dimension output in terms of the design matrix. Because the dichotomous predictors are treated as dummy variables (with the reference category coded 0), as in a regression model, the number of levels for each predictor now will be one (instead of two), and the resulting number of fixed and random effects is five, which is consistent with Equation 4.13. We summarize this difference in model parameterization in Table 4.8.

Examining Whether a Level-1 Slope Varies Between Schools

As a third step, we might wish to examine whether a Level-1 slope varies across schools. Alternatively, we could add Level-2 predictors at this step. In this case, in our third model, we examine whether a Level-1 slope varies across schools. We could actually investigate any one of the individual-level predictors first because all three of them were significant in explaining likelihood to be proficient in reading. For demonstration purposes, we will examine whether the relationship between the student race/ethnicity and likelihood to be proficient in reading varies across schools. This might help us identify schools that are more (or less) equitable in producing outcomes for students of varying racial/ethnic backgrounds. In this case, we might ask: Does the slope vary randomly across schools, and, if it does, is there a relationship between features of schools' environments and the strength of the slope relationship? The within-school model remains the same as Equation 4.11.

First, the step is to specify the Level-1 slope as randomly varying (because it has been previously fixed as in Equation 4.12). To examine whether the slope varies, we need to make a couple of changes in the model. First, we need to estimate the slope's variance at Level 2. To do so, we need to change the Level-2 minority by race/ethnicity slope from fixed ($\beta_{3j} = \gamma_{30}$) to randomly varying:

$$\beta_{3j} = \gamma_{30} + u_{3j}. \tag{4.15}$$

TABLE 4.8 Alternative Model 1.2: Covariance Parameters (Predictors Defined as Scale)

Covariance parameters	
Residual effect	0
Random effects	1
Design matrix columns	
Fixed effects	4
Random effects	1
Common subjects	122

The combined school-level model changes from the one presented in Equation 4.13 because of the presence of the random slope coefficient for race/ethnicity (u_{3j}) at Level 2. Through substitution, the combined, single-equation model is now as follows:

$$\eta_{ij} = \gamma_{00} + \gamma_{10}lowses_{ij} + \gamma_{20}female_{ij} + \gamma_{30}minor_{ij} + u_{3j}minor_{ij} + u_{0j} \qquad (4.16)$$

Second, we need to note that the covariance matrix of random effects also changes to accommodate the random slope. Adding the randomly varying slope changes the number of random effects in the model from one (the intercept) to two (intercept, slope). If we add the randomly varying slope, we will have a default diagonal covariance structure at Level 2. We can estimate this structure using the default *Variance component* specification, which is the same as specifying the covariance structure at Level 2 as *Diagonal*. As Equation 4.16 suggests, this provides estimates of the intercept and slope variances, but not their covariance:

$$\begin{bmatrix} \sigma_I^2 & 0 \\ 0 & \sigma_S^2 \end{bmatrix}. \qquad (4.17)$$

Previously, we estimated five total parameters in the model specification summarized in Equation 4.13 (i.e., four fixed effects, one random Level-2 variance). With the added random slope parameter, we will now be estimating six parameters (i.e., four fixed effects and two Level-2 random effects).

In contrast, if we specified the Level-2 covariance structure as *Unstructured*, it indicates that we are also estimating the covariance between the intercept and slope. If we also estimate the covariance (σ_{IS}), an additional estimated parameter is added to the model:

$$\begin{bmatrix} \sigma_I^2 & \sigma_{IS} \\ \sigma_{IS} & \sigma_S^2 \end{bmatrix}. \qquad (4.18)$$

Although it is often of interest to obtain an estimate of the covariance between the intercept and slope, we note that sometimes a warning is issued when an unstructured covariance structure is specified. For example, if the covariance or slope estimate is near 0, the program may not be able to estimate the parameter accurately. Therefore, it is often desirable to begin with a more simplified covariance structure, using the default specification (as in Equation 4.17).

Defining Model 1.3 with IBM SPSS Menu Commands
Note: IBM SPSS settings default to those used in Model 1.2.

1. Go to the tool-
bar and select
ANALYZE,
MIXED MODELS,
GENERALIZED
LINEAR. This
command opens
the *Generalized
Linear Mixed Models*
(GENLIN MIXED)
main dialog box.

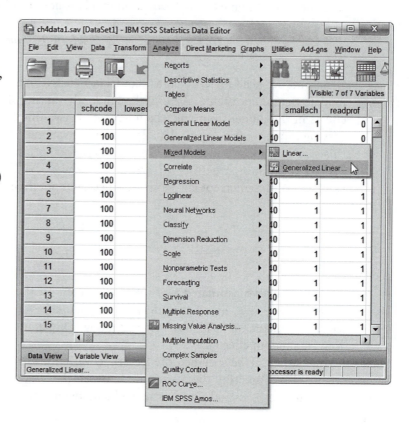

2a. The *Generalized
Linear Mixed
Models* dialog box
displays the *Fields
& Effects* com-
mand tab and
Fixed Effects item.

Note: The *Fixed
Effects* item
is preselected
(represented in
the illustration
by the dashed-

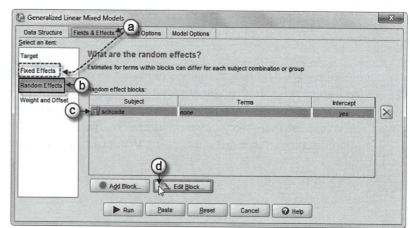

line box but would appear as grayed-out on the computer screen) as it was the last setting
used from the prior model.

b. We are interested in examining whether the relationship between student race/ethnic-
ity and likelihood to be proficient in reading varies across schools, so we will now add a
Level-2 randomly varying effect to the model. First, click to select the *Random Effects* item
to display the main screen.

c. Click on *schcode* to select and activate the random effect block to enable further editing
changes.

d. Click the EDIT BLOCK button to access the *Random Effect Block* screen.

e. We will add a random effect (*minor*) to the model. Click to select *minor* and then drag the variable to the *Main* section of the *Effect builder* box.

f. Note that we continue using the default settings for *Include intercept*, *schcode*, and *variance components*.

Click the OK button to close the *Random Effect Block* dialog box and return to the *Random Effects* main screen.

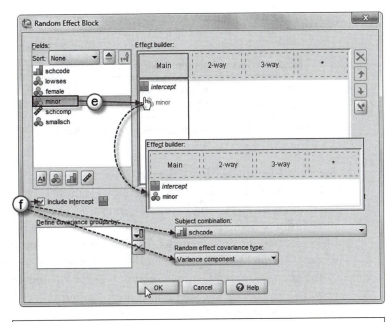

$$\beta_{3j} = \gamma_{30} + u_{3j} \quad (\text{Eq. 4.15})$$

g. In the main screen notice that the *Random effect blocks* displays a summary of the block with *minor* added to the model.

Click the RUN button to generate the model results.

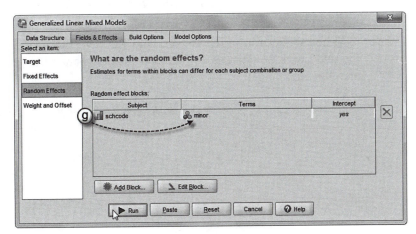

Interpreting the Output of Model 1.3

The fixed effects are similar to the previous model (not tabled). The relevant covariance parameters are summarized in Table 4.9. The variance in the minority-proficiency slope ($\sigma_S^2 = 0.096$) is significant across schools ($z = 2.300$, $p < .05$). These results suggest that we could develop a between-school model to explain this variability in slopes.

Adding Level-2 Predictors to Explain Variability in Intercepts

Our next step is to add the Level-2 predictors that might explain variability in school intercepts. In Model 4, we will use a measure of student composition (i.e., a continuous variable describing average income, standardized with mean = 0 and SD = 1) and school size (a dichotomous variable coded 1 = <600 students and 0 = >599 students). We note that with continuous variables, it is important to consider how they are measured so that 0 has some type of meaning within

TABLE 4.9   Model 1.3: Variance Components

					95% Confidence interval	
Random effect	Estimate	Std. error	z-Test	Sig.	Lower	Upper
Var(Intercept)	0.226	0.063	3.576	0.000	0.130	0.390
Var(minor)	0.096	0.042	2.300	0.021	0.041	0.225

Note: Covariance structure: variance components; subject specification: schcode.

the multilevel model. For example, if we grand-mean center a continuous variable, the mean will then be 0. If school composition were family income measured in dollars, however, leaving it in dollar units would create an intercept (i.e., the point on the Y axis where X is 0) as the probability of being proficient in reading when family income is 0. This coefficient would have little practical use in our model. In that case, grand-mean centering would create an intercept where 0 is equal to the *average* family income for the sample. This would be much more useful in the proposed model. In our model, because *schcomp* has been standardized already, there is very little difference in either grand-mean centering it or leaving it as standardized (because the mean for the sample is 0.01).

We will first attempt to explain variability in the intercepts. The Level-2 intercept model is as follows:

$$\beta_{0j} = \gamma_{00} + \gamma_{01}schcomp_j + \gamma_{02}smallsch_j + u_{0j}. \tag{4.19}$$

When we substitute the Level-1 and Level-2 equations into one combined model (adding the two Level-2 predictors will make eight total parameters to estimate), we have the following one-equation model:

$$\eta_{ij} = \gamma_{00} + \gamma_{01}schcomp_j + \gamma_{02}smallsch_j + \gamma_{10}lowses_{ij} + \gamma_{20}female_{ij} + \gamma_{30}minor_{ij} + u_{3j}minor_{ij} + u_{0j}. \tag{4.20}$$

Defining Model 1.4 With IBM SPSS Menu Commands
Note: IBM SPSS settings default to those used in Model 1.3.

1. Go to the toolbar and select ANALYZE, MIXED MODELS, GENERALIZED LINEAR. This command opens the *Generalized Linear Models* (GENLIN MIXED) main dialog box.

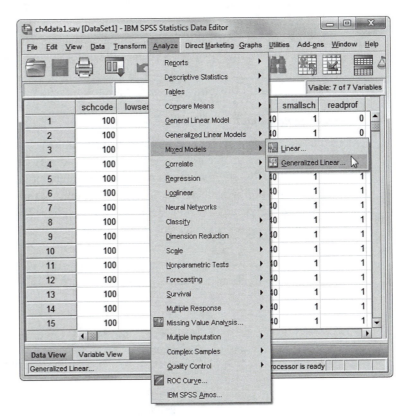

2a. The *Generalized Linear Mixed Models* dialog box displays the *Fields & Effects* command tab and *Random Effects* main screen (not illustrated).

   Note: The *Random Effects* item is preselected (represented in the illustration by the dashed-line box but would appear in blue on the computer screen) as it was the last setting used from the prior model.

   b. We will add two Level-2 variables (*schcomp, smallsch*) to see whether they may help explain variability

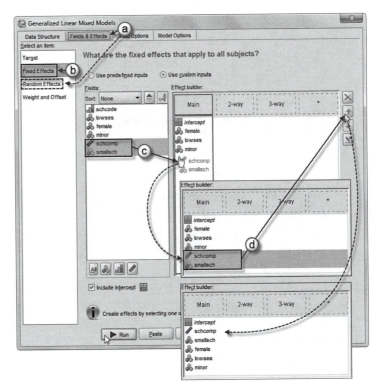

$$\beta_{0j} = \gamma_{00} + \gamma_{01}\textit{schcomp}_j + \gamma_{02}\textit{smallsch}_j + u_{0j} \quad \text{(Eq. 4.19)}$$

in school intercepts. To add the variables to the model, click the *Fixed Effects* item to access the main screen.

c. Now click to select *schcomp* and *smallsch* and then drag the variables onto the *Main* section.

d. To facilitate reading of the output tables we will rearrange the variable order. Click to select *schcomp* and *smallsch* and then click the up-arrow button on the right to place them in first and second position. The variable order is now *schcomp, smallsch, lowses, female, minor.*

Click the RUN button to generate the model results

Interpreting the Output of Model 1.4

The model dimensions for this model are given in Table 4.10. There are eight total parameters to be estimated (i.e., six fixed effects—the intercept, low SES, female, minority, school composition, and small school—and two random effects: the intercept and minority-reading proficiency slope). The table lists 10 fixed effects. Because we have defined the dichotomous predictors as factors (i.e., low SES, female, minority, small school), they each have two levels (total = 8). School composition is continuous (1) and the intercept (1) comprise the remainder of the 10 fixed effects specified in the table.

The fixed-effect output is presented in Table 4.11. Regarding intercepts, the within-school estimates remain about the same. Between schools, school composition is negatively related to

TABLE 4.10 Model 1.4: Covariance Parameters

Covariance parameters	
Residual effect	0
Random effects	2
Design matrix columns	
Fixed effects	10
Random effects	3
Common subjects	122

TABLE 4.11 Model 1.4: Fixed Effects

Model term	Coefficient	Std. error	T	Sig.	Exp(coefficient)	95% Confidence interval for exp(coefficient)	
						Lower	Upper
Intercept	1.289	0.091	14.125	0.000	3.631	3.036	4.343
schcomp	−0.389	0.043	−9.023	0.000	0.678	0.623	0.737
smallsch = 1	0.144	0.087	1.641	0.101	1.154	0.972	1.370
smallsch = 0	0[a]						
female = 1	0.479	0.059	8.110	0.000	1.615	1.438	1.814
female = 0	0[a]						
lowses = 1	−0.775	0.065	−11.888	0.000	0.461	0.405	0.523
lowes = 0	0[a]						
minor = 1	−0.595	0.078	−7.577	0.000	0.552	0.473	0.643
minor = 0	0[a]						

Note: Probability distribution: binomial; link function: logit.

[a] This coefficient is set to 0 because it is redundant.

TABLE 4.12 Model 1.4: Variance Components

Random effect	Estimate	Std. error	z-Test	Sig.	95% Confidence interval Lower	Upper
Var(Intercept)	0.058	0.039	1.492	0.136	0.016	0.216
Var(minor)	0.097	0.041	2.350	0.019	0.042	0.224

Notes: Covariance structure: variance components; subject specification: schcode.

the probability of being proficient in reading ($\gamma_{01} = -0.389$, $p < .001$). This suggests that increasing one standard deviation (SD) in terms of student composition (e.g., increasing the percentages of students requiring targeted resource support) decreases the probability of being proficient in reading by 0.389 log odds units compared with schools at the grand mean of student composition (holding other effects constant). In terms of odds ratios, as school composition is increased by one standard deviation, the odds that $Y = 1$ are multiplied by 0.678 (or reduced by 32.2%).

In other words, the odds of being proficient are decreased 0.678 times for every 1-SD increase in student composition. Note that for a 2-SD increase in student composition, the odds ratios of being proficient are multiplied ($0.678*0.678 = 0.46$), as opposed to being added together as in the case of log odds coefficients. To illustrate, the log odds for student composition are -0.389. If we wish to know the effect on log odds units for a 2-SD increase, we obtain -0.778 [$-0.389 + (-0.389)$]. If we now exponentiate it ($e^{-0.778}$), the resulting odds ratio is 0.46. School enrollment size is unrelated to proficiency in reading ($\gamma_{02} = .144$, $p > .05$).

The variance components table (assuming a diagonal covariance matrix) is presented in Table 4.12. We will now concentrate on the slope model at Level 2. We can see in the table that the slope variance between schools is significant at $p < .05$, suggesting we might build a Level 2 model to explain this variance.

Adding Level-2 Variables to Explain Variation in Level-1 Slopes (Cross-Level Interaction)

We can build a similar model to explain variability in Level-1 minority-reading proficiency slopes across schools (β_{3j} in Eq. 4.12). Because the intercept and slope models can be correlated, it is often useful to develop each model using the same set of predictors, and only if some predictors are found to be nonsignificant, can they then be dropped from the final models (Raudenbush & Bryk, 2002). We can add the same two school-level variables to explain variability in school proficiency intercepts. The slope model will therefore be as follows:

$$\beta_{3j} = \gamma_{30} + \gamma_{31}schcomp_j + \gamma_{32}smallsch_j + u_{3j}. \tag{4.21}$$

Through substitution, Level-2 variables that explain Level-1 relationships will appear in the combined model as cross-level interactions. A cross-level interaction tests whether a variable measured at a higher level of the data hierarchy moderates (enhances or diminishes) a relationship observed at a lower level of the hierarchy. When we substitute Equation 4.21 and Equation 4.19 into Equation 4.11, the combined model will be the following:

$$\eta_{ij} = \gamma_{00} + \gamma_{01}schcomp_j + \gamma_{02}smallsch_j + \gamma_{10}lowses_{ij} + \gamma_{20}female_{ij} + \gamma_{30}minor_{ij}$$
$$+ \gamma_{31}minor_{ij}*schcomp_j + \gamma_{32}minor_{ij}*smallsch_j + u_{3j}minor_{ij} + u_{0j}. \tag{4.22}$$

In this case, the random coefficient for the slope (u_{3j}) takes into consideration the cross-level interactions involving the effect of minority by race/ethnicity on reading proficiency.

Interactions are typically moderators, which suggest that a relationship under consideration between X and Y depends on a third variable. The interaction could be at the same level (e.g., minority*female) or it might exist across levels (e.g., minority*student composition). When there are two categorical variables in an interaction, the meaning of the interaction term is that these two variables exert joint effects on the outcome beyond the main effects of each variable individually. For example, the cross-level interaction might indicate that the effect of minority by race/ethnicity on outcomes differs due to levels of school size. The number of interaction terms needed to represent the interaction between the two predictors is formed by multiplying the number of indicators needed for each variable involved in the interaction. In this case, minority by race/ethnicity is coded 0, 1 and school size is also coded 0, 1, so only one category indicator is needed to represent each variable as a main effect. Therefore, the number of interaction terms needed is 1 (1*1).

We note that when the explanatory variables are both nominal (and coded as descending so the reference group is 0), the cross-level interaction will represent the additional effect of categories coded 1 on the outcomes. This is because multiplication of the other categories will equal 0. For example, the reference category for small school (i.e., average and large schools are coded 0) when multiplied by the reference category for minority (coded 0) will equal 0 (0*0 = 0).

Similarly, a reference category multiplied by a category coded 1 will also equal 0 (0*1 = 0; 1*0 = 0). These redundant categories are shown in the output. This leaves only the two categories coded 1 (i.e., small schools, minority by race/ethnicity), which when multiplied (1*1 = 1) will not equal 0. Hence, the weight assigned to the interaction will represent the moderating effect of the school-level variable on the within-school slope relationship between ethnicity and likelihood to be proficient.

We can also examine interactions between continuous and categorical predictors. If the interaction is significant, it can be considered as a moderation effect; that is, the coefficient indicates that the effect of the continuous variable (e.g., student composition) varies with levels of the categorical predictor (Azen & Walker, 2011). We will discuss the substantive interpretation of these two types of interactions subsequently. We also note in passing that we could investigate a mediating effect, which implies that an added predictor is intermediate in the causal chain between X and Y. The mediating relationship implies a temporal relationship, as X occurs before the mediator M and M occurs before Y. We provide an investigation of one such relationship in Chapter 6.

Defining Model 1.5 With IBM SPSS Menu Commands

Note: IBM SPSS settings default to those used in Model 1.4.

1. Go to the tool-bar and select ANALYZE, MIXED MODELS, GENERALIZED LINEAR. This command opens the *Generalized Linear Mixed Models* (GENLIN MIXED) main dialog box.

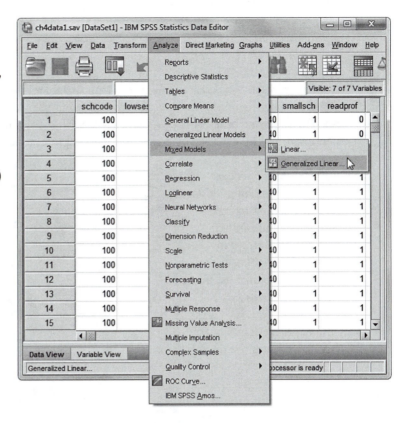

2a. The *Generalized Linear Mixed Models* dialog box displays the *Fields & Effects* command tab and *Fixed Effects* item. The *Fixed Effects* item (represented in the illustration by the dashed-line box) is the preselected default as it was the last setting used in the prior model.

b. Click to select *minor* and *schcomp* and then drag both variables to the *2-way* column to create *minor*schcomp*.

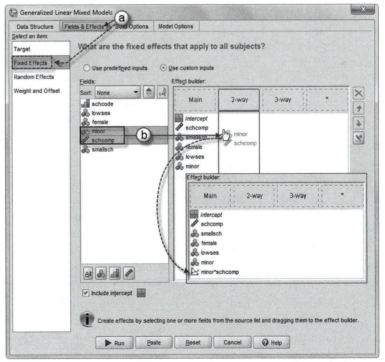

$$\eta_{ij} = \gamma_{00} + \gamma_{01}schcomp_j + \gamma_{02}smallsch_j + \gamma_{10}lowses_{ij} + \gamma_{20}female_{ij} + \gamma_{30}minor_{ij}$$
$$+ \gamma_{31}\textbf{\textit{minor}}_{ij}\textbf{*}\textbf{\textit{schcomp}}_j + \gamma_{32}minor_{ij}*smallsch_j + u_{3j}minor_{ij} + u_{0j}. \quad \text{(Eq. 4.22)}$$

We will add two cross-level interactions into the model to estimate the interactions of (1) minority students by school student composition, and (2) minority students by school size. These interactions are coded as

minor*schcomp
minor*smallsch

Interactions may be created by (1) selecting variables and then dragging them onto the *2-way* or *3-way* columns on the *Effect builder*, or (2) using the *Add a Custom Term* feature, which controls the sequence order of the variables. We will use the first method of simply selecting and then dragging the variables to create a cross-level interaction:

c. Create the second cross-level interaction by clicking to select *minor* and *smallsch* and then dragging both variables to the *2-way* column to create *minor*smallsch*.

Click the RUN button to generate the model results.

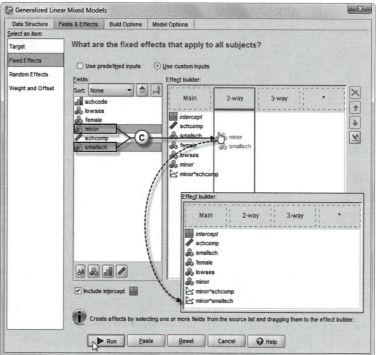

$$\eta_{ij} = \gamma_{00} + \gamma_{01}schcomp_j + \gamma_{02}smallsch_j + \gamma_{10}lowses_{ij} + \gamma_{20}female_{ij} + \gamma_{30}minor_{ij}$$
$$+ \gamma_{31}minor_{ij}*schcomp_j + \gamma_{32}\textbf{\textit{minor}}_{ij}*\textbf{\textit{smallsch}}_j + u_{3j}minor_{ij} + u_{0j}. \qquad \text{(Eq. 4.22)}$$

Interpreting the Output of Model 1.5

The model dimensions are summarized in Table 4.13. It suggests that 10 total parameters are being estimated (i.e., eight fixed effects and two random effects). We note that the variance parameters are very similar, so we do not report them here.

The fixed effects of this model with the slope interactions added are presented in Table 4.14. The results for the intercept model are similar to the last model, so we focus primarily on interpreting the cross-level interactions in the slope model. Regarding the nominal interaction (*minority*smallsch*), it appears that school enrollment is unrelated to the relationship between minority status and probability of being proficient in reading ($\gamma_{32} = -0.026$, $p > .10$). We can interpret the main effects of each relationship (i.e., school size and minority by race/ethnicity) separately at their respective levels of the data hierarchy. We also might consider removing the interaction subsequently because it is not statistically significant, unless we intend it to be in the model as a particular hypothesis we are investigating in our work. For example, if the proposed model were primarily about school size and its possible moderating effect on within-school

TABLE 4.13   Model 1.5: Covariance Parameters

Covariance parameters	
Residual effect	0
Random effects	2
Design matrix columns	
Fixed effects	16
Random effects	3
Common subjects	122

processes, it might be important to leave it in the model so that readers can see that it was tested specifically and found not to be relevant to the set of relationships included in the model.

If, on the other hand, the *minority*smallsch* interaction were significant, it would be important to discuss the interpretation of all the relevant parameters (Azen & Walker, 2011). We would note the main effect of minority by race/ethnicity (coded 1) in terms of the reference category of the school size variable (0 = large schools). In terms of odds ratios, for example, this would be the ratio of the probability that minority versus nonminority students are proficient in large schools (i.e., γ_{30}). The interaction coefficient represents the difference between the effect of race/ethnicity in small schools versus the effect of race/ethnicity in the reference group (large schools). The interaction odds ratio coefficient (γ_{32}) is interpreted as how many times larger (or smaller) the odds ratio (OR) for race/ethnicity is in small schools versus large schools. We can summarize this as follows:

$$e^{\gamma_{32}} = \frac{OR_{\text{(small schools)}}}{OR_{\text{(large schools)}}}.$$

TABLE 4.14   Model 1.5: Fixed Effects

Model term	Coefficient	Std. error	T	Sig.	Exp(coefficient)	95% Confidence interval for exp(coefficient)	
						Lower	Upper
Intercept	1.283	0.103	12.413	0.000	3.606	2.945	4.416
schcomp	−0.457	0.063	−7.726	0.000	0.633	0.560	0.716
smallsch = 1	0.149	0.127	1.170	0.242	1.161	0.904	1.490
smallsch = 0	0[a]						
female =1	0.479	0.059	8.098	0.000	1.615	1.438	1.813
female = 0	0[a]						
lowses = 1	−0.772	0.065	−11.832	0.000	0.462	0.407	0.525
lowes = 0	0[a]						
minor = 1	−0.601	0.105	−5.734	0.000	0.548	0.446	0.673
minor = 0	0[a]						
schcomp*[minor = 1]	0.131	0.074	1.782	0.075	1.140	0.987	1.317
schcomp*[minor = 0]	0[a]						
[smallsch = 1]*[minor = 1]	−0.026	0.150	−0.173	0.863	0.974	0.727	1.306
[smallsch = 1]*[minor = 0]	0[a]						
[smallsch = 0]*[minor = 1]	0[a]						
[smallsch = 0]*[minor = 0]	0[a]						

Note: Probability distribution: binomial; link function: logit.

[a] This coefficient is set to 0 because it is redundant.

We can use the log odds coefficients to calculate the relevant odds ratios in each school setting because they can be added. For example, the log odds of a student who is minority by race/ethnicity being proficient in a small school will be $1.283 - 0.601 + (-0.026) = 0.656$, with the odds ratio then $e^{0.656}$ or 1.927. The log odds of a student who is minority by race/ethnicity being proficient in a large school will be $1.283 - 0.601 = 0.682$ and the odds ratio ($e^{0.682}$) will be 1.978. The odds ratio in Table 4.14 for the interaction is then 0.974:

$$0.974 = \frac{1.927_{\text{(small schools)}}}{1.978_{\text{(large schools)}}}.$$

Turning to the *student composition*minority* interaction, if we interpret the coefficient as not significant, we would say the effect of race/ethnicity on probability of being proficient is the same across levels of student composition. For purposes of demonstration, however, we will suggest that there is preliminary evidence (at $p < .10$) that students who are minority by race/ethnicity receive some advantage from being in schools with higher percentages of students requiring special academic services compared with their peers in schools at the grand mean of student composition.

We note that the average effect of minority by race/ethnicity (coded 1) on probability of a student being proficient in reading is negative and significant (log odds $= -0.601$, $p < .001$), holding the other predictors in the model constant. In this case, the significance of the interaction term ($\gamma_{31} = 0.131$, $p < .10$) indicates that increases in student composition (i.e., percentages of students requiring special academic services) reduce the negative effect of race/ethnicity on the probability of being proficient. More specifically, for a 1-SD increase in student composition above the grand mean (coded 0), the average log odds of being proficient (-0.601) are reduced by 0.131 units ($-0.601 + 0.131 = -0.470$). If we exponentiate this log odds coefficient ($e^{-0.470}$), the odds ratio is 0.625.

From Table 4.14, the odds ratio for minority by race/ethnicity on probability of being proficient is 0.548 at the grand mean for student composition. Therefore, the ratio of the odds of a student who is minority by race/ethnicity being proficient in a school one standard deviation above the grand mean in student composition versus the ratio of the odds of that student being proficient in a school at the grand mean of student composition (adjusted for other variables) is $0.625/0.548 = 1.14$, which is the odds ratio coefficient in Table 4.14. This suggests that, for students who are minority by race/ethnicity in schools one standard deviation above the grand mean in student composition, the odds are multiplied by 1.14 (or a 14% increase) and indicate higher odds of being proficient for students who are minority by race/ethnicity compared to their peers in schools that are more average in student composition, adjusted for other predictors in the model.

The output also produces the percentage of cases correctly classified. Alternatively, one can also save classifications and use *Crosstabs* to determine whether many cases were correctly

TABLE 4.15 readprof*Predicted Value Cross Tabulation

			Predicted value		
			0	1	Total
readprof	0	Count	676.0	1343.0	2019.0
		Within readprof	33.5%	66.5%	100.0%
	1	Count	442.0	4067.0	4509.0
		Within readprof	9.8%	90.2%	100.0%
Total		Count	1118.0	5410.0	6528.0
		Within readprof	17.1%	82.9%	100.0%

Warnings

> glmm: The final Hessian matrix is not positive definite although all convergence criteria are satisfied. The procedure continues despite this warning. Subsequent results produced are based on the last iteration. Validity of the model fit is uncertain.

FIGURE 4.11 Warning message generated when using an unstructured covariance matrix.

classified. These are summarized in Table 4.15. The percentage correctly classified was almost 73% (676 + 4067 = 4743/6528 = 0.727). The model did a better job of correctly classifying students who were proficient (90.2%) than students who were not proficient (33.5%).

Of course, adding other variables to the model would improve the correct classification of individuals. For example, knowing the student's math proficiency status boosts the correct classification percentage to over 81%.

Finally, when we attempted to rerun this model with an *Unstructured* covariance matrix, in order to obtain a covariance between the intercept and slope, we obtained the following warning message shown in Figure 4.11.

The introduction of a more complex covariance matrix results in a situation where the matrix cannot be inverted. A common cause of this is redundancy in covariance parameters. This is a good example of why we place a great deal of emphasis on the theoretical underpinnings in model specification. Less thoughtful specifications are more likely to lead to problems like this; that is, additional complexity is sometimes incorporated into the model without a clear rationale for why it is necessary. But even in instances where the specification is defensible, these problems will sometimes arise. Although there is some guidance on how to proceed in the particular instance of a nonpositive definite Hessian matrix (see Gill & King, 2004), oftentimes these barriers force a reconsideration of model specification.

Estimating Means

We also provide a simple illustration of how we can obtain means for particular variables of interest in our estimated models. To illustrate this feature, we use two of the Level-1 factors (i.e., under-represented minority coded 1, low socioeconomic status coded 1) and their interaction on reading proficiency (*minority*lowses*). We built a simple two-level model with two random effects at the school level (i.e., intercept and minority slope). We summarize this simple model using the combined-equation model:

$$\eta_{ij} = \gamma_{00} + \gamma_{10}lowses_{ij} + \gamma_{20}minor_{ij} + \gamma_{30}lowses * minor_{ij} + u_{3j}minor_{ij} + u_{0j}. \tag{4.23}$$

For the model, we found that under-represented minority (log odds = −0.995) and low SES (log odds = −0.780) had negative main effects on probability of being proficient in reading. In addition, the interaction was also significant and positive (log odds = 0.271, $p < .05$), suggesting that the gap in probability of being proficient was smaller for low SES, under-represented minority students compared against the other groups. We can investigate this further using the "compare means" options. In Table 4.16 we provide the means, standard errors, and confidence levels for the four cells. (We provide the model syntax in Appendix A.)

The significance tests for the simple contrasts are next displayed. They have been adjusted for the number of tests conducted (using the Bonferroni adjusted significance level). They suggest that both contrasts are significant ($p < .01$). (See Table 4.17.)

Finally, we illustrate a typical chart (Figure 4.12) that can be constructed showing these relationships. We note that the mean gap proficiency status is actually a bit smaller for the low SES group, which is apparent in the significant interaction effect for the log odds coefficients in the table.

TABLE 4.16   Cell Means

	Mean	Std. error	95% Confidence interval for exp(coefficient)	
			Lower	Upper
Lowses (1)				
Minor = 1	0.549	0.017	0.516	0.583
Minor = 0	0.670	0.019	0.631	0.707
Lowses (0)				
Minor = 1	0.716	0.016	0.684	0.745
Minor = 0	0.846	0.012	0.820	0.869

TABLE 4.17   Tests of Simple Contrasts

Minor	Lowses simple contrasts	Contrast estimates	Std. error	t	df	Adj. sig.[a]	95% Confidence interval for exp(coefficient)	
							Lower	Upper
1	1 – 0	–0.166	0.019	–8.838	6,524	0.000	–0.203	–0.129
0	1 – 0	–0.176	0.018	–9.717	6.524	0.000	–0.212	–0.141

Note: Confidence interval bounds are approximate.

[a] The 0.000 values indicate significant contrasts. In the actual IBM SPSS output, significant values are shaded gold. The sequential Bonferroni adjusted significance level is .05.

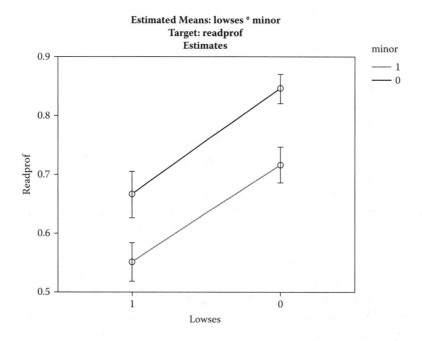

FIGURE 4.12   Illustrating the variables and interaction relationship.

TABLE 4.18 Examining the Pattern of Residuals in Explaining Reading Proficiency

Pearson residual

readprof	N	Mean	Std. dev.	Range	Skewness	Kurtosis
0	2019	−1.32722	0.531908	3.008	−1.472	1.978
1	4509	0.60742	0.258691	1.267	0.738	−0.282
Total	6528	0.00907	0.966119	5.185	−1.018	0.020

Saving Output

We can also save various types of output. For example, we can save the predicted category classifications based on the independent variables in the model. We might examine the pattern of residuals around predicting students' probability of being proficient. Table 4.18 suggests considerably more variability in residuals explaining predictions of not proficient (coded 0) versus proficient (coded 1). More specifically, the standard deviation for not proficient is about twice the size (0.53) of the standard deviation for proficient (0.26). This suggests that the model is better at predicting probability of being proficient as opposed to not proficient.

Probit Link Function

Although the logit link function is an appropriate choice for examining dichotomous outcomes, other choices are available (e.g., probit). The probit model is another type of generalized linear model that extends the linear regression model by linking the range of real numbers to the 0–1 range. It is often used in the context of a dose–response experiment. The probit model makes use of the natural response rate, which is the probability of getting a response with no dose. For example, if we had a treatment to improve students' proficiency in reading, the natural response rate would be the proportion of students who would be proficient without the treatment being introduced. We note that a logistic regression model is generally more useful than a probit model if the study is not a dose–response experiment.

In our extended example, the assumption is that the observed values of the dependent variable reflect an underlying continuous latent variable (η_{ij}) that is assumed to be normally distributed. This is easy to argue in the case of reading proficiency because, generally, "proficiency" represents a somewhat arbitrary point along an underlying continuum of reading performance. As in the logistic GLMM, the underlying latent outcome can be predicted by a set of explanatory variables. Because the relationship between the observed values of Y (0, 1) and X is not linear, however, we must again employ some type of linking function. Whereas the logit link function for a dichotomous outcome is based on the binomial version of the cumulative distribution function (CDF), which describes the distribution of the residuals, the probit link function is based on the normal CDF. The probit model is defined as

$$P(y = 1 | x) = \Phi(\eta_{ij}), \tag{4.24}$$

where the capital Greek letter phi (Φ) is the CDF of the normal distribution and η_{ij} is an underlying continuous variable, which might be thought of as a student's likelihood to be proficient, with larger values of η_{ij} corresponding to greater probabilities of being proficient. We note that "normal distribution" is also referred to as a Gaussian distribution. The Gaussian distribution is a continuous function that approximates the exact binomial distribution of events and whose domain is the real line and range is (0, 1).

The probit link function $[\eta_{ij} = \Phi^{-1}(\pi_{ij})]$, where Φ^{-1} is the inverse standard normal cumulative distribution function, is used to translate the probability that $Y = 1$ from Equation 4.24 into a standard normal (or z) score of the underlying continuous predictor (η_{ij}) of Y, instead of Y itself. The Level-1 structural model considers the following:

$$\eta_{ij} = \Phi^{-1}(\pi_{ij}) = \beta_{0j} + \beta_{1j}X_{ij} + \beta_{2j} + X_{ij} + \cdots + \beta_{qj}X_{qij}, \qquad (4.25)$$

where, similarly to a logistic regression model, a set of linear relations can be used to predict the underlying latent variable representing the probability that $Y = 1$.

In Equation 4.25, the βx term is called the probit score or index. A probit (probability unit) coefficient (β) describes a one-unit increase in a predictor (X) corresponding with a β standard deviation increase in the probit score. The Gaussian distribution is normalized so that the sum over all values of X gives a probability of $Y = 1$. Visually, the probit of a proportion represents a point on the normal curve that has a specific proportion (or area under the normal curve) to the left of it. This means that it outputs the probability of the events that will fall between 0 and 1 (0% to 100%). Some researchers prefer this standardized score to the odds ratio associated with a logistic regression model. We note that the unstandardized coefficients are not as easily interpreted because they represent the probit-transformed change in the predicted score for every unit change in the predictor.

As an example, consider the fact that a normal distribution (i.e., with mean = 0 and $\sigma = 1$) places approximately 68.3% of the area under the curve between z-scores of -1.00 and 1.00, symmetric around 0. From a normal curve table, we can find that at $z = -1.00$, the proportion to the left under the curve is 0.1587 (see Table 4.19). This can be translated as follows:

$$\Phi(-1.00) = .1587 = 1 - \Phi(1.00).$$

This implies that the probit function provides an "inverse" computation, generating a z-score value of a normally distributed underlying variable associated with a specified cumulative probability. In order to obtain the cumulative proportion of the probablity that a proportion $\pi = 1$, in the probit link function we make use of the inverse of the CDF, which can be denoted $\eta = \Phi^{-1}(\pi)$. In the preceding example, then, probit $(0.1587) = -1.00$, which is equal to a negative probit of 0.8413, or $-$probit(0.8413) because a z-score of $1 = 0.8413$. Given a set of independent variables, therefore, one can predict the probability of the event occurring.

Readers will note the similarity of the two link approaches, differing by the distribution function used for examining the conditional probability of $\pi = 1$, given x. Both represent nonlinear functions of x that vary between 0 and 1. The choice of either link function will result in slight differences in model parameters, but when modeled proportions are all between 0.1 and 0.9, the differences are generally negligible (Hox, 2002). When there are proportions close to 0 or 1, the logit transformation does a better job of spreading them over a wider range of the transformed scale than the probit transformation. If the proportions are close to 0 or 1 and the interest is in these proportions, the logistic transformation will be a better approach.

Let us return to our previous multilevel analysis (within cross-level interactions), but this time using a probit link function.

Defining Model 1.6 With IBM SPSS Menu Commands

Note: IBM SPSS settings default to those used in Model 1.5.

1. Go to the toolbar and select ANALYZE, MIXED MODELS, GENERALIZED LINEAR. This command opens the *Generalized Linear Mixed Models* (GENLIN MIXED) main dialog box.

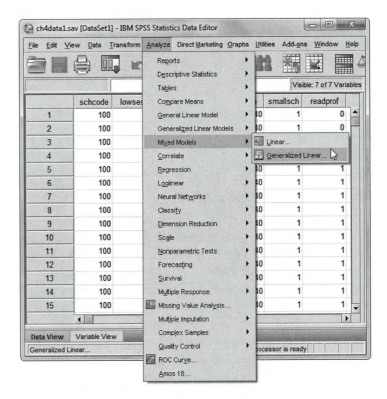

2a. The *Generalized Linear Mixed Models* dialog box displays the *Fields & Effects* command tab and *Fixed Effects* item. The *Fixed Effects* item (represented in the illustration by the dashed-line box) is the preselected default as it was the last setting used in the prior model.

b. We will change the model's distribution to a binary probit, which specifies a binomial distribution with a probit link and is used when the target is a binary response with an underlying normal distribution. First, click the *Target* item to access the target and distribution main screen.

c. Now click the *Binary probit* option.

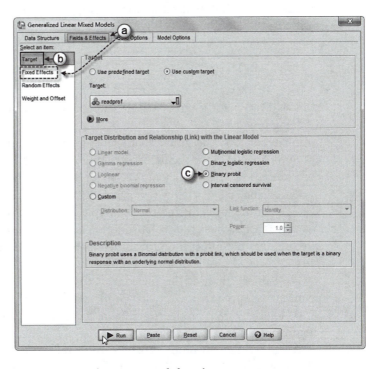

Click the RUN button to generate the model results.

TABLE 4.19   z-Scores and Associated Probabilities

z-Score	Probability
–2.0	0.0228
–1.0	0.1587
–0.5	0.3085
0.0	0.5000
0.5	0.6915
1.0	0.8413
2.0	0.9772

Interpreting Probit Coefficients

Before we present the results, we briefly discuss the interpretation of probit coefficients. A positive probit coefficient means that an increase in the predictor leads to an increase in the predicted probability. A negative coefficient means that an increase in the predictor leads to a decrease in the predicted probability. We can interpret the predicted probit score for η_{ij} (0.774) in Table 4.20 as a z-score of a student who has values of 0 on the independent variables—that is, a male (coded 0) of average/high SES (coded 0) and not a minority (coded 0) by race/ethnicity, who attends an average/large school (coded 0) at the grand mean for student SES composition (coded 0). This means that the predicted probability $Y = 1$ is $\Phi(\eta_{ij})$.

If we consult a table of proportions under the curve, a z-score of 0.774 takes in 0.781 of the area under the curve, suggesting that the average probability of being proficient for students who were 0 on all predictors would be 78.1%. This coefficient is consistent with the logit intercept of 1.283 reported in Table 4.14. The logit score can be translated into a probability of $\pi = 1$:

$$1/(1+e^{-(\beta_0)}) = 1/(1+2.71828^{-(1.283)}) = 1/(1+0.277) = 1/1.277 = 0.783$$

This suggests that the probit and logistic models provide very similar estimates of the probability of being proficient.

For predictors, the probit coefficients give the change in the probit index, or z-score, for a one-unit change in the predictor. One might say that, for each unit of X_{ij}, the z-score is increased (or reduced) by the coefficient. We note, however, that although it is possible to interpret the probit coefficients as changes in z-scores (i.e., similarly to how log odds coefficients represent changes in the log odds metric), we typically convert the z-scores to probabilities that $Y_{ij} = 1$. We note that because the probit function is based on the normal distribution, an increase in the z-score does not affect the probability that $Y = 1$ in a uniform way. Consider the z-scores and their associated probabilities in Table 4.19.

We can see that, at the mean ($z = 0$), a variable that produces a z-score increase of 1.0 increases the probability of being proficient from 0.5000 to 0.8413. In contrast, at a z-score of –2.0, a variable that produces a z-score increase of 1.0 increases the probability of being proficient from about 0.0228 to about 0.1587.

When we convert probit scores to probabilities, therefore, we must take into consideration that a specific change in X_{ij} cannot tell us the predicted change in the probability that $\pi_{ij} = 1$ because the amount of change depends upon where we are beginning on the curve. The increase in the probability of $\pi = 1$ attributed to a one-unit increase in a given X (e.g., school SES) is therefore dependent on both the values of the other predictors and the intervals on the normal curve we are trying to predict. For example, increasing the probability $Y_{ij} = 1$ for SES increases of one unit (regardless of whether or not they are in standard deviation units) does not affect the probability $Y_{ij} = 1$ *uniformly* across all school SES levels.

In contrast, when we work with odds ratios, the increase of one unit in X_{ij} results in the same increase in predicted odds because, as X increases by one unit, the expected odds of proficiency are multiplied e^β times. Of course, when we consider the relationship between the probability $Y = 1$ in logistic regression, we also find that the rate of change in the predicted probability $Y = 1$ varies depending on the value of X (Azen & Walker, 2011); however, we note that model results are not typically presented in this fashion for models using the logit link function.

Interpreting the Output of Model 1.6

We present the fixed effects in Table 4.20. In this case, the adjusted coefficient intercept is 0.774. If we look at the coefficients in the table, we see that being female (rather than male) increases the predicted probit index by 0.289 of a standard deviation. Being in a small school (rather than a large school) increases the predicted probit index by 0.09 of a standard deviation. Conversely, being in a school one standard deviation above the grand mean in student composition (which, because of reverse coding, indicates lower average community SES) decreases that probit index by 0.26 of a standard deviation. We note that because logit coefficients are about 1.6–1.8 times larger than probit coefficients, we can see that the intercepts differ, respectively, by a factor of about 1.66 (1.283/.774) = 1.658). We can also note that the probit coefficient for school composition shown in Table 4.20 is −0.263, which is also smaller than the corresponding logit coefficient from Table 4.14 by 1.74 (−0.457/−0.263 = 1.74).

Examining the Effects of Predictors on Probability of Being Proficient

The simplest technique used to present probit estimates is to set each independent variable to its mean (or to select the category with the largest number of members for a categorical variable).

TABLE 4.20 Model 1.6: Fixed Effects

Model term	Coefficient	Std. error	*t*	Sig.	95% Confidence interval for exp(coefficient) Lower	Upper
Intercept	0.774	0.060	12.946	0.000	0.657	0.892
schcomp	−0.263	0.036	−7.329	0.000	−0.334	−0.193
smallsch = 1	0.089	0.073	1.218	0.223	−0.054	0.232
smallsch = 0	0[a]					
female = 1	0.289	0.035	8.211	0.000	0.220	0.358
female = 0	0[a]					
lowses = 1	−0.467	0.039	−12.128	0.000	−0.543	−0.392
lowes = 0	0[a]					
minor = 1	−0.358	0.062	−5.791	0.000	−0.479	−0.237
minor = 0	0[a]					
schcomp*[minor = 1]	0.063	0.043	1.465	0.143	−0.021	0.147
schcomp*[minor = 0]	0[a]					
[smallsch = 1]*[minor = 1]	−0.013	0.088	−0.143	0.886	−0.184	0.159
[smallsch = 1]*[minor = 0]	0[a]					
[smallsch = 0]*[minor = 1]	0[a]					
[smallsch = 0]*[minor = 0]	0[a]					

Note: Probability distribution: binomial; link function: probit.
[a] This coefficient is set to 0 because it is redundant.

Then one can show the effect of $\pi_{ij} = 1$ for each independent variable. This is done by calculating $\Phi\beta X$ for a selected level of the independent variable. To illustrate this approach, for a continuous variable, if the mean of school SES composition is 0.01 in the sample, one can calculate the probability of being proficient when school SES composition is at the grand mean.

We will use 0.0 to simplify the calculation [0.774 − 0.263(0) = 0.774]. The resulting proportion under the curve when $\beta_{0j} = 0.774$ is 0.781. Keeping in mind that SES composition is negatively related to likelihood to be proficient, we can recalculate the probability of $Y = 1$ when school SES is increased by one unit [0.774 − 0.263(1) = 0.511]. The resulting proportion under the curve for a z-score of $\beta_{0j} = 0.511$ is 0.643. Similarly, one can calculate the likelihood of $Y = 1$ for males (female = 0) [0.774 + 0.289(0) = 0.774, with corresponding area under the curve = 0.781]. Next, one can recalculate the probability $Y = 1$ when changing by one unit—that is, from male to female [0.774 + 0.289(1) = 1.063, and the corresponding area under the curve is 0.856].

A second way to deal with the situation is to choose values of the X_{ij} variables where the probability $\pi_{ij} = 1$ is near 0.50 (i.e., right in the middle). This maximizes the effect because the slope on the S-curve will be the steepest at this point. In this example, however, the mean proportion of proficient students is actually about 0.691, so we know the outcome does not occur in equal likelihood.

A third way is to take several values on the curve and demonstrate what the effect of X_{ij} on the probability $Y_{ij} = 1$ would be at the bottom of the S-curve (nearer $Y_{ij} = 0$), near the middle (closer to the .50/.50 case), and then near the top of the S-curve (nearer to $Y_{ij} = 1$). This has the advantage of seeing the change over a wide range of the probability curve. We return to the relevant part of the probit equation to illustrate this approach:

$$\eta_{ij} = 0.774 - 0.263 \text{ (School SES)}$$

In Table 4.21, we provide several changes in predicted probability $Y_{ij} = 1$ for one-unit increases in school SES composition. As the table suggests, the greater the probability $\pi_{ij} = 1$ is, the less an SES increase of +1 unit changes the predicted probability of being proficient.

We show a plot of the predicted probability that $\pi_{ij} = 1$ for different values of school SES in Figure 4.13. This approach is one good way to summarize the implied relationship between X and Y.

Extending the Two-Level Model to Three Levels

We can also extend this type of two-level proficiency model to a three-level model by adding a classroom level. We use a data set very similar to the last data set but with a classroom level including a standardized measure of teacher effectiveness, a class size measure, and a classroom composition measure. This data set is *ch4data2.sav*. We provide a list of the variables in

TABLE 4.21 Predicted Probability $\pi = 1$, Given a One-Unit Change in School Composition

School SES	Probit core(η_{ij})	$P(\pi = 1)$	Change
−4	1.826	0.966	NA
−3	1.563	0.941	−0.025
−2	1.300	0.903	−0.038
−1	1.037	0.850	−0.053
0	0.774	0.781	−0.069
1	0.551	0.709	−0.072
2	0.248	0.597	−0.112
3	−0.015	0.494	−0.103
4	−0.278	0.391	−0.103

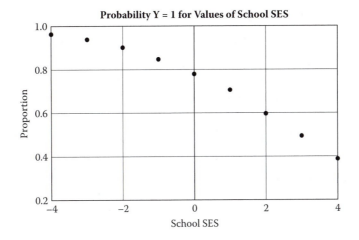

FIGURE 4.13   Predicted probabilities for reverse-coded values of school SES.

Table 4.22. At the classroom level (Level 2), we may wish to determine specifically whether teacher effectiveness (i.e., a z-score describing the relative impact of the teacher's skill in contributing to student literacy outcomes) increases the probability that a student is proficient in reading. We also add classroom controls for class size and student composition.

The Unconditional Model

At Level 1, we have the following model for individual i in classroom j and school k:

$$\eta_{ijk} = \log\left(\frac{\pi_{ijk}}{1-\pi_{ijk}}\right) = \beta_{0jk}. \tag{4.26}$$

At Level 2, we can have the intercepts vary across classrooms:

$$\beta_{0jk} = \gamma_{00k} + u_{0jk}, \tag{4.27}$$

and at Level 3, the model will be as follows:

$$\gamma_{00k} = \gamma_{000} + u_{00k}. \tag{4.28}$$

We then specify the single-equation unconditional model as follows (adding a random effect for the intercept at Level 2 and Level 3):

$$\eta_{ijk} = \gamma_{000} + u_{00k} + u_{0jk}, \tag{4.29}$$

where γ_{000} represents the intercept at the school level, u_{00k} represents the school level random effect for the intercept, and u_{0jk} represents the random effect for intercepts at the classroom level. We note that some users might prefer to use a different letter to refer to errors at either Level 2 or Level 3. This suggests three effects to estimate (one fixed effect for the intercept and random effects at Levels 2 and 3).

TABLE 4.22 Data Definition of *ch4data2.sav* (N = 9,068)

Variable	Level[a]	Description	Values	Measurement
schcode	School	School identifier (122 schools)	Integer	Ordinal
rteachid	Classroom	Recoded teacher identifier numbered 1 to 10 with each school	Integer	Ordinal
lowses	Individual	Predictor dichotomous variable representing student's social economic status (SES)	0 = not low socioeconomic status 1 = low socioeconomic status	Nominal
female	Individual	Demographic predictor variable representing student's gender	0 = male 1 = female	Nominal
minor	Individual	Demographic predictor variable representing student's ethnicity	0 = nonminority 1 = minority	Nominal
readprof	Individual	Two-category dependent variable referring to students' reading proficiency	0 = not proficient 1 = proficient	Nominal
zteacheff	Classroom	Predictor variable (standardized factor score) representing teachers' relative contribution in producing higher or lower student outcomes[b]	−2.03 to 1.96	Scale
schcomp	School	Weighted composite representing the percentage of students receiving additional support services related to economic status, language background, or other special learning needs	−1.67 to 2.32	Scale
smallsch	School	Predictor dichotomous variable representing school size based on the number of students	0 = not small school (>599) 1 = small school (<600)	Nominal
gmclasscomp	Classroom	Predictor variable (grand-mean centered) representing the percentage of students receiving additional support services related to economic status, language background, or other special learning needs	−0.41 to 0.59	Scale
gmclass_size	Classroom	Predictor variable (grand-mean centered) representing the number of students in the class	−9.85 to 10.15	Scale

[a] Individual = Level 1; school = Level 2; classroom = Level 3.

[b] Higher positive scores suggest performance considerably above the average teacher's contribution, and higher negative scores suggest performance considerably below average.

Defining Model 2.1 With IBM SPSS Menu Commands

Note: Refer to Chapter 3 for discussion concerning settings and modeling options available in the IBM SPSS GENLIN MIXED routine.

Launch the IBM SPSS application program and select the *ch4data2.sav* data file.

1. Go to the tool-
bar and select
ANALYZE,
MIXED MODELS,
GENERALIZED
LINEAR. This
command opens
the *Generalized
Linear Mixed Models*
(GENLINMIXED)
main dialog box.

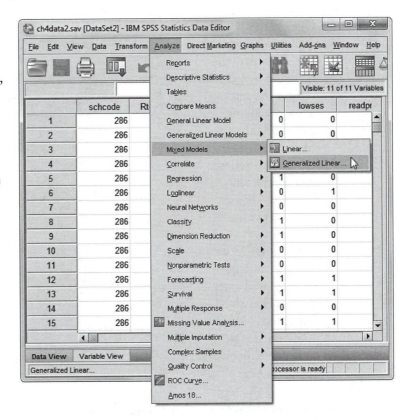

2a. The *Generalized
Linear Mixed Models*
display screen shows
four command tabs:
*Data Structure, Fields
& Effects, Build
Options*, and *Model
Options*.

b. The *Data Structure*
command tab is
the default screen
display when
creating a model
for the first time.
The options enable
specifying struc-
tural relationships
between data
set records when
observations are
correlated.

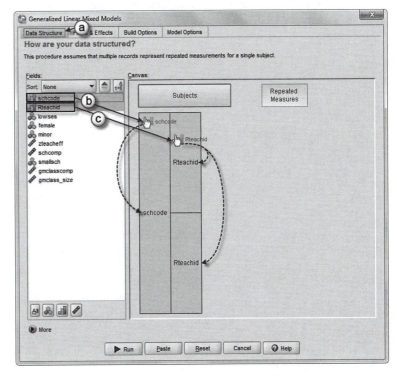

c. To build our model, begin by defining the subject variable. A "subject" is an observational unit that can be considered independent of other subjects. Click to select the *schcode* variable from the left column and then drag and drop *schcode* onto the canvas area beneath the *Subjects* box.

3a. Click the *Fields & Effects* command tab, which displays the default *Target* display screen. The settings allow specifying a model's target (dependent variable), distribution, and relationship to model predictors through the link functions.

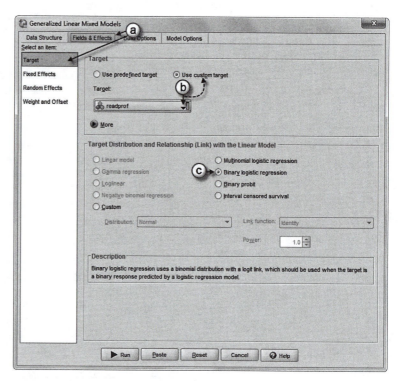

b. For this model we will use student reading proficiency (*readprof*) as the dependent variable. Click the *Target* pull-down menu and select *readprof. Note:* Selecting the target variable will automatically change the setting to *Use custom input.*

c. Next we will designate a binomial distribution and logit link function for the model. Click to select *Binary logistic regression.*

4a. Click the *Fixed Effects* item to access the screen for adding fixed effects such as predictor variables and interactions into the model.

b. Note that the *Include intercept in model* option is preselected. This is because a Level-1 model is intercept only (default), so other model effects must be specified.

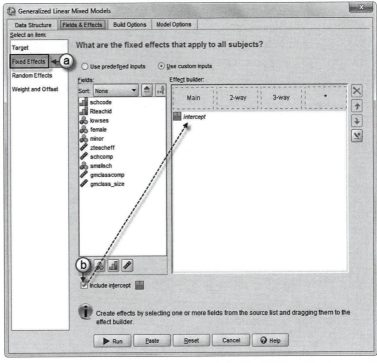

$$\eta_{ijk} = \log\left(\frac{\pi_{ijk}}{1 - \pi_{ijk}}\right) = \beta_{0jk} \quad \text{(Eq. 4.26)}$$

5a. Because the unconditional model does not have any fixed effects, we will skip this setting. The next step is to add a random effect (intercept) to the model. Click on the *Random Effects* item. The *Random Effects* command is used to specify which variables are to be treated as randomly varying across groups (random effects).

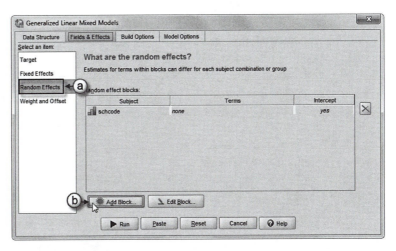

b. Click the ADD BLOCK button to access the *Random Effect Block* main dialog box.

c. In this model only the intercept is going to be randomly varying, so click to select the *Include intercept* option, which changes the setting within the *Effect builder* to *intercept* (from *no intercept*). Because there is only one random effect (the intercept), we can use the default variance component (VC) setting. This specifies a diagonal covariance matrix for the random effects; that is, it provides a separate

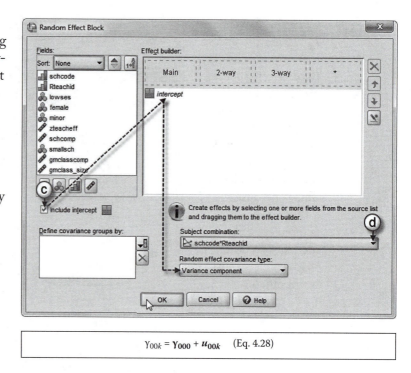

$$Y_{00k} = Y_{000} + u_{00k} \quad (Eq. 4.28)$$

variance estimate for each random effect, but not covariances between random effect, VC defaults to an identity covariance matrix.

d. Select the *Subject combination* by clicking the pull-down menu and selecting *schcode*.

Click the OK button to close the *Random Effect Block* dialog box and return to the *Random Effects* main screen.

e. In the main screen notice that the *Random effect blocks* displays a summary of the block just created: *schcode* (Subject) and *intercept* (Yes).

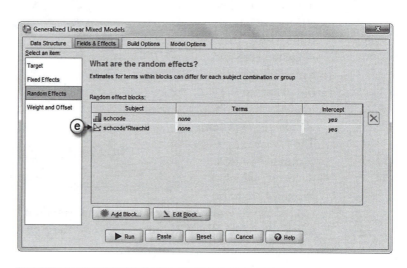

$$\beta_{0jk} = Y_{00k} + u_{0jk} \quad (Eq. 4.28)$$

6a. Finally, click the *Build Options* command tab located at the top of the screen to access advanced criteria for building the model.

b. Change the sorting order for both categorical targets and categorical predictors. Click the pull-down menu for each option and select *Descending*. We will retain the default settings for *Stopping Rules, Post Estimation,* and *Degrees of Freedom.*

c. Change the *Tests of fixed effects and coefficients.* Click to select *Use robust estimation to handle violations of model assumptions.*

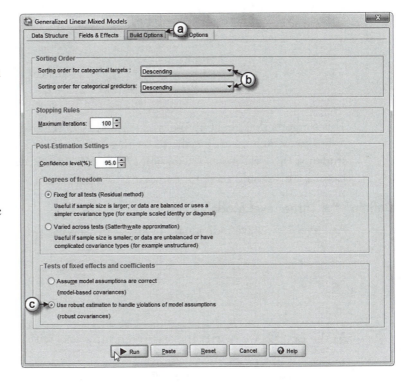

Click the RUN button to generate the model results.

Interpreting the Output of Model 2.1

The intercept (not tabled) for this model is 0.937 ($t = 14.956$, $p < .001$). Table 4.23 provides the relevant variance components, with the Level-1 residual scaled to 1.0. We can use the variance components to calculate estimates of the variability in probability of being proficient at Levels 2 and 3. Once again, for a logistic distribution, the variance is $\pi^2/3$, or approximately 3.29. For Level 3, the variability will be as follows:

$$\left[\frac{\sigma^2_{School}}{\sigma^2_{School} + \sigma^2_{Class} + 3.29} \right] = 0.378/3.784 = 0.100$$

TABLE 4.23   Model 2.1: Variance Components

Random and residual effects	Estimate	Std. error	z-Test	Sig.	95% Confidence interval for exp(coefficient) Lower	Upper
Random[a]						
Level-3 Var(Intercept)	0.378	0.071	5.322	0.000	0.262	0.547
Level-2 Var(Intercept)	0.116	0.045	2.604	0.009	0.055	0.246
Residual[b]	1.00					

[a] Covariance structure: variance components; subject specification: id.
[b] Covariance structure: scaled identity; subject specification: id.

For Level 2, the variability will be

$$\left[\frac{\sigma^2_{Class}}{\sigma^2_{School} + \sigma^2_{Class} + 3.29}\right] = 0.116/3.784 = 0.031$$

This suggests that about 10% of the variability in reading proficiency lies between schools and only about 3% between teachers. As these are reading proficiency scores of upper elementary students, these estimates are feasible. Although, ideally, we would like a bit more variability at Level 2, this should serve for demonstration purposes.

Defining the Three-Level Model

At Level 1, we have individual i in class j in school k as follows:

$$\eta_{ijk} = \log\left[\frac{\pi_{ijk}}{1 - \pi_{ijk}}\right] = \beta_{0jk} + \beta_{1jk}lowses_{ij} + \beta_{2jk}female_{ij} + \beta_{3jk}minor_{ij}. \qquad (4.30)$$

At the Level 2, we specify the classroom variables j within schools k:

$$\beta_{0jk} = \gamma_{00k} + \gamma_{01k}zteacheffect_{jk} + \gamma_{02k}gmclasscomp_{jk} + \gamma_{03k}gmclass\_size_{jk} + u_{0jk}. \qquad (4.31)$$

At Level 3, we specify the school covariates:

$$\gamma_{00k} = \gamma_{000} + \gamma_{001}schcomp_k + \gamma_{002}smallsch_k + u_{00k}. \qquad (4.32)$$

We also specify the individual-level predictors as fixed at Level 2 ($\beta_{1jk} = \gamma_{10k}$; $\beta_{2jk} = \gamma_{20k}$; $\beta_{3jk} = \gamma_{30k}$). Similarly, we specify the class-level predictors to be fixed at Level 3 ($\gamma_{01k} = \gamma_{010}$; $\gamma_{02k} = \gamma_{020}$; $\gamma_{03k} = \gamma_{030}$).

If we combine the information into a single equation (see Chapter 1), it implies that there are 11 parameters to estimate. The nine fixed effects are the intercept and eight slope coefficients. The two random effects include the Level-3 intercept and the Level-2 intercept:

$$\begin{aligned}\gamma_{00k} = {} & \gamma_{000} + \gamma_{001}schcomp_k + \gamma_{002}smallsch_k + \gamma_{010}zteacheffect_{jk} + \gamma_{020}gmclasscomp_{jk} \\ & + \gamma_{030}gmclass\_size_{jk} + \gamma_{100}lowses_{ij} + \gamma_{200}female_{ij} + \gamma_{300}minor_{ij} + u_{00k} + u_{0jk}. \end{aligned} \qquad (4.33)$$

Defining Model 2.2 With IBM SPSS Menu Commands

Note: Settings default to those used in Model 2.1.

1. Go to the tool-
bar and select
ANALYZE,
MIXED MODELS,
GENERALIZED
LINEAR. This
command opens
the *Generalized
Linear Mixed Models*
(GENLINMIXED)
main dialog box.

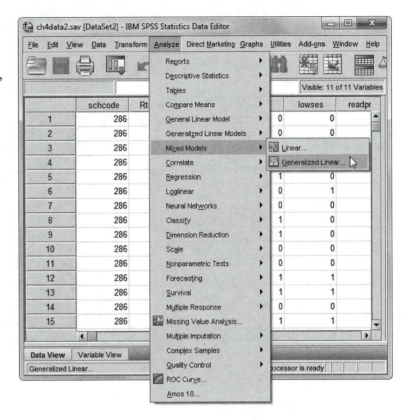

2a. The *Generalized Linear Mixed Models* dialog box displays the *Build Options* command tab, which was the last setting used in the prior model.

b. We will begin to build the model by adding predictor variables. First, click to select the *Fields & Effects* command tab, which displays the *Random Effects* main screen (not illustrated).

Note: The *Random Effects* item is preselected (represented in the illustration by the dashed-line box but would appear as grayed-out on the computer screen) as it was the last setting used from the prior model.

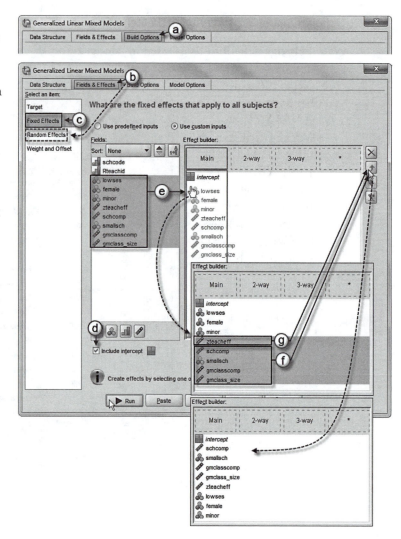

c. Click to select the *Fixed Effects* item to display the main screen, where we may add variables to the model.

$$\gamma_{00k} = \gamma_{000} + \gamma_{001}schcomp_k + \gamma_{002}smallsch_k + \gamma_{010}zteacheffect_{jk} + \gamma_{020}gmclasscomp_{jk} + \gamma_{030}gmclass\_size_{jk} + \gamma_{100}lowses_{ij} + \gamma_{200}female_{ij} + \gamma_{300}minor_{ij} + u_{00k} + u_{0jk}. \quad \text{(Eq. 4.33)}$$

d. Note that the *Include intercept in model* option is preselected and displayed in the *Effect builder* as *intercept*. This is because a Level-1 model is intercept only (default), so other model effects must be specified.

e. We will add eight predictors to the model (*lowses, female, minor, zteacheff, schcomp, smallsch, gmclasscomp, gmclass_size*). Click to select the eight variables and then drag and drop them into the *Main* (effect) column of the *Effect builder*.

f. To facilitate reading of the output tables, we will rearrange the variable order. Click to select the last four variables (*schcomp, smallsch, gmclasscomp, gmclass_size*) and then click the up-arrow button on the right so that *schcomp* is the first variable listed.

g. Next click to select *zteacheff* and then click the up-arrow button to place the variable below *gmclass_size*. The variable sequence is now *schcomp smallsch, gmclasscomp, gmclass_size, zteacheff, lowses, female, minor*.

Note: The order may also be controlled by entering the variables into the *Effect builder* beginning with *female*, followed by the remainder (*lowses, minor*).

Click the RUN button to generate the model results.

TABLE 4.24   Model 2.2: Covariance Parameters

Covariance parameters	
Residual effect	0
Random effects	2
Design matrix columns	
Fixed effects	13
Random effects	11
Common subjects	158

Interpreting the Output of Model 2.2

First, we can check to see if the model is specified as we intended. We note in Table 4.24 that there are two random effects as specified in the null model. The fixed effects correspond to the output in Table 4.25. There are 13 fixed effects in the table (but 4 are redundant because they were defined as factors). This leaves nine fixed effects as specified in the model (Equation 4.33).

The fixed-effect estimates are presented in Table 4.25. The effects of the predictors are similar to those of other models we have presented in this chapter. We focus in particular on the classroom-level effects. We can see that the adjusted odds ratio for teacher effectiveness is 1.069 ($p < .05$), holding the other variables constant. This suggests that the predicted odds of proficiency are multiplied by 1.069 (or increased by about 7%) for a 1-SD increase in teacher effectiveness. For a 2-SD increase in effectiveness, note that the odds would be multiplied by a factor of 1.143 (1.069*1.069), or about 14.3%. Class size and student composition do not appear to affect the predicted odds of being proficient in reading ($p > .05$).

We provide the variance estimates in Table 4.26. We note that the estimate of the teacher-level variance is slightly larger than in the unconditional model. This is likely due to the rescaling that takes place at Level 1 when variables are added to the model or to other errors in their

TABLE 4.25   Model 2.2: Fixed Effects

Model term	Coefficient	Std. error	t	Sig.	Exp(coefficient)	95% Confidence interval for exp(coefficient) Lower	Upper
Intercept	1.091	0.084	12.997	0.000	2.978	2.526	3.511
schcomp	−0.372	0.068	−5.456	0.000	0.689	0.603	0.788
smallsch = 1	0.046	0.086	0.534	0.593	1.047	0.884	1.240
smallsch = 0	0[a]						
gmclasscomp	−0.434	0.262	−1.659	0.097	0.648	0.388	1.082
gmclass_size	0.004	0.008	−0.517	0.605	1.004	0.989	1.019
zteacheff	0.067	0.033	1.991	0.047	1.069	1.001	1.141
lowses = 1	−0.424	0.071	−6.598	0.000	0.624	0.543	0.718
lowses = 0	0[a]						
female = 1	0.503	0.066	7.602	0.000	1.654	1.453	1.884
female = 0	0[a]						
minor = 1	−0.471	0.071	−6.598	0.000	0.624	0.543	0.718
minor = 0	0[a]						

Note: Probability distribution: binomial; link function: logit.

[a] This coefficient is set to 0 because it is redundant.

TABLE 4.26   Model 2.2: Variance Components

Random and residual effects	Estimate	Std. error	z-Test	Sig.	95% Confidence interval for exp(coefficient)	
					Lower	Upper
Random						
Level-3 Var(Intercept)[a]	0.089	0.037	2.387	0.000	0.039	0.201
Level-2 Var(Intercept)[b]	0.122	0.046	2.640	0.008	0.058	0.257
Residual[c]	1.00					

[a] Covariance structure: variance components; subject specification: schcode.
[b] Covariance structure: variance components; subject specification: schcode*Rteachid.
[c] Covariance structure: scaled identity; subject specification: none.

initial estimation that often occurs in multilevel modeling (e.g., see Hox, 2010). This can affect the estimate of variance components at higher levels of the data hierarchy. We could, of course, further investigate whether teacher effectiveness varies across schools and, if so, whether we can explain this variation due to school context or process variables. These could imply more complex model formulations about how school contexts or process variables might affect classroom mediators such as teacher effectiveness and its significant relationship to student outcomes. We refer readers to MacKinnon (2008) for further discussion about how multilevel mediation models can be specified and tested.

Summary

In this chapter we introduced the basics of developing a two-level model with a dichotomous outcome, primarily using a logit link function. We demonstrated that the same model can also be developed using a probit link function. We illustrated how the basic two-level model can easily be extended to three-level models. We can also extend this basic dichotomous cross-sectional model to a longitudinal model. We develop this longitudinal approach for single-level and multilevel models in the next chapter.

CHAPTER 5

Multilevel Models With a Categorical Repeated Measures Outcome

Introduction

In the previous chapter, the outcome and explanatory variables were only measured on one occasion. One of the limitations of cross-sectional data is that they are not well suited to studying processes that are assumed to be developmental—more specifically, when time is a key factor in understanding how such processes unfold within individuals and groups. Longitudinal studies have become increasingly important in understanding how social, behavioral, and health-related processes may change over time. In such studies, individuals are observed over several occasions, often to determine whether a particular intervention may produce changes in their status or behavior.

When an outcome is measured several times for an individual, we have a type of repeated-measures design (RMD), perhaps comparing change in behavior over several occasions between participants in an intervention group versus a control group. The RMD can be conceptualized as another type of hierarchical data structure, with measurement occasions nested within individuals. When data are collected on multiple occasions on individuals, the responses will not be independent of each other; that is, prior responses will typically be correlated with subsequent responses.

Repeated measures data can be continuous (as in a typical growth model), binary, ordinal, counts, or events in trials. In our first workbook, we devoted two chapters to the study of individual-level (two levels) and organizational-level (three levels) change for continuous outcomes. In the past, changes in individuals were often examined using repeated measures ANOVA (analysis of variance). This type of analysis, however, is limited to within- and between-subject effects (i.e., as a single-level, multivariate design), requires assumptions about the similarity and spacing of measurements over time, and can be considerably influenced by missing data. Increasingly, however, methods are available that can provide a much more rigorous and thorough examination of longitudinal data. Generalized linear mixed models (GLMMs) also provide a means of incorporating longitudinal designs with categorical outcomes into situations where there are clustered data structures.

One of the attractive properties of the GLMM is that it allows for linear as well as nonlinear models under a single framework, which will address issues of clustering. It is possible to fit models with outcomes resulting from various probability distributions including normal (or Gaussian), inverse Gaussian, gamma, Poisson, multinomial, binomial, and negative binomial through an appropriate link function $g(\cdot)$. At Level 1, repeated observations (e.g., students' proficiency status in math, students' enrollment over successive semesters in college, changes in clinical or health status) are nested within individuals, perhaps with additional time-varying covariates. At Level 2, we can define variables describing differences between individuals (e.g., treatment groups, participation status, subject background variables and attitudes). At Level 3, individuals might be nested in various organizational units that differ by relevant contextual and process variables.

The focus of these types of models is generally on randomly varying parameters at the individual level (i.e., between individuals) and often between groups. When we suspect variability across subjects, we assume that the sampled subjects differ in ways that are not completely captured by the observed covariates in the model; that is, we assume a degree of heterogeneity of subjects with respect to the measured outcome process. For this reason, GLMMs are often referred to as conditional models because they focus on the conditional distribution of the outcome given the random effects (Hedeker, 2005). As Hedeker notes, one needs to decide on a reasonable model for both the fixed and the random effects.

Alternatively, using the generalized estimated equations (GEE) approach, we can examine a number of categorical measurements nested within individuals (i.e., individuals represent the clusters), but where individuals themselves are considered to be independent and randomly sampled from a population of interest; that is, the pairs of dependent and independent variables $(Y_i; X_i)$ for individuals are assumed to be independent and identically distributed (Ziegler, Kastner, & Blettner, 1998). GEE is used to characterize the marginal expectation of a set of repeated measures (i.e., average response for observations sharing the same covariates) as a function of a set of study variables.

As a result, the growth parameters are not assumed to vary randomly across individuals (or higher groups) as in a typical random-coefficients (or mixed) model. This is an important distinction between the two types of models to keep in mind: Although random-coefficient models explicitly address variation across individuals as well as clustering among subjects in higher order groups, GEE models assume simple random sampling of subjects representing a population as opposed to a set of higher order groups. Hence, GEE models provide "population average" results; that is, they model the marginal expectation as a function of the explanatory variables.

We note that the coefficients and interpretation of population-average models and unit-specific models are slightly different, given the same set of predictors and a dichotomous (or other categorical) outcome. Regression coefficients based on population averages (GEE) will be generally similar to unit-specific (random-effect models) coefficients but smaller in size (Raudenbush & Bryk, 2002). This distinction does not arise in models with continuous outcomes and identity link functions. For example, for a GEE model, the odds ratio is the average estimate in the population—that is, the expected increase for a unit change in X in the population. In contrast, in random-effect (unit-specific) models, the odds ratio will be the subject-specific effect for a particular level of clustering (i.e., the person or unit of clustering) given a unit change in X. As we have noted, this latter approach models the expected outcome for a Level-2 unit conditional on a given set of random effects.

There are currently a number of options for modeling longitudinal categorical data. For example, one common situation using ordinal data might be to model changes in individuals' perceptions of particular organizational processes, which can be formulated as a proportional odds model. We could also have count data collected on subjects over several different time periods. We can capture the change in a categorical outcome through specifying a single, time-related indicator (e.g., coding successive, repeated-measures intervals (i.e., days, weeks, or years) or by defining separate contrasts over time (e.g., changes between the baseline, separate immediate intervals, and end of a treatment) using dummy coding, effect coding, or reference cell coding. In this chapter, we demonstrate some possibilities for examining longitudinal categorical data through an extended example examining change in a dichotomous outcome over time. The dichotomous example can be easily extended to other types of longitudinal categorical outcomes (Hedeker, 2005).

Generalized Estimating Equations

We first begin with a within- and between-subjects model estimated using the GEE (or fixed-effect) approach. GEE was developed to extend a generalized linear model (GLM) further by accommodating repeated categorical measures, logistic regression, and various other models for time series or other correlated data where relationships between successive measurements on the same individual are assumed to influence the estimation of model parameters (Horton & Lipsitz, 1999; Liang & Zeger, 1986; Zeger, Liang, & Albert, 1988). The GEE analytic approach handles a number of different types of categorical outcomes, their associated sampling distributions, and corresponding link functions. It is suitable to use when the repeated observations are nested within individuals over time, but the individuals are considered to be a random sample of a population.

In one scenario, individuals are randomly assigned to treatment conditions that unfold over time. When the response variable consists of a number of events occurring within a set of trials, the dependent variable contains the number of events that occur over particular time periods. For this type of longitudinal data structure, the analyst can select an additional variable containing the number of trials. An offset term, which is a *structural* predictor, can be added to the model. Its coefficient is not estimated by the model but is assumed to have the value 1; thus, the values of the offset are simply added to the linear predictor of the dependent variable. This extra parameter can be especially useful in Poisson regression models, where each case may have different levels of exposure to the event of interest. The GEE approach, therefore, is quite flexible in accommodating a wide variety of longitudinal models.

At present in IBM SPSS, the GEE approach only accommodates a two-level data hierarchy (measurements nested in individuals). In cases where individuals are also nested in groups, we can make use of GENLIN MIXED to specify the group structure. We develop this type of three-level model using a random-coefficients approach in the second part of the chapter. We reiterate Zeger and colleagues' (1988) point that the primary distinction between GEE (or population-average) models versus random-coefficients (subject- or unit-specific) models is whether the regression coefficients describe the average population or an individual's (or group's) response to changing levels of *X*. A secondary distinction is in the nature of the assumed time dependence. GEE models only describe the covariance among repeated observations (*Y*) for each subject, whereas random coefficients attempt to model and explain the source of this covariance.

GEE Model Estimation

IBM SPSS supports several estimation methods for GEE models. A set of estimating equations is solved through an iterative process to find the value of the regression parameters. For the estimation of these parameters, the analyst can choose from a number of well-established maximum likelihood (ML) approaches including Newton-Raphson, Fisher scoring, or the default hybrid approach, which uses Fisher scoring iterations before switching to the Newton-Raphson method. If convergence is reached during the Fisher scoring phase of estimation before the maximum number of Fisher iterations is reached, the algorithm continues with the Newton-Raphson method. One way of estimating the variance available from GEE models is the *model-based* (or naive) estimate, which is consistent when both the mean model and the covariance model are correctly specified. However, the association between observations within clusters is treated as a nuisance parameter (Ziegler et al., 1998).

The model-based estimator (designated MODEL) is the negative of the generalized inverse of the Hessian matrix. If misspecification of the association structure is suspected, using *robust* estimates of variance parameters is recommended. The robust estimator (also referred to as the Huber/White/sandwich estimator) is a corrected model-based estimator that provides a consistent estimate of the covariance, even when the specification of the variance and link functions is incorrect (Horton & Lipsitz, 1999). Because the analyst will generally not know the correct covariance structure, the empirical variance estimate will be preferred when the number of clusters is large. With a small number of clustered occasions, the model-based variance estimator may have better properties even if the working variance is wrong (Prentice, 1988). This is because the robust variance estimator is asymptotically unbiased, but could be highly biased when the number of clusters is small.

Users have flexibility to set the maximum number of iterations and maximum step-halving (i.e., at each iteration the step size is reduced by a factor of 0.5 until the log likelihood increases or maximum step-halving is reached), and the routine provides a check for separation of data points (i.e., with multinomial and binary responses, this algorithm performs tests to ensure that the parameter estimates have unique values, with separation occurring when the procedure can produce a model that correctly classifies every case). Users can also set convergence criteria (i.e., parameter convergence, log-likelihood convergence, or Hessian convergence). Finally, it is also important to note that the GEE approach assumes that data are missing completely at random (MCAR). This remains a considerable limitation; however, it is easier to fit models under this type of assumption because they require fewer computational restrictions on the data. Often, it may be necessary to impute plausible values to improve model efficiency in providing optimal estimates of parameters.

An Example Study

Consider a study to examine students' likelihood to be proficient in reading over time and to assess whether their background might affect their varying patterns of meeting proficiency or not. We may first be interested in answering whether a change takes place over time in students' likelihood to be proficient. This concern addresses whether the probability of a student being proficient is the same or different over the occasions of measurement. The assumption is that if we can reject the hypothesis that the likelihood of being proficient is the same over time, it implies that a change in individuals has taken place.

In this situation, occasions of measurement are assumed to be nested within subjects but independent between subjects. The GEE approach therefore, as conceptualized in IBM SPSS, results in a population-average model—that is, with measurements nested within individuals and the explanatory variables indicating the expected change in the population for a unit change in one of the explanatory variables (Hox, 2010). In this way, the effect of the clustering of the data within individuals is treated as a type of "noise" that can be removed by accounting for the correlation between observations in a GLM. As we have noted previously, this assumption is different from the random-coefficients approach, which partitions the variance at each level of the data hierarchy to represent the clustering effects. A second interest might be whether there are differences in likelihood to be proficient across groups of individuals. These differences could be due to an experimental treatment or other types of factors (e.g., background).

Research Questions

In our extended example, we may have a number of research questions we are interested in examining, such as the following:

- What is the probability of students being proficient in reading over time?
- Do probabilities of being proficient change over time?
- What do students' trends look like over time?
- Are there between-individual variables that explain students' likelihood to be proficient over time?

In the second part of our analyses, we can use GLMM to investigate ways in which a random-coefficients approach can enhance our consideration of both individual- and group-level predictors that might influence likelihood to be proficient in the presence of missing data, varied occasions of measurement, and more complex error structures.

The Data

The data in this study consist of 2,228 individuals who were measured on four occasions regarding their proficiency in reading (Table 5.1). To examine growth within and between individuals in GENLIN MIXED, the data must first be organized differently (see Chapter 2). The time observations must be organized vertically, which requires multiple lines for each subject. As explained in Chapter 2, we use *Data* and *Restructure* menu commands to restructure the data vertically.

Closer inspection shows that we have four observations on each subject; that is, the reading proficiency scores (coded 1 = proficient, 0 = not proficient) are observed over 4 years on 2,228 students. We can observe that the repeated observations are nested within individual identification (*id*) numbers, and student IDs are nested within school identifiers (*nschode*). The grouping variables (*id, nschcode*) are used to identify each predictor as belonging to a particular level of the data hierarchy (although we will assume that school structure does not exist for our GEE example).

Readers will recall that an intercept is defined as the level of Y when X (*time*) is 0. For categorical outcomes, the time variable functions to separate contrasts between time—for example, between a baseline measurement and end of a treatment intervention—or to examine change over a particular time period. This coding pattern for *time* (0, 1, 2, 3) identifies the intercept in the model as students' *initial* (time1) proficiency status (because it is coded 0 and the intercept represents the individual's status when the other predictors are 0). This is the most common type of coding for models involving individual change.

We could code in reverse order (e.g., placing 0 last) if we wished to examine students' likelihood to be proficient at the *end* of the trend. If we desired, we could also place the 0 point at some other relevant place within the repeated measurements (e.g., Interval 3). Alternatively, it is often useful to treat the time-related variable as categorical (rather than continuous) and examine different effects for the different intervals (evenly spaced or not) using various types of indicator coding.

Defining the Model

Several important steps must be specified in conducting the analysis. Users identify the type of outcome and appropriate link function, define the regression model, select the correlation structure between repeated measures, and select either model-based or robust standard errors. There are a number of different ways to notate the models. We will let Y_{ti} be the dichotomous response at time t ($t = 1, 2,..., T$) for individual i ($i = 1, 2,..., N$), where we assume that observations of different individuals are independent but we allow for an association between the repeated measures for the same subject. This will allow us later in the chapter to add the subscript j to define random effects of individuals nested within groups such as classrooms or schools. We assume the following marginal regression model for the expected value of Y_{ti}:

$$g(E[Y_{ti}]) = x'_{ti}\beta, \qquad (5.1)$$

where x'_{ti} is a $(p + 1)*1$ vector (prime designates a vector) of covariates for the ith subject on the tth measurement occasion ($t = 1, 2,..., T$), β represents the corresponding regression parameters, and $g(\cdot)$ refers to one of several corresponding link functions, depending on the measurement of Y_{ti}.

This suggests that the data can be summarized to the vector Y_i and the matrix X_i (Horton & Lipsitz, 1999; Ziegler et al., 1998). The slope β can be interpreted as the rate of change in the population-averaged Y_i associated with a unit change in X_i (Zeger et al., 1986). Typically, the β

TABLE 5.1   Data Definition of *ch5data.sav* (N = 9,068)

Variable	Level[a]	Description	Values	Measurement
id	Individual	Student identifier (2,228 students across four time occasions)	Integer	Nominal
nschcode	School	School identifier (122 schools)	Integer	Ordinal
female	Individual	Demographic predictor variable representing student's gender	0 = male 1 = female	Nominal
zses_mean	School	Predictor interval variable converted to standardized factor score representing average student socioeconomic status composition within the schools	−2.43 to 2.50	Scale
index1	Within individual	Identifier variable resulting in reindexing the *id* individual identifier (1 to 2,267) to create new identifier with a number sequence (1, 2, 3, 4) corresponding to the four time occasions measuring students' reading proficiency	1 = first time 2 = second time 3 = third time 4 = fourth time	Nominal
time	Within individual	Variable representing four *time* occasions in time measuring students' reading proficiency	0 = first time 1 = second time 2 = third time 3 = fourth time	Scale
time1	Within individual	Recoded *time* variable from four yearly occasions in time (0, 1, 2, 3) into a sequence that encompasses the whole 4-year period of time (0.00, 0.33, 0.67, 1.00) and interpreted as the change in log odds between Year 1 and Year 4	0.00 = first time 0.33 = second time 0.67 = third time 1.00 = fourth time	Nominal
timeend	Within individual	Recoded *time* to represent students' ending reading proficiency status	−3 = first time −2 = second time −1 = third time 0 = fourth time	Nominal
readprof	Individual	Two-category dependent variable referring to student's reading proficiency	0 = not proficient 1 = proficient	Nominal
lowses	Individual	Predictor dichotomous variable representing student's social economic status (SES)	0 = not low socioeconomic status 1 = low socioeconomic status	Nominal
smallsch	School	Predictor dichotomous variable representing school size based on the number of students	0 = not small (≥600) 1 = small (<600)	Nominal
Rid	Individual	A within-group level identifier[b] representing a sequential identifier for each student within each school	1 to 15	Ordinal

[a] Individual = Level 1; school = Level 2; within individual = repeated measure lowest level.
[b] Results from ranking student cases (*id*) with the school group identifier (*nschcode*).

parameters are constant for all t (Ziegler et al., 1998); however, the GEE model can be extended to include time-dependent predictors. This extension can then accommodate longitudinal studies where the influence of a covariate (e.g., motivation) changes over time. For example, X_{ti} might be a measure of student motivation observed before the corresponding yearly reading test. It should be noted that a time-varying covariate must be arranged vertically like the response variable. A similar approach can be used to support analyses that jointly estimate time-varying and time-constant parameters (Park, 1994; Ziegler et al., 1998).

When the data are dichotomous, the marginal mean—a probability—is most commonly modeled via the logit link function (i.e., whether a child is proficient or not at time t). The coefficients are then interpreted as log odds. As we noted in the last chapter, the probit model, which is based on the standard normal distribution, as opposed to the logistic distribution, is another alternative to the logistic model for investigating dichotomous and ordinal outcomes. For the Bernoulli case (i.e., where the number of trials is 1), Y_{ti} has a binomial distribution with probability of success π_{ti} and variance of $\pi_{ti}(1 - \pi_{ti})$. For binary data with the logit link function, we have the familiar

$$\eta_{ti} = \log(\pi_{ti}/(1-\pi_{ti})) = x'_{ti}\beta, \tag{5.2}$$

where η_{ti} is the underlying transformed predictor of Y_{ti}—in this case, the log of the odds of $\pi_{ti}(1-\pi_{ti})$.

We again draw readers' attention to the fact that the model represents a ratio of the probability of the event coded 1 occurring versus the probability of the event coded 0 occurring at a particular time point. There is no residual variance parameter (ε_i), as the variance is related to the expected value of π_{ti} and therefore cannot be uniquely defined. The corresponding regression model can be expressed in terms of the predicted probability as

$$\pi_{ti} = \frac{\exp(x'_{ti}\beta)}{1 + \exp(x'_{ti}\beta)}. \tag{5.3}$$

Alternatively, the predicted log odds can also be converted back to a predicted probability by computing

$$\pi_{ti} = \frac{1}{1 + e^{-(\eta_{ti})}}. \tag{5.4}$$

Model Specifying the Intercept and Time

In the first model, we specify the repeated-measures outcomes in two parameters that describe the intercept and time-related slope as follows:

$$\eta_{ti} = \log(\pi_{ti}/(1-\pi_{ti})) = \beta_0 + \beta_1 time, \tag{5.5}$$

where *time* is coded to indicate the interval between successive measurements, β_0 is an intercept, and β_1 describes the rate of change on a logit scale in the fraction of positive responses in the population of subjects per unit time, rather than the typical change for an individual subject.

As Equation 5.5 suggests, β_0 is the log odds of response when time is 0 (i.e., initial status). In this case, β_1 is the log odds associated with a 1-year interval. The model assumes that there are no between-subject random effects; therefore, there are two parameters to estimate. Because this is a population-average model, we will drop the subscripts referring to the predictors.

Through various coding schemes, we can change the meaning of the time-related slope parameter. In this case, the measurements are evenly spaced (one per year). We could instead code the successive measurements to describe the change in likelihood to be proficient over the whole period of time by coding in the following manner: (0.0, 0.33, 0.67, 1.0). Because the slope represents the change in Y for a unit change in X, in the latter example, the interpretation would now be the change in proficiency for the *entire* trend studied (i.e., from the interval coded 0 to the interval coded 1). We demonstrate this difference later in the chapter.

When measurements are not evenly spaced, we might be able to make them approximate a linear trend, for example, by taking the square root of the time measurements. A simple example would be when observations were taken at 1 month, 4 months, and 9 months. We can also model the change in the categorical outcome through defining separate time-related contrasts over time (e.g., changes between the baseline, separate immediate intervals, and end of a treatment). These various types of coding can help accommodate different spacing of measurements within individuals.

Correlation and Covariance Matrices

As noted, in a GEE model we have a vector of repeated measures $(y_1, y_2, ..., y_j)'$. The GEE approach allows for describing the correlation between the repeated-measures observations without specifically addressing the origin of the dependency; therefore, it is most suitable when possible random effects at Level 2 and their variances are not of direct interest. As we have indicated, the focus is on estimating the average response over the population ("population-averaged" effects), rather than the regression parameters that would enable prediction of the effect of changing one or more covariates on a given individual. Zeger and Liang (1986) referred to a *working* correlation matrix, which is not required to be correctly specified for the parameter estimates and the estimated variance of the parameter estimates in the model to be consistent, as long as the mean model itself is correct and no data are missing. Users must specify the structure of the working correlation matrix a priori.

It is possible to specify several different types of correlation structures to describe the within-subject dependencies over time. However, because one does not often know what the correct structure is ahead of time, different choices can make some difference in the model's parameter estimates; therefore, the structure is chosen to improve efficiency. It often does take a bit of preliminary work to determine the optimal working correlation matrix for a particular data structure. Examples of correlation/covariance structure specifications include independent, exchangeable, autoregressive, stationary *m*-dependent, and unstructured.

The *independent* matrix assumes that the repeated measurements are uncorrelated; however, this will not be the case in most instances. Generally, in longitudinal models, the successive measurements are correlated at least to some extent. An *exchangeable* (or *compound symmetry*) covariance (or correlation) matrix assumes homogenous correlations between elements (which is sometimes difficult to assume in longitudinal studies); that is, the correlations are assumed to be the same over time. This is often a reasonable place to start in a longitudinal study, however. The *autoregressive*, or AR(1), matrix assumes that the repeated measures have a first-order autoregressive structure. This implies that the correlation between any two adjacent elements is equal to ρ (rho), to ρ^2 for elements separated by a third, and so on, with ρ constrained such that $-1 < \rho < 1$.

An *m*-dependent matrix assumes that consecutive measurements have a common correlation coefficient, pairs of measurements separated by a third have a common correlation coefficient, and so on, through pairs of measurements separated by $m - 1$ other measurements. When measurements are not evenly spaced, it may be reasonable to consider a model where the correlation is a function of the time between observations (i.e., *m*-dependent or autoregressive). Measurements with greater separation are assumed to be uncorrelated. When choosing this structure, specify a value of m less than the order of the working correlation matrix. Finally, an *unstructured* correlation (or covariance) matrix provides a separate coefficient for each covariance.

As with cross-sectional models, we have found that model estimates can vary slightly according to the matrix structure specified.

By default, the GEE procedure will adjust correlation estimates by the number of nonredundant parameters. Removing this adjustment may be desirable if the analyst wishes the estimates to be invariant to subject-level replication changes in the data.

Standard Errors

Model-based standard errors are based on the correlational structure chosen. Hence, they may be inconsistent if the correlation structure is incorrectly specified. They are usually a little smaller than robust standard errors (SEs). For smaller numbers of clusters, model-based SEs are generally preferred over robust SEs. In contrast, robust standard errors vary only slightly depending on the choice of hypothesized correlational structure among the repeated measures; that is, the estimates are consistent even if the correlational structure is specified incorrectly. As we noted, the robust SE approach uses a "sandwich" estimator based on an approximation to maximum likelihood.

Because of this, there can be occasions that occur when one approach will converge and the other may not. Robust standard errors are often preferred when the number of clustered observations is large. We will estimate our models in this example using robust standard errors. Once again, we note that users should keep in mind that GEE uses a type of quasilikelihood estimation (as opposed to full information ML), which can make direct model comparison based on fit statistics that depend on the real likelihood (e.g., deviance, AIC, BIC) not very accurate (Hox, 2010).

Defining Model 1.1 With IBM SPSS Menu Commands

Launch the IBM SPSS application program and select the *ch5data1.sav* data file.

1. Go to the toolbar and select ANALYZE, GENERALIZED LINEAR MODELS, GENERALIZED ESTIMATING EQUATIONS. This command allows access to the *Generalized Estimating Equations* (GEE) main dialog box.

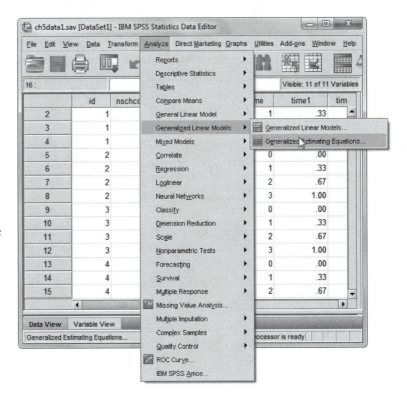

2a. The *Generalized Estimating Equations* screen displays 10 command tabs, each having specific features: *Repeated, Type of Model, Response, Predictors, Model, Estimation, Statistics, EM Means, Save, and Export*. We begin with the *Repeated* tab, which is the default setting when building a model for the first time. The *Repeated* tab enables specifying subject and within-subject variables along with covariance and correlation matrix settings.

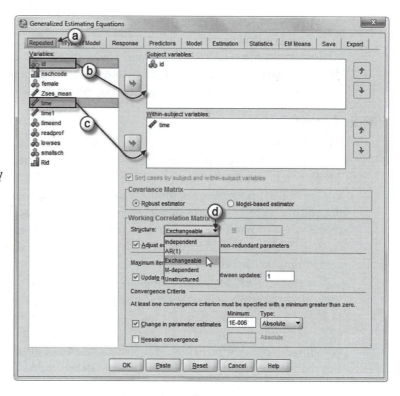

b. For this model we will designate *id* (individual identification numbers) as the subject variable. Click to select *id* from the *Variables* column and then click the right-arrow button to move the variable into the *Subject variables* box.

c. Now click to select *time* from the *Variables* column and then click the right-arrow button to move the variable into the *Within-subject variables* box. The combination of values for *time* and *id* defines a particular time period and student for each case.

d. Change the *Structure* option of the *Working Correlation Matrix*. This correlation matrix represents the within-subject dependencies. Its size is determined by the number of measurements and thus the combination of values of within-subject variables. Click the pull-down menu and select *Exchangeable*. This structure has homogenous correlations between elements and is also known as a compound symmetry structure.

Note: We will use the default robust SEs.

3a. Click the *Type of Model* command tab; this enables specifying model types, distribution, and link functions.

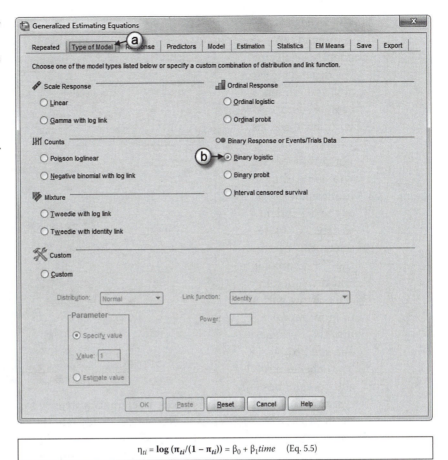

$$\eta_{ti} = \log\left(\pi_{ti}/(1 - \pi_{ti})\right) = \beta_0 + \beta_1 time \quad \text{(Eq. 5.5)}$$

b. For this model we will specify a binomial distribution with a logistic link function. Click to select *Binary logistic* within the *Binary Response or Events/Trials Data* section.

4a. Click the *Response* command tab; this enables specifying the dependent variable and reference category for variables with dichotomous values.

b. For this model we will use reading proficiency scores as the dependent variable. Click to select *readprof* from the *Variables* column and then click the right-arrow button to move *readprof* into the *Dependent Variable* box.

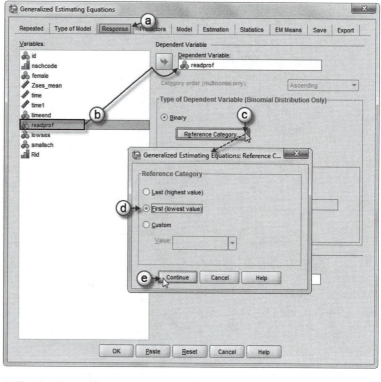

c. The dependent variable *readprof* has only two values (0, 1), so we will specify a reference category for parameter estimation. Click the REFERENCE CATEGORY button to activate the pop-up dialog box.

d. Change the default reference category from *Last (highest value)* by clicking to select *First (lowest value)*. Interpretation of the resulting parameter estimates will now be relative to the likelihood of the value "0" category.

e. Click the CONTINUE button to close the *Reference Category* pop-up box.

5a. Click the *Predictors* command tab; this allows specifying factors and covariates in predicting the dependent variable. Factors are categorical predictors that may be numeric or string. Covariates are scale predictors that must be numeric.

b. To examine changes within and between students in their reading proficiency over time, we will use the *time* variable. The variable's coding pattern (0, 1, 2, 3) identifies the intercept

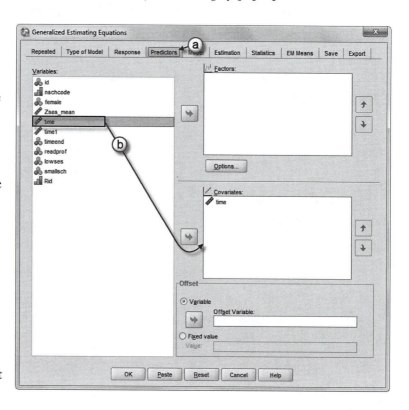

in the model as students' initial reading proficiency status (coded 0). Click to select *time* from the *Variables* list and then click the right-arrow button to move *time* into the *Covariates* box.

6a. Click the *Model* command tab; this enables specifying model effects based on the covariate variable (*time*).

b. Note that the *Include intercept in model* option is preselected. This is because a Level-1 model is intercept only (default), so other model effects must be specified.

c. We will use the default *Main effects* setting, which creates a main-effects term for each variable selected for the model, which in this case is 1: *time*.

d. Click to select *time* and then click the right-arrow button to move the variable into the *Model* box. Note at the bottom of the box that one effect is now identified in the model (*Number of Effects in Model: 1*).

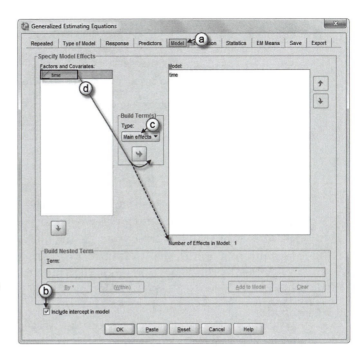

$$\eta_{ti} = \log (\pi_{ti}/(1 - \pi_{ti})) = \beta_0 + \beta_1 time \quad (Eq. 5.5)$$

7a. We will retain the default settings used in the *Estimation* command tab, so skip over to click the *Statistics* command tab; this enables specifying model effects and model output (print) options.

b. We will retain the *Model Effects* settings but add three additional print options for the output results by clicking to select Go to the *Print* section and select two additional options: *Include exponential parameter estimates* and *Correlation matrix for parameter estimates*.

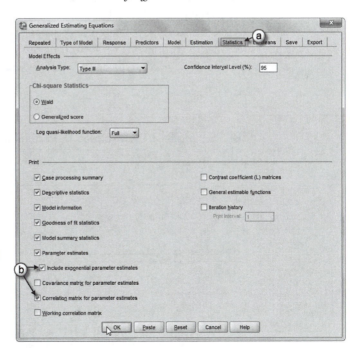

Note: We included the option *Correlation matrix for parameter estimates* to generate the results shown later in Table 5.8.

Click the OK button to generate the model results.

TABLE 5.2   Model 1.1: Model Information

Dependent variable		readprof[a]
Probability distribution		Binomial
Link function		Logit
Subject effect	1	id
Within-subject effect	1	Time
Working correlation matrix structure		Exchangeable

[a] The procedure models 1 as the response, treating 0 as the reference category.

Interpreting the Output of Model 1.1

Table 5.2 provides information about how the model is defined (e.g., probability distribution and link function, number of effects in the model, type of correlation matrix used to describe within-subject structure). As the output shows, the distribution is binomial and a logit link function is used to transform Y_{ti}. The working correlation structure is exchangeable, which is the same as compound symmetry. This implies that the correlations are the same over each time interval. We can subsequently investigate whether this is a viable assumption for these data.

Table 5.3 provides information about the overall data set. As shown, there are 2,228 individuals and four repeated measures within subjects. The table also shows whether there are fewer measurements for some individuals. In this case, all subjects have the same number of measurements, suggesting that the dimension of the correlation matrix is 4 by 4.

We can next note from the case processing summary (Table 5.4) that no cases were removed for missing data. Readers may note that there are 8,912 processed cases because each subject takes four lines (i.e., 2,228*4 = 8,912).

Next, we can observe how many of the total cases for the dependent variable (reading proficiency) are coded 1 (proficient) versus 0 (not proficient). As Table 5.5 suggests, across the four time periods, an average 68% of the individuals were proficient and 32% were not.

TABLE 5.3   Model 1.1: Correlated Data Summary

Number of levels	Subject effect	id	2228
	Within-subject effect	Time	4
Number of subjects			2228
Number of measurements per subject	Minimum		4
	Maximum		4
Correlation matrix dimension			4

TABLE 5.4   Model 1.1: Case Processing Summary

	N	Percent
Included	8912	100.0
Excluded	0	0.0
Total	8912	100.0

TABLE 5.5 Model 1.1: Categorical Variable Information

			N	Percent
Dependent variable	readprof	0	2850	32.0
		1	6062	68.0
		Total	8912	100.0

If we did not include the time variable, the log odds intercept would be 0.755 (not tabled), which would be the grand mean log odds coefficient across the four time periods. We can translate the odds ratio back to the predicted population probability of $\pi = 1$. For this model, the predicted probability of proficiency can then be calculated from Equation 5.4 as

$$\pi = \frac{1}{1 + e^{-(\beta_0)}},\qquad(5.6)$$

where β_0 is the log odds of the intercept. In this first case, where the log odds are 0.755, the predicted probability will then be $1/[1 + (2.71828)^{-(0.755)}]$, which reduces to 1/1.47 or 0.680 (i.e., 68.0%, as noted in Table 5.5).

In Table 5.6 we present the parameter estimates for the model including the time-related variable. The estimated intercept log odds coefficient is 0.838, which, because of the coding of the time variable (i.e., 0, 1, 2, 3), can be interpreted as the percentage of individuals who are proficient at the start of the study. The intercept represents the predicted log odds when any variables in the model are 0. If we exponentiate the log odds, we obtain the corresponding odds ratio of 2.311. This suggests that individuals are almost 2.3 times more likely to be proficient than nonproficient at the beginning of the study (0.70/0.30 ~ 2.3). We note that, consistent with our previous models in Chapter 4, a scale factor is fixed to 1.0 (with the variance in a logistic distribution equal to $\pi^2/3$).

The log odds can be changed back to a predicted probability as $1/[1 + (2.71828)^{-(0.838)}]$, which reduces to 1/1.4326 or 0.702. This suggests that the initial proportion of students who were proficient at the beginning of the trend was about 0.70, or 70%. This is consistent with the proportion of students who were proficient at Time 0 (0.6984) in Table 5.7. From the predicted probability, we can also confirm that the odds ratio is 0.702/0.300, which equals approximately 2.33. The Wald chi-square test suggests that the intercept is not 0 (Wald chi-square = 306.910, $p < .001$).

Regarding the *time* variable, the coefficient suggests that, over each interval, students' likelihood of being proficient decreases significantly (log odds = −0.055, $p < .001$). We can translate this into a predicted probability by adding it to the intercept. Initially (i.e., at time = 0), the log odds of being proficient are 0.838. For the second interval (time = 1), the estimated log odds are then 0.783 [0.838 + (−0.055) = 0.783]. This new probability is then estimated to

TABLE 5.6 Model 1.1: Parameter Estimates

Parameter	β	Std. error	95% Wald confidence interval		Hypothesis test			Exp(β)	95% Wald confidence interval for exp(β)	
			Lower	Upper	Wald χ²	df	Sig.		Lower	Upper
(Intercept)	0.838	0.0478	0.744	0.931	306.910	1	0.000	2.311	2.104	2.538
Time	−0.055	0.0165	−0.087	−0.022	11.021	1	0.000	0.947	0.916	0.978
(Scale)	1									

Note: Dependent variable: readprof; model: (intercept), time.

TABLE 5.7   Proportions of Proficient Students Over Time

Readprof

Time	Mean	N	Std. dev.
0	0.6984	2228	0.459
1	0.6988	2228	0.459
2	0.6481	2228	0.478
3	0.6755	2228	0.468
Total	0.6802	8912	0.466

TABLE 5.8   Model 1.1: Correlations of Parameter Estimates

	(Intercept)	Time
(Intercept)	1.000	−0.611
Time	−0.611	1.000

Note: Dependent variable: readprof; model: (intercept), time.

be $1/[1 + (2.71828)^{-(0.783)}]$, which reduces to $1/1.457$ or 0.69. Note that this estimate is slightly different from the actual observed probability in Table 5.7 because no actual change took place between Time 0 and Time 1.

The odds ratio suggests that the odds of being proficient are multiplied by 0.947 (or reduced by 5.3%) over the first interval. We can estimate the change from Time 0 to Time 2 by again adding the intercept and log odds for Time covering two intervals $[.838 + (−.055) + (−.055) = 0.728]$. We can again use Equation 5.3 to calculate the probability of being proficient at Time 2; that is, $1/1.483 = 0.674$. This estimate for Time 2 is obviously closer to describing the change in proficiency that occurred between Time 1 and Time 3 in Table 5.7. We would estimate the proficiency level for the last interval as approximately 0.66, but this estimate will also be off, since Table 5.7 shows that the level of proficiency did not continue to drop over time. We can see, in this particular situation, an assumed negative linear time trend in reduced probability of being proficient does not quite fit the data optimally.

We next present the correlation between the time-related slope and the initial status intercept. Table 5.8 suggests that they are moderately and negatively correlated, as is common in growth models (i.e., individuals who are initially low will tend to grow more over time and vice versa).

Alternative Coding of the Time Variable

We could decide to code the data somewhat differently. We might wish to treat the time-related variable as ordinal $(1, 2..., C)$ rather than scale. If we make this change, we will have $C − 1$ estimates because one category will serve as the reference group. In this case, we will specify "descending" for the factor category order so that the first category (time = 0) will serve as the reference group. This is the same as creating a series of $C − 1$ dummy variables for a categorical factor and specifying them in the model.

Defining Model 1.2 With IBM SPSS Menu Commands

Note: To facilitate illustrating this model, we will use a different data set (*ch5data1-ord.sav*) that has the variable *time* changed to an ordinal measure. Refer to Chapter 3 and Figure 3.9 for further information on changing a variable's measurement setting.

Locate and select the *ch5data1-ord.sav*.

1. Go to the toolbar and select ANALYZE, GENERALIZED LINEAR MODELS, GENERALIZED ESTIMATING EQUATIONS. This command allows access to the *Generalized Estimating Equations* (GEE) main dialog box.

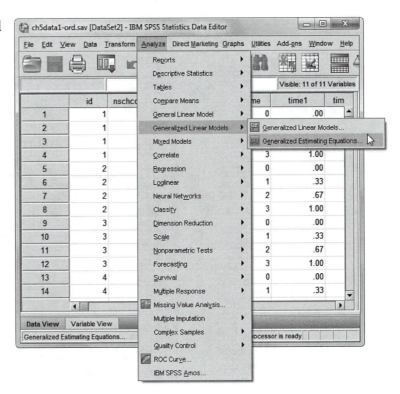

2a. The *Generalized Estimating Equations* screen displays ten command tabs, each having specific features: *Repeated, Type of Model, Response, Predictors, Model, Estimation, Statistics, EM Means, Save,* and *Export*. We begin with the *Repeated* tab, which is the default setting when building a model for the first time. The *Repeated* tab enables specifying subject and within-subject variables along with covariance and correlation matrix settings.

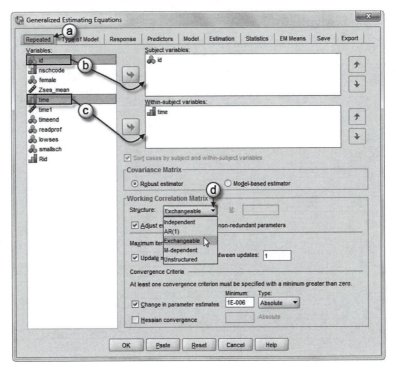

b. For this model we will designate *id* (individual identification numbers) as the subject variable. Click to select *id* from the *Variables* column and then click the right-arrow button to move the variable into the *Subject variables* box.

c. Now click to select *time* from the *Variables* column and then click the right-arrow button to move the variable into the *Within-subject variables* box.

Note: Notice *time*'s "bars" icon, which identifies the variable's ordinal measurement. The combination of values for *time* and *id* defines a particular time period and student for each case.

d. Change the *Structure* option of the *Working Correlation Matrix*. This correlation matrix represents the within-subject dependencies. Its size is determined by the number of measurements and thus the combination of values of within-subject variables. Click the pulldown menu and select *Exchangeable*. This structure has homogenous correlations between elements and is also known as a compound symmetry structure.

3a. Click the *Type of Model* command tab; this enables specifying model types, distribution, and link functions.

b. For this model we will specify a binomial distribution with a logistic link function. Click to select *Binary logistic* within the *Binary Response or Events/ Trials Data* section.

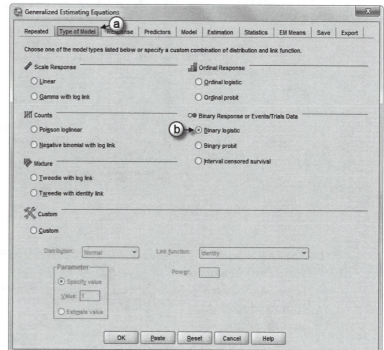

4a. Click the *Response* command tab; this enables specifying the dependent variable and reference category for variables with dichotomous values.

b. For this model we will use reading proficiency scores as the dependent variable. Click to select *readprof* from the *Variables* column and then click the right-arrow button to move *readprof* into the *Dependent Variable* box.

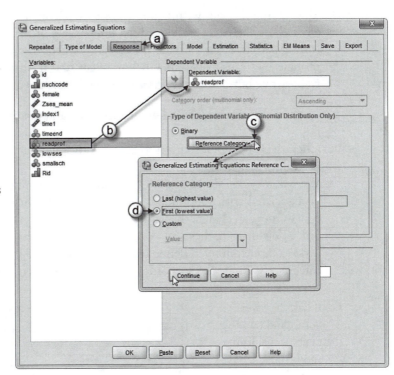

c. The dependent variable *readprof* has only two values (0, 1), so we will specify a reference category for parameter estimation. Click the REFERENCE CATEGORY button to activate the pop-up dialog box.

d. Change the default reference category from *Last (highest value)* by clicking to select *First (lowest value)*. Interpretation of the resulting parameter estimates will now be relative to the likelihood of the value "0" category.

e. Click the CONTINUE button to close the *Reference Category* pop-up box.

5a. Click the *Predictors* command tab; this allows specifying factors and covariates in predicting the dependent variable. Factors are categorical predictors that may be numeric or string. Covariates are scale predictors that must be numeric.

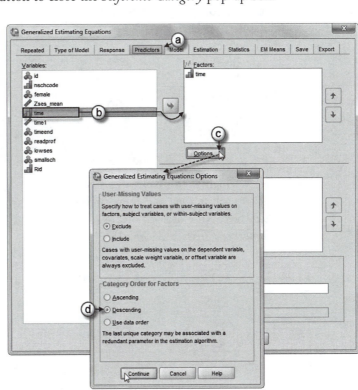

b. To examine changes within and between students in their reading proficiency over time, we will use the *time* variable, which now has an ordinal measurement. The variable's coding pattern (0, 1, 2, 3) identifies the intercept in the model as students' initial reading proficiency status (coded 0). Click to select *time* from the *Variables* list and then click the right-arrow button to move *time* into the *Factors* box.

Note: Factors are categorical predictors that can be numeric or string.

c. Click the OPTIONS button to access additional factor-related settings concerning missing values and categorical order factors.

d. We will focus on the *Category Order* option, which determines a factor's last level that may be associated with a redundant parameter in the estimation algorithm. Changing the category order can change the values of factor-level effects. Factors may be sorted in ascending order (lowest to highest value), descending order (highest to lowest value), or data order (first value encountered defines the first category and last value encountered defines the last category). Change the sort order by clicking to select *Descending*.

6a. Click the *Model* command tab; this enables specifying model effects based on the covariate variable (*time*).

b. Note that the *Include intercept in model* option is preselected. This is because a Level-1 model is intercept only (default), so other model effects must be specified.

c. We will use the default *Main effects* setting, which creates a main-effects term for each variable selected for the model, which in this case is 1: *time*.

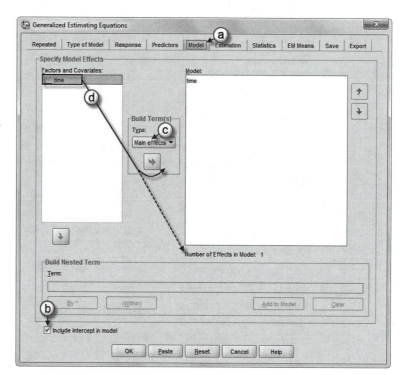

d. Click to select *time* and then click the right-arrow button to move the variable into the *Model* box. Note at the bottom of the box that one effect is now identified in the model (*Number of Effects in Model: 1*).

7a. We will retain the default settings used in the *Estimation* command tab, so skip over to click the *Statistics* command tab; this enables specifying model effects and model output (print) options.

b. We will retain the *Model Effects* settings but add three additional print options for the output results by clicking to select Go to the *Print* section and select two additional options:

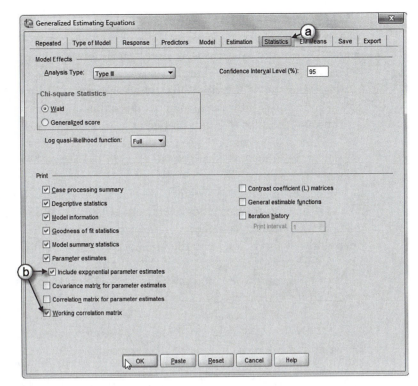

Include exponential parameter estimates and *Working correlation matrix*.

Note: We included the option *Working correlation matrix* to generate the results shown later in Table 5.11.

Click the OK button to generate the model results.

Interpreting the Output of Model 1.2

We provide results of this model in Table 5.9. The intercept log odds are now 0.840. This is only slightly different from the last table. If we calculate the predicted probability of being proficient initially (time = 0), we see that it will be $1/(1 + e^{-(0.840)})$ or $1/1.432 = 0.698$. This is consistent with the observed probability of 0.6984 in Table 5.7. We can see further that at time = 1, there

TABLE 5.9 Model 1.2: Parameter Estimates

Parameter	β	Std. error	Hypothesis test Wald χ²	df	Sig.	Exp(β)	95% Wald confidence interval for exp(β) Lower	Upper
(Intercept)	0.840	0.0462	330.845	1	0.000	2.315	2.115	2.535
[time = 3]	−0.106	0.0457	5.438	1	0.020	0.899	0.822	0.983
[time = 2]	−0.229	0.0445	26.440	1	0.000	0.795	0.729	0.868
[time = 1]	0.002	0.0177	0.014	1	0.904	1.002	0.968	1.038
[time = 0]	0[a]					1		
(Scale)	1							

Note: Dependent variable: readprof; model: (intercept), time (ordinal).

[a] Set to 0 because this parameter is redundant.

was little change in log odds units regarding students' probability of being proficient (log odds = 0.002, p = .904). At Time 2 (log odds = −0.229, p < .001) and Time 3 (log odds = −0.106, p < .001), however, students were significantly lower in probability of being proficient relative to their proficiency status at Time 0.

Regarding the odds ratios (ORs), we can interpret the nonsignificant relationship at time = 1 as indicating no significant change in odds of being proficient at Time 1 (OR = 1.002, p = .904). In contrast, the odds of being proficient at time = 2 versus Time 0 are multiplied by 0.795 (or reduced by 20.5%) compared to the initial level. At time = 3, it suggests that the odds of being proficient at Time 3 versus Time 0 (i.e., initial status intercept) are multiplied by 0.899 (or reduced by 10%). (See Table 5.9)

We can estimate the probability of being proficient at Time 3 versus Time 0 in several ways. We can add the two log odds coefficients (0.840 − 0.106 = 0.734). This will provide the log odds of being proficient at Time 3. The exponentiated slope can be interpreted as the change in the odds that Y = 1 relative to the reference category (Time 0). If we exponentiate the log odds ($e^{0.734}$), we obtain the odds ratio of 2.08. From Equation 5.3, we can then calculate the probability of being proficient at Time 3 as 2.08/3.08 = 0.675, which is consistent with the Time 3 mean of 0.6755 in Table 5.7.

Alternatively, we can also represent the new odds ratio as the product of the two odds ratios (2.315*0.899) = 2.08; that is, we multiply the odds ratio for time = 0 by the difference in odds between Time 0 and Time 3 (0.899), which provides the new odds ratio (2.08). We can again calculate the probability of being proficient as 0.675 (2.08/3.08). Applying this approach for time = 2, we have 2.315*0.795 = 1.840, which is then 0.648 (1.84/2.84 = 0.648). This estimate of the probability Y = 1 is consistent with the observed mean proportion of 0.6481 in Table 5.7. We can see that defining the time trend as categorical in this instance provides some benefits in representing the change probability of being proficient that takes place between each measurement more accurately.

Another possibility would be to estimate the change in proficiency across the whole trend (4 years) instead of at each yearly interval. In this case, we can code the data such that Time 1 is coded 0 and Time 4 (the end year) is coded 1. The time variable would then be interpreted as the change in log odds between Year 1 and Year 4 because this will represent a one-unit change in the time variable from 0 to 1. We might code Time 2 as 0.33 and Time 3 as 0.67. Readers can observe this in the data set for the variable *time1*. In this case, the log odds coefficient for *time1* will be −0.165 (i.e., 0.055*3). The odds ratio is then 0.848, suggesting that the odds of being proficient are multiplied by 0.848 (or reduced by 15.2%) over the course of the study.

Finally, we note that we could also decide to code the time-related variable to represent students' ending proficiency status. In this case, the correlation would be positive, suggesting that ending proficiency status is positively related to change in likelihood to be proficient over the course of years studied. To illustrate, if we change the coding to represent the intercept as students' end status (i.e., −3, −2, −1, 0), the correlation between the intercept and slope will now be positive, as summarized in Table 5.10. This variable is *timeend* in the data set. This reflects that a student's probability of being proficient at the end of the study is in part related to her or his change in proficiency status over time. (We provide the model syntax for this table in Appendix A.)

As this discussion and our two initial analyses in Tables 5.6 and 5.9 suggest, the analyst has considerable flexibility in how she or he defines the key parameters in the longitudinal model (i.e., intercept and slope), and the choice of coding pattern will influence both the strength and

TABLE 5.10 Correlations of Parameter Estimates (Recoded *time* Variable)

	(Intercept)	Timeend
(Intercept)	1.000	0.475
timeend	0.475	1.000

Note: Dependent variable: readprof; model: (intercept), timeend.

TABLE 5.11 Model 1.2: Working Correlation Matrix

Measurement	Measurement			
	[time = 0]	[time = 1]	[time = 2]	[time = 3]
[time = 0]	1.000	0.605	0.605	0.605
[time = 1]	0.605	1.000	0.605	0.605
[time = 2]	0.605	0.605	1.000	0.605
[time = 3]	0.605	0.605	0.605	1.000

Note: Dependent variable: readprof; model: (intercept), time (ordinal).

TABLE 5.12 Model 1.2: Goodness of Fit[a]

	Value
Quasilikelihood under independence model criterion (QIC)[b]	11160.933
Corrected quasilikelihood under independence model criterion (QICC)[b]	11160.933

Notes: Dependent variable: readprof; model: (intercept), time (ordinal).
[a] Information criteria are in small-is-better form.
[b] Computed using the full log quasilikelihood function.

direction of the correlation observed between them (which, of course, can affect their substantive interpretation as well).

We next show the working correlation matrix for Model 1.2, presented in Table 5.11. This matrix describes the within-subject correlations over time. As we noted previously, we assumed compound symmetry (i.e., equal correlations between time intervals) in this first example. The estimates suggest a moderate correlation between likelihood to be proficient at each interval.

Finally, we present information about the model's fit in Table 5.12. For the GEE approach, the estimator is a quasilikelihood estimator, which makes direct model comparisons tenuous (Hox, 2010) because they are only approximate.

We could also explore some different types of working correlation matrices. For example, an autoregressive matrix was very similar. We could also try a completely unstructured correlation matrix, as shown in Model 1.3.

Defining Model 1.3 With IBM SPSS Menu Commands

Note: From this model forward, we will resume using the *ch5data1.sav* file that was last used for Model 1.1. If you are continuing from Model 1.1, the program settings will default to those used for that model.

Resume using *ch5data1.sav*.

1. Go to the toolbar and select ANALYZE, GENERALIZED LINEAR MODELS, GENERALIZED ESTIMATING EQUATIONS. This command allows access to the *Generalized Estimating Equations* (GEE) main dialog box.

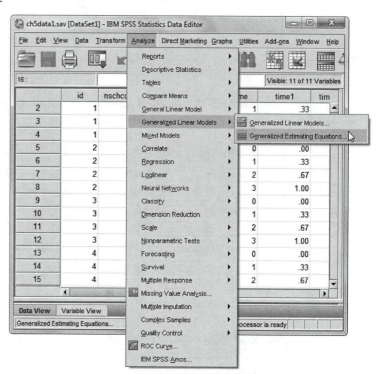

2a. The *Generalized Estimating Equations* display screen highlights the *Statistics* tab, which was the last command window used for Model 1.1.

b. Click the *Repeated* command tab to change the correlation matrix option.

c. We will change the correlation matrix by clicking the pull-down menu and selecting *Unstructured*.

Note: Unstructured is a general correlation matrix.

Click the OK button to generate the model results.

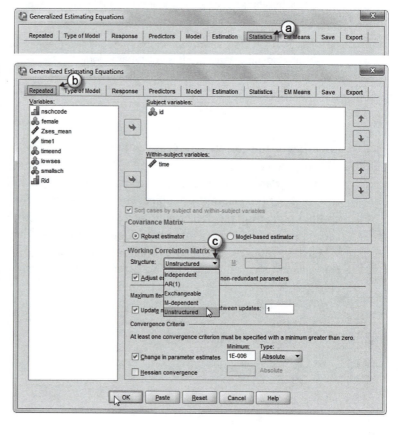

TABLE 5.13 Model 1.3: Working Correlation Matrix

Measurement	Measurement			
	[time = 0]	[time = 1]	[time = 2]	[time = 3]
[time = 0]	1.000	0.906	0.522	0.496
[time = 1]	0.906	1.000	0.520	0.505
[time = 2]	0.522	0.520	1.000	0.671
[time = 3]	0.496	0.505	0.671	1.000

Note: Dependent variable: readprof; model: (intercept), time.

Interpreting the Output of Model 1.3

We provide the unstructured working correlation matrix in Table 5.13. It does show some variability in the correlations between measures for each interval (primarily between time interval 0 and 1, $r = 0.91$). Generally, however, a completely unstructured matrix is not required for within-subject relationships because it produces a more complex model formulation that, in our experience, does not appreciably affect the fixed effects in the model. Especially in the multilevel formulation, it may be much harder to fit such a model because of its greater complexity in estimating each time-related relationship.

Adding a Predictor

We can next add between-subjects predictors, but the outcome parameters are treated as fixed; that is, the slopes cannot vary across individuals in the sample. We provide an example where we add gender (female coded 1; male coded 0) to the model. We can define this model as follows:

$$\eta_{ti} = \log[\pi_{ti}/(1 - \pi_{ti})] = \beta_0 + \beta_1 time + \beta_2 female. \tag{5.7}$$

Defining Model 1.4 With IBM SPSS Menu Commands

Note: Settings default to those used for Model 1.3.

1. Go to the toolbar and select ANALYZE, GENERALIZED LINEAR MODELS, GENERALIZED ESTIMATING EQUATIONS. This command allows access to the *Generalized Estimating Equations* (GEE) main dialog box.

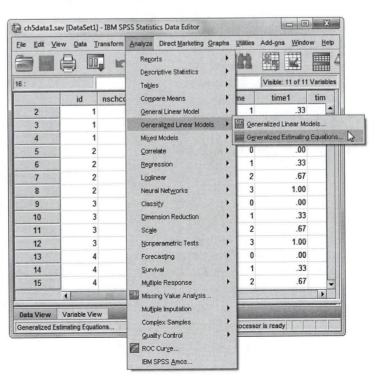

2a. The *Generalized Estimating Equations* display screen highlights the *Repeated* tab, which was the last command window used for Model 1.3.

b. For this model we will change the correlation matrix. Click the pull-down menu and select *Exchangeable*. This structure has homogenous correlations between elements and is also known as a compound symmetry structure.

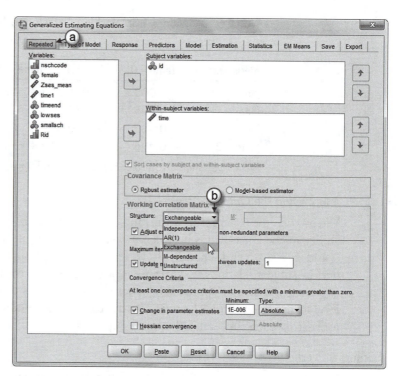

3a. Click the *Predictors* command tab.

b. We will add *female* to the model. First, click to select *female* and then click the right-arrow button to move the variable into the *Factors* box.

c. We will change the category order of *female*, which was coded 1 (female) or 0 (male). Factors may be sorted in ascending order (lowest to highest value) or descending order (highest to lowest value). Because "1" represents female, we will select a descend-

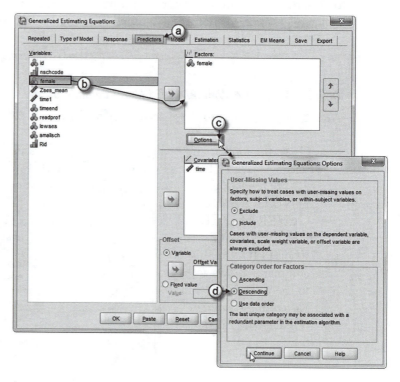

ing sort order. First, click the OPTIONS button, which will display the *Generalized Estimating Equations: Options* dialog box.

d. Now click to select *Descending*. Click the CONTINUE button to close the *Generalized Estimating Equations: Options* dialog box.

4a. Now click the *Model* command tab.

b. Note that the *Include intercept in model* option is the default setting.

c. *Main effects* remains as the default setting.

d. We will add *female* to the model. First, click to select *female* and then click the right-arrow button to move the variable into the *Model* box. Note at the bottom of the box that one effect is identified in the model (*Number of Effects in Model: 1*).

e. To facilitate reading of the output table, we will rearrange

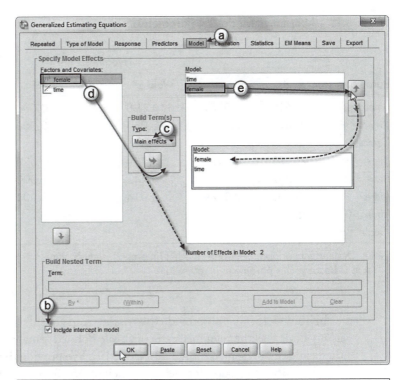

$$\eta_{ti} = \log[\pi_{ti}/1 - \pi_{ti})] = \beta_0 + \beta_1 time + \beta_2 female \quad \text{(Eq. 5.7)}$$

the variables. Click to select *female* and then click the up-arrow button to move the variable above *time*. The variable sequence is now: *female, time*.

Click the OK button to generate the model results.

Interpreting the Output of Model 1.4

In Table 5.14 we see that females are more likely to be proficient in reading than males at the start of the study (log odds = 0.502, $p < .001$). We can calculate the likelihood of males being proficient as $0.607 + 0.502(0)$, which is the intercept log odds (0.607). For females, the likelihood of being proficient at initial status is $0.607 + 0.502(1)$ or 1.109. We can then calculate the predicted probability as $1/(1 + e^{-(0.607)})$, which will be 1/1.545 or 0.647 for males. For females it will be $1/1.330 = 0.752$.

TABLE 5.14 Model 1.4: Parameter Estimates

| Parameter | β | Std. Error | 95% Wald confidence interval | | Hypothesis test | | | Exp(β) | 95% Wald confidence interval for exp(β) | |
			Lower	Upper	Wald χ²	df	Sig.		Lower	Upper
(Intercept)	0.607	0.0586	0.492	0.722	107.499	1	0.000	1.835	1.636	2.058
[female = 1]	0.502	0.0769	0.352	0.653	42.694	1	0.000	1.652	1.421	1.921
[female = 0]	0[a]							1		
Time	−0.056	0.0168	−0.089	−0.023	11.019	1	0.001	0.946	0.915	0.977
(Scale)	1									

Note: Dependent variable: readprof; model: (intercept), female, time.

[a] Set to 0 because this parameter is redundant.

TABLE 5.15 Alternative Model 1.4 Parameter Estimates

Parameter	β	Std. error	Hypothesis test			Exp(β)	95% Wald confidence interval for exp(β)	
			Wald χ^2	df	Sig.		Lower	Upper
(Intercept)	0.610	0.0571	114.198	1	0.000	1.841	1.646	2.059
[female = 1]	0.498	0.0768	42.054	1	0.000	1.645	1.415	1.913
[female = 0]	0ᵃ					1		
[time = 3]	−0.108	0.0463	5.439	1	0.020	0.898	0.820	0.983
[time = 2]	−0.232	0.0451	26.442	1	0.000	0.793	0.726	0.866
[time = 1]	0.002	0.0179	0.014	1	0.904	1.002	0.968	1.038
[time = 0]	0ᵃ					1		
(Scale)	1							

Note: Dependent variable: readprof; model: (intercept), female, time.
ᵃ Set to 0 because this parameter is redundant.

In Table 5.15, we also provide the results from considering the time variable as ordinal (again using Time 0 as the reference category). We defined female as nominal and "descending" so that the results are presented for females (coded 1). We provide the syntax for this model in Appendix A. The initial log odds for being proficient are 0.610 for males (0.610 + 0.448(0) = 0.610). For females it will be 1.108 (0.610 + 0.498 = 1.108). These results are consistent with the results in Table 5.14.

Adding an Interaction Between Female and the Time Parameter

We can create an interaction to test whether likelihood of change in proficiency status over time is contingent on gender. This effect can be added as an interaction term in the model (*time*female*):

$$\eta_{ti} = \log[\pi_{ti}/(1-\pi_{ti})] = \beta_0 + \beta_1 time + \beta_2 female + \beta_3 time * female. \qquad (5.8)$$

We note that if we define time as categorical, there will be an interaction associated with each time interval. There are several ways to create interaction terms with the IBM SPSS GUI. We have chosen the simple method in GEE of using the *Interaction* setting and then "dragging" variables onto the *Model* canvas, as shown in Step 2 of the modeling instructions.

Defining Model 1.5 With IBM SPSS Menu Commands
Note: Settings default to those used for Model 1.4.

1. Go to the toolbar and select ANALYZE, GENERALIZED LINEAR MODELS, GENERALIZED ESTIMATING EQUATIONS. This command allows access to the *Generalized Estimating Equations* (GEE) main dialog box.

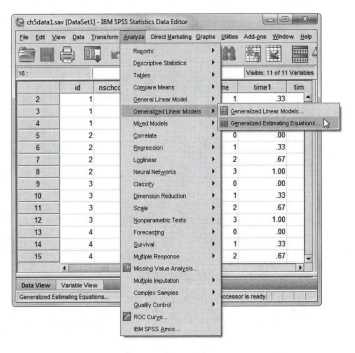

Adding an Interaction to Model 1.5

2a. The *Generalized Estimating Equations* display screen highlights the *Model* tab, which was the last command window used for Model 1.4.

b. We will add an interaction term to the model: *female*time*. First, change the model *Type* by clicking on the pull-down menu to select *Interaction*.

c. Interactions may be added to the model by two different methods. First, click to select both *female* and *time* and then (1) drag the variables into the *Model* box, or (2) click the right-arrow button to move the variables into the *Model* box. The interaction *female*time* is added to the

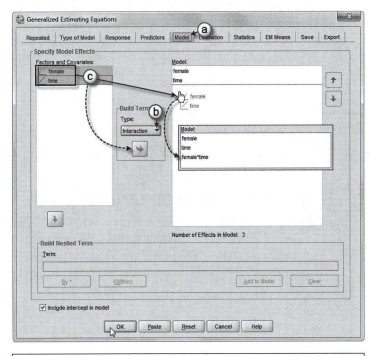

$$\eta_{ti} = \log[\pi_{ti}/(1 - \pi_{ti})] = \beta_0 + \beta_1 time + \beta_2 female + \beta_3 time * female \qquad (Eq. 5.8)$$

model and increases the number of model effects to three.

Click the OK button to generate the model results.

TABLE 5.16   Model 1.5: Parameter Estimates

Parameter	β	Std. error	95% Wald confidence interval		Hypothesis test			Exp(β)	95% Wald confidence interval for exp(β)	
			Lower	Upper	Wald χ²	df	Sig.		Lower	Upper
(Intercept)	0.626	0.0647	0.500	0.753	93.768	1	0.000	1.871	1.648	2.124
[female = 1]	0.458	0.0968	0.268	0.647	22.357	1	0.000	1.580	1.307	1.910
[female = 0]	0ᵃ							1		
Time	−0.068	0.0229	−0.113	−0.023	8.799	1	0.003	0.934	0.893	0.977
[female = 1]*time	0.028	0.0335	−0.038	0.094	0.692	1	0.405	1.028	0.963	1.098
[female = 0]*time	0ᵃ							1		
(Scale)	1									

Note: Dependent variable: readprof; model: (intercept), female, time, female*time.
ᵃ Set to 0 because this parameter is redundant.

Interpreting the Output of Model 1.5

Regarding the interaction term we tested, the results in Table 5.16 suggest that males and females do not differ in their probability of being proficient over time (log odds = 0.028, $p > .05$).

In Table 5.17, we also provide alternative results for Model 1.5 with time defined as an ordinal indicator. We provide syntax specification of the formulation in Appendix A. The positive interactions of the dummy time-related variables (time = 3, time = 2, time = 1) with female provides tests of the difference in mean log odds for females versus males at each time interval (with time = 0 serving as the reference group). The results suggest that females are significantly more likely to be proficient than males at the beginning of the study (i.e., there is main effect of female on the intercept, which represents initial status) and at time Interval 1 (OR = 1.117, $p < .05$), but not at Intervals 2 and 3. As the table implies, this can be an advantageous way to present time-related results in some situations.

Categorical Longitudinal Models Using GENLIN MIXED

In this section we develop several single-level and multilevel longitudinal models using GENLIN MIXED.

Specifying a GEE Model Within GENLIN MIXED

We can also specify the previous model (from Table 5.16, where *time* is defined as scale instead of ordinal) in GENLIN MIXED. To illustrate consistency with the single-level GEE approach, we will first estimate the model with fixed effects only (i.e., with no randomly varying intercept across individuals). The single-level model with fixed effects is the same as Equation 5.8.

TABLE 5.17 Alternative Model 1.5: Parameter Estimates

Parameter	β	Std. error	95% Wald confidence interval		Hypothesis test			Exp(β)	95% Wald confidence interval for exp(β)	
			Lower	Upper	Wald χ^2	df	Sig.		Lower	Upper
(Intercept)	0.634	0.0620	0.512	0.755	104.391	1	0.000	1.884	1.669	2.128
[female = 1]	0.445	0.0935	0.262	0.629	22.701	1	0.000	1.561	1.300	1.875
[female = 0]	0[a]							1		
[time = 3]	–0.169	0.0631	–0.293	–0.045	7.183	1	0.007	0.845	0.746	0.956
[time = 2]	–0.220	0.0603	–0.338	–0.102	13.336	1	0.000	0.802	0.713	0.903
[time = 1]	–0.046	0.0229	–0.091	–0.001	4.013	1	0.045	0.955	0.913	0.999
[time = 0]	0[a]							1		
[female = 1]*[time=3]	0.140	0.0927	–0.042	0.322	2.275	1	0.131	1.150	0.959	1.379
[female = 1]*[time=2]	–0.025	0.0905	–0.202	0.153	0.074	1	0.785	0.976	0.817	1.165
[female = 1]*[time=1]	0.110	0.0365	0.039	0.182	9.137	1	0.003	1.117	1.040	1.200
[female = 1]*[time=0]	0[a]									
[female = 0]*[time=3]	0[a]									
[female = 0]*[time=2]	0[a]									
[female = 0]*[time=1]	0[a]									
[female = 0]*[time=0]	0[a]									
(Scale)	1									

Note: Dependent variable: readprof; model: (intercept), female, time, female*time.

[a] Set to 0 because this parameter is redundant.

Defining Model 2.1 With IBM SPSS Menu Commands

Note: Refer to Chapter 4 for discussion concerning settings and modeling options available in the IBM SPSS GENLIN MIXED routine.

Use the *ch5data1.sav* data file.

1. Go to the toolbar and select ANALYZE, MIXED MODELS, GENERALIZED LINEAR. This command opens the *Generalized Linear Mixed Models* (GENLINMIXED) main dialog box.

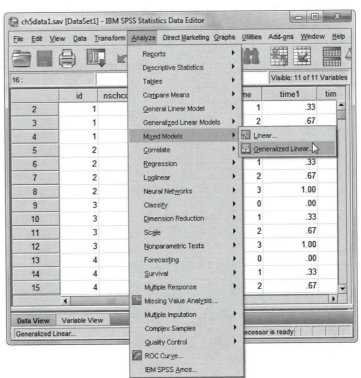

2a. The *Generalized Linear Mixed Models* display screen shows four command tabs: *Data Structure, Fields & Effects, Build Options,* and *Model Options.*
The *Data Structure* command tab is the default screen when creating a model for the first time. The options enable specifying structural relationships between data set records when observations are correlated.

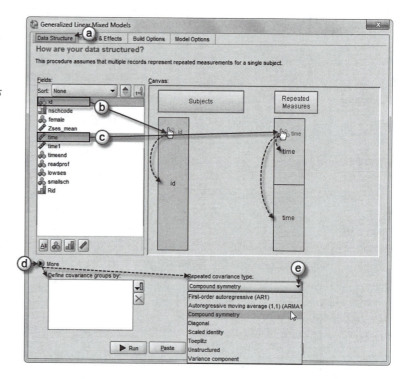

b. To build our model, begin by defining the subject variable.
A "subject" is an observational unit that can be considered independent of other subjects. Click to select the *id* variable from the left column and then drag and drop *id* onto the canvas area beneath the *Subjects* box.

c. We will designate *time* as a repeated measure. Click to select *time* and then drag the variable beneath the *Repeated Measures* column.

d. We will select the covariance type by clicking the *More* button to reveal additional options.

e. Now click the *Repeated covariance type* pull-down menu and select *Compound symmetry.* The *Repeated covariance type* provides eight options for specifying the covariance structure for residuals.

Note: The compound symmetry structure has constant variance and constant covariance.

3a. Click the *Fields &
Effects* command tab,
which displays the
default *Target* display
screen when creating
a model for the first
time. The settings
allow specifying
a model's target
(dependent variable),
distribution, and
relationship to model
predictors through
the link functions.

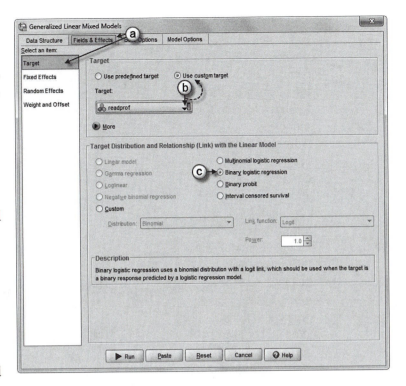

b. For this model we
will use student
reading proficiency
(*readprof*) as the
dependent variable.
Click the *Target*
pull-down menu and
select *readprof.*

Note: Selecting the target variable will automatically change the target setting to *Use
custom input.*

c. Next, we will designate a binomial distribution and logit link function for the model. Click
to select *Binary logistic regression.*

Note: The binary logistic regression uses a binomial distribution with a logit link and is used
due to the binary aspect of the *readprof* dependent variable.

4a. Now click the *Fixed Effects* item to access the main dialog box; this enables introducing predictor variables into the model.

b. Note that the *Include intercept in model* option is preselected and displayed in the *Effect builder* as *intercept*. This is because a Level-1 model is intercept only (default), so other model effects must be specified.

c. We will begin to build the model by adding two predictor variables (*female, time*). Click to select both variables and then drag them into the *Main* (effect) column of the *Effect builder*.

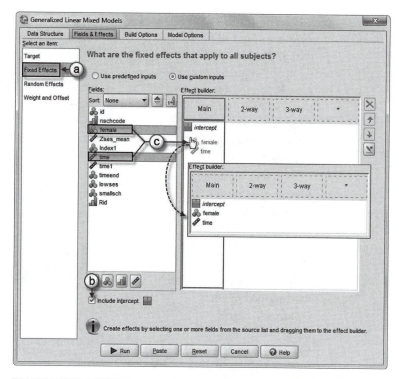

$$\eta_{ti} = \log[\pi_{ti}/(1 - \pi_{ti})] = \beta_0 + \beta_1 time + \beta_2 female \quad \text{(Eq. 5.7)}$$

5a. Finally, click the *Build Options* command tab located at the top of the screen to access advanced criteria options for building the model.

b. Change the sorting order for both categorical targets and categorical predictors. Click the pull-down menu for each option and select *Descending*. We will retain most of the remaining default settings.

c. We will change one setting. Click to select *Use robust estimation to handle violations of model assumptions (robust covariances)*.

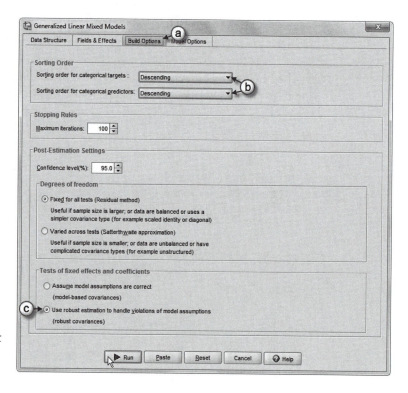

Click the RUN button to generate the model results.

TABLE 5.18 Comparing Models 2.1 and 1.4: Parameter Estimates

Parameter estimates	Coefficient	Std. error	t	Sig.	Exp(coefficient)	95% Confidence interval for exp(coefficient)	
						Lower	Upper
Model 2.1 (GENLIN MIXED)							
Intercept	0.607	0.059	10.368	0.000	1.835	1.636	2.058
Female = 1	0.502	0.077	6.534	0.000	1.652	1.421	1.921
Female = 0	0[a]						
Time	–0.056	0.017	–3.319	0.001	0.946	0.915	0.977
Model 1.4 (GEE)							
Intercept	0.607	0.059	10.368	0.000	1.835	1.636	2.058
Female = 1	0.502	0.077	6.534	0.000	1.652	1.421	1.921
Female = 0	0[a]						
Time	–0.056	0.017	–3.319	0.001	0.946	0.915	0.977

Note: Probability distribution: binomial; logit link function: logit.
[a] This parameter is set to 0 because it is redundant.

Interpreting the Output of Model 2.1

As shown in Table 5.18, the results are identical with either program, as we would expect. In Table 5.19, the compound symmetry structure is presented as a covariance. The covariance estimate (0.600) of the repeated measures, which is adjusted for gender, is only slightly different from those presented in Table 5.9 for the exchangeable covariance structure (which is compound symmetry) in the GEE model (0.605). The diagonal offset is 0.401 (i.e., the variance estimate for the repeated measures).

In Table 5.20, we can verify that this particular GENLIN MIXED formulation for longitudinal data provides results consistent with the GEE formulation (i.e., population-average estimates) because it does not have any random effects (i.e., subject-specific results). The four fixed effects are the intercept, gender (reflecting specification as a factor with two categories), and time. The two residual effects are the covariance estimates. As this table indicates, there are no random effects at the between-subject level. The "common subjects" line provides information about the number of Level-2 units in the analysis. In this case, individuals are the Level-2 units because the repeated observations are nested within individuals. We do not report the information about common subjects appearing below the table in subsequent tables in this chapter.

Examining a Random Intercept at the Between-Student Level

Up to this point in the chapter, we have treated the clustering of repeated measurements within individuals as "nuisance" parameters, consistent with a GEE formulation. In this next section, however, our goal shifts to building a two-level model to examine possible variation in the random parameters between subjects. In this type of "random coefficients" formulation, at Level

TABLE 5.19 Model 2.1: Variance Components

Residual effect	Estimate	Std. error	z-Test	Sig.	95% Confidence interval	
					Lower	Upper
CS diagonal offset	0.401	0.007	57.806	0.000	0.388	0.415
CS covariance	0.600	0.021	28.484	0.000	0.558	0.641

[a] Covariance structure: compound symmetry; subject specification: id.

TABLE 5.20 Model 2.1: Covariance Parameters

Covariance Parameters	
Residual effect	2
Random effects	0
Design Matrix Columns	
Fixed effects	4
Random effects	0[a]
Common subjects	2,228

Note: Common subjects are based on the subject specifications for the residual and random effects and are used to chunk the data for better performance.

[a] This is the number of columns per common subject.

1 we define the time-related measurements as nested within individuals. This type of two-level model is often formulated from the fixed-effects formulation (Equation 5.1) as follows:

$$\log\left[\frac{P(Y_{ti}=1)}{1-P(Y_{ti}=1)}\right] = x'_{ti}\beta_i + u_i, \tag{5.9}$$

where, in the basic formulation, a random effect u_i for the intercept is added at the between-subjects level, which is normally distributed with mean of 0 and some variance (Hedeker, 2005).

The model can be extended to include a vector of random effects—for example, to include the time trend as randomly varying. The vector of random effects is assumed to follow a multivariate normal distribution with means of 0 and a variance–covariance matrix describing the variances and covariances between the random effects (Hedeker, 2005).

Within subjects (Level 1), we have t measurements nested within individual students i as follows:

$$\eta_{ti} = \log[\pi_{ti}/(1-\pi_{ti})] = \beta_{0i} + \beta_{1i}time_{ti}. \tag{5.10}$$

In this type of random-coefficients model, we may expect that a subset of the Level-1 regression coefficients may vary across individuals (i) while others may remained fixed. The model differs from the previous GEE formulation by adding a random intercept (β_{0i}) between subjects (Level 2):

$$\beta_{0i} = \gamma_{00} + u_{0i}, \tag{5.11}$$

where u_{0i} is a random subject effect distributed normally and independently $(0, \sigma^2)$. We will treat the time-related variable as fixed between subjects:

$$\beta_{1i} = \gamma_{10}. \tag{5.12}$$

The combined model will therefore be the following:

$$\eta_{ti} = \gamma_{00} + \gamma_{10}time_{ti} + u_{0i}, \tag{5.13}$$

where η_{ti} is the estimated log odds.

This suggests three parameters to estimate. The two fixed effects include the intercept and fixed slope for time, and the random parameter is the slope variance between individuals at

Level 2. We will use an autoregressive covariance matrix at Level 1 (because we obtained an error message that the compound symmetry covariance did not produce model convergence).

Defining Model 2.2 With IBM SPSS Menu Commands

Note: Settings default to those used for Model 2.2.

1. Go to the tool-bar and select ANALYZE, MIXED MODELS, GENERALIZED LINEAR. This command opens the *Generalized Linear Mixed Models* (GENLINMIXED) main dialog box.

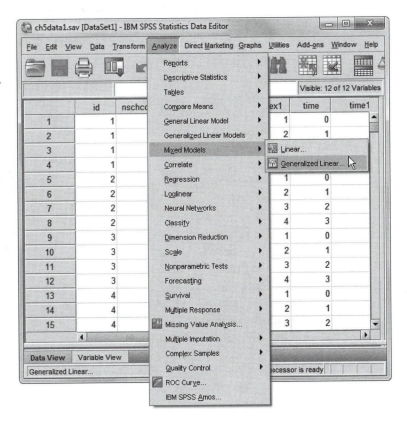

2a. *The Generalized Linear Mixed Models* dialog box displays the *Build Options* command tab, which was the last setting used in the prior model.

b. We will change the covariance type for this model. First, click to select the *Data Structure* command tab.

c. Due to a warning message generated when running the complete Model 2.2, we will change the compound symmetry covariance matrix used in the prior model. Click the *Repeated covariance type* pull-down menu to select *First-order autoregressive (AR1)*.

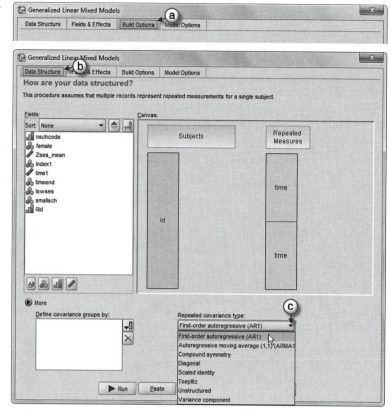

Note: The AR(1) is a first-order autoregressive structure with homogenous variances.

3a. Click the *Fields & Effects* command tab. The *Fixed Effects* screen is displayed as it was last used in the prior model.

b. We will change the model by removing the *female* predictor. Click to select *female* and then click the red "X" button to delete the variable from the *Effect builder*.

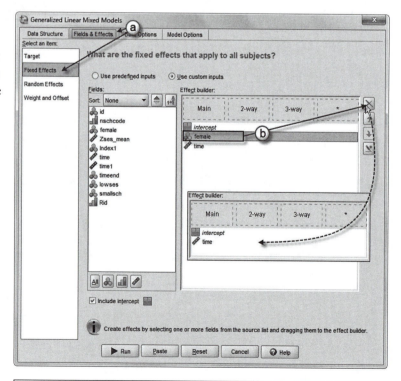

$$\eta_{ti} = \log[\pi_{ti}/(1 - \pi_{ti})] = \beta_{0i} + \beta_{1i}time_{ti} \quad \text{(Eq. 5.10)}$$

4a. The next step is to add a random effect (intercept) to the model. Click on the *Random Effects* item. The *Random Effects* command is used to specify which variables are to be treated as randomly varying across groups (random effects).

b. Click the ADD BLOCK button to access the *Random Effect Block* main dialog box.

c. In this model only the intercept is going to be randomly varying, so click to select the *Include intercept* option; this changes the setting within the *Effect builder* to *intercept* (from *no intercept*). Because there is only one random effect (the intercept), we can use the default variance component (VC) setting. This specifies a diagonal covariance matrix for the random effects; that is, it provides a separate variance estimate for each random effect, but not covariances between random effects.

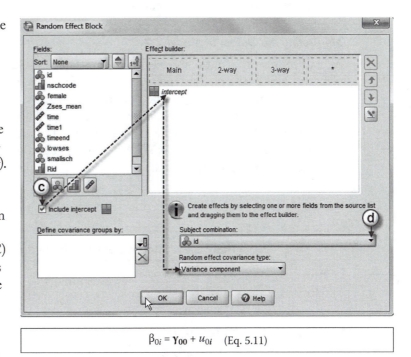

$$\beta_{0i} = \gamma_{00} + u_{0i} \quad \text{(Eq. 5.11)}$$

d. Select the *Subject combination* by clicking the pull-down menu and selecting *id*. Click the OK button to close the *Random Effect Block* dialog box and return to the *Random Effects* main screen.

e. In the main screen notice that the *Random effect blocks* displays a summary of the block just created: *id* (Subject) and *intercept* (Yes).

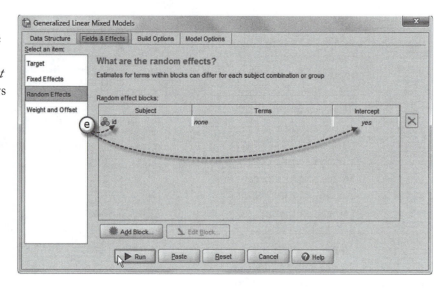

Click the RUN button to generate the model results.

Interpreting the Output of Model 2.2

We can confirm this in the model dimensions table (Table 5.21). Here we can see one random effect at Level 2 (the random intercept) and two residual effects (for the diagonal estimate and the off-diagonal correlation parameter of the autoregressive covariance matrix). We can also see in the subject specification that 2,228 individuals are defined at Level 2.

In Table 5.22, we next provide the student-level intercept (log odds = 1.220) and the time slope (log odds = −0.080, $p < .001$). We note that these subject-specific results are different from the population-average results in Table 5.6 (i.e., intercept log odds = 0.838; time log odds = −0.055).

TABLE 5.21　Model 2.2: Covariance Parameters

Covariance Parameters	
Residual effect	2
Random effects	1
Design Matrix Columns	
Fixed effects	2
Random effects	1
Common subjects	2,228

TABLE 5.22　Model 2.2: Fixed Effects

Parameter	Coefficient	Std. Error	t	Sig.	Exp(coefficient)	95% Confidence interval for exp(coefficient)	
						Lower	Upper
(Intercept)	1.220	0.069	17.612	0.000	3.387	2.957	3.880
Time	−0.080	0.027	−2.973	0.003	0.923	0.876	0.973

Probability distribution: binomial; link function: logit.

It turns out that the absolute value of the random-coefficients model is larger than the population average model by the square root of the sampling design effect (Hedeker, 2005; Zeger et al., 1988). We can estimate this difference as

$$\sqrt{d} = \sqrt{\frac{\sigma_u^2 + \sigma^2}{\sigma^2}} = \sqrt{\frac{RE\sigma^2}{FE\sigma^2}}, \tag{5.14}$$

where *RE* is the random-effects model variance and *FE* is the fixed-effects model variance (Hedeker, 2005).

In Table 5.23 we provide the model covariance parameters. As readers will recall, for a logistic distribution the variance is estimated as $\pi^2/3$, or approximately 3.29. In this case, this would be as follows:

$$\sqrt{d} = \sqrt{\frac{4.161 + 3.29}{3.29}} \sim 1.5$$

The actual difference in the two intercepts 1.22/0.838 is estimated as 1.46 (which is close to 1.5). For the odds ratio, the difference in size is also 1.46 (−0.08/−0.055). We note that Zeger and colleagues (1988) suggest a slightly different estimate of the variance for the logistic distribution. However, $\pi^2/3$ will probably suffice in most instances.

We can also use the between-subject level (Level 2) variance to estimate an intraperson correlation, which is defined as the variance between individuals (Level 2) divided by the total variance (Level-2 variance + $\pi^2/3$). This provides an estimate of how much variance in reading proficiency lies between individuals. We can calculate the between-subject variance as 4.161/(4.161 + 3.29), which is approximately 0.558. This suggests that about 55.8% of the variance in initial reading proficiency lies between subjects. At Level 2, the results suggest that likelihood to be proficient varies significantly across individuals in the study ($z = 17.369$, $p < .001$). (See Table 5.23.)

What Variables Affect Differences in Proficiency Across Individuals?

We will define female and low SES as two possible predictors at Level 2 (between students). The Level-1 model is the same as Equation 5.11. We will also include two cross-level interactions between the predictors and time. At Level 2, we have the following model for the intercept (β_{0i}) and time slope (β_{1i}):

$$\beta_{0i} = \gamma_{00} + \gamma_{01} female_i + \gamma_{02} lowses_i + u_{0i}$$

$$\beta_{1i} = \gamma_{10} + \gamma_{11} female_i + \gamma_{12} lowses \tag{5.15}$$

TABLE 5.23   Model 2.2: Variance Components

Random and Residual Effects	Estimate	Std. Error	z-Test	Sig.	95% Confidence interval	
					Lower	Upper
Random[a]						
Var(intercept)	4.161	0.240	17.369	0.000	3.717	4.659
Residual						
AR1 diagonal	0.521	0.018	29.412	0.000	0.487	0.557
AR1 rho	0.395	0.021	19.122	0.000	0.353	0.434

[a] Covariance structure: first-order autoregressive; subject specification: schcode.

We note that we keep the time random effect fixed as in Equation 5.12. The combined model will then be as follows:

$$\eta_{ti} = \gamma_{00} + \gamma_{01} female_i + \gamma_{02} lowses_i + \gamma_{10} time_{ti} + \gamma_{11} time_{ti} * female_i + \gamma_{12} time_{ti} * lowses_i + u_{0i}. \quad (5.16)$$

Defining Model 2.3 With IBM SPSS Menu Commands

Note: Settings default to those used for Model 2.2.

1. Go to the toolbar, select ANALYZE, MIXED MODELS, GENERALIZED LINEAR. This command opens the *Generalized Linear Mixed Models* (GENLINMIXED) main dialog box.

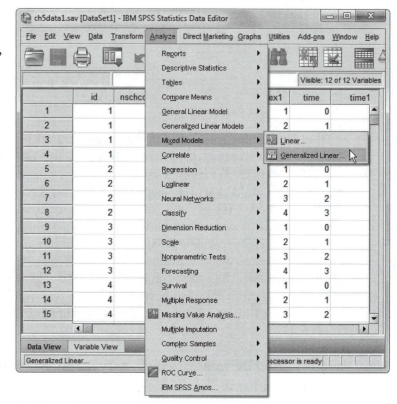

2a. The *Generalized Linear Mixed Models* dialog box displays the *Random Effects* command tab, which was last used in the prior model.
Note: The *Random Effects* item is preselected (represented in the illustration by the dashed-line box but appears on screen colored blue) as it was the last setting used from the prior model.

b. We will add two predictors and two interactions to the model. First, click to select the *Fixed Effects* item to access the main screen.

c. We will add two predictors (*female, lowses*) to the model. Click to select both variables and then drag both variables onto the *Main* column of the *Effect builder.*

$$\eta_{ti} = \gamma_{00} + \gamma_{01}\textbf{\textit{female}}_i + \gamma_{02}\textbf{\textit{lowses}}_i + \gamma_{10}\textbf{\textit{time}}_{ti} + \gamma_{11}\textbf{\textit{time}}_a * \textbf{\textit{female}}_i + \gamma_{12}\textbf{\textit{time}}_{ti} * \textbf{\textit{lowses}}_i + u_{0i} \quad \text{(Eq. 5.16)}$$

Adding Two Interactions to Model 2.3

d. We will now add the first interaction to the model. Click to select *female* and *time* and then drag both variables to the *2-way* column to create *female*time.*

e. To add the second interaction to the model, click to select *time* and *lowses* and then drag both variables to the *2-way* column to create *time*lowses.*

f. To facilitate reading of the output tables, we will rearrange the variable order. Click to select *time* and then click the down-arrow button on the right to place *time* below *lowses.* The variable order is now *female, lowses, time, lowses*time, female*lowses.*

Note: Further adjustment was made to the model syntax to reverse-order the two interactions *time*female, time*lowses* (as shown in Table 5.27).
Click the OK button to generate the model results.

Interpreting the Output of Model 2.3

We can confirm the model specifications in Table 5.24. The 10 fixed effects are composed of the intercept, time slope, female (two categories), low SES (two categories), interaction (two categories), and interaction (two categories). There are one random effect (the randomly varying subject intercept) and two residual effects (i.e., the structure of the within-subjects covariance matrix).

TABLE 5.24 Model 2.3: Covariance Parameters

Covariance Parameters	
Residual effect	2
Random effects	1
Design Matrix Columns	
Fixed effects	10
Random effects	1
Common subjects	2,228

In Table 5.25 we provide the fixed effects. We can see that after adding the covariate interactions with time, the main effect of *time* is not significant ($p = .224$). However, there is an interaction with low SES—that is, change in student proficiency status over time appears to be linked to socioeconomic status. More specifically, over time, the probability of low SES students being proficient declines more sharply than among their average and high SES peers ($p < .05$). In particular, the odds of being proficient for low SES students are multiplied by 0.868 (or decline by 13.2%) for each yearly interval. The interaction between female and time is not significant ($p > .05$). Table 5.26 presents the covariance parameters for this model.

We also provide the GEE estimates for this same model in Table 5.27. Readers will note the same pattern of results (main effects for female and low SES are significant; only the interaction between *time*lowSES* is significant) but, again, the absolute value of the population average coefficients will be smaller by $\sqrt{d}$ (see Equation 5.14). For example, in the subject-specific results (Table 5.25), the log odds for the intercept are 1.359 and the coefficient for female is 0.608 (OR = 1.836). In Table 5.27, for the population average model, the log odds coefficient for the intercept is 0.937, and the log odds for female are 0.447 (OR = 1.563). Similarly, for low SES students in the subject-specific results, the log odds are −0.996 (OR = 0.369), and for the population average model they are −0.732 (OR = 0.481).

We may wish to save predicted probabilities $Y = 1$ over time and plot the significant SES–time relationship. We first removed the nonsignificant interaction (*female*time*) and then reestimated the model saving the predicted probabilities. We can use those to plot the relationship over time. Fixed-effect estimates were quite similar, so we do not show them. Figure 5.1 illustrates

TABLE 5.25 Model 2.3: Fixed Effects

Model term	Coefficient	Std. error	t	Sig.	Exp(coefficient)	95% Confidence interval for exp(coefficient)	
						Lower	Upper
Intercept	1.359	0.116	11.760	0.000	3.894	3.104	4.884
female = 1	0.608	0.143	4.250	0.000	1.836	1.387	2.430
female = 0	0ᵃ						
lowses = 1	−0.996	0.145	−6.862	0.000	0.369	0.278	0.491
lowses = 0	0ᵃ						
time	−0.057	0.047	−1.215	0.224	0.945	0.862	1.035
time*[female = 1]	0.076	0.057	1.331	0.183	1.079	0.965	1.208
time*[female = 0]	0ᵃ						
time*[lowses = 1]	−0.142	0.058	−2.456	0.014	0.868	0.775	0.972
time*[lowses = 0]	0ᵃ						

Note: Probability distribution: binomial; link function: logit.
ᵃ This coefficient is set to 0 because it is redundant.

TABLE 5.26 Model 2.3: Variance Components

Random and residual effects	Estimate	Std. error	z-Test	Sig.	95% Confidence interval for exp(coefficient)	
					Lower	Upper
Random[a]						
Var(intercept)	4.294	0.232	18.540	0.000	3.863	4.773
Residual[b]						
AR1 diagonal	0.500	0.015	33.024	0.000	0.471	0.531
AR1 rho	0.363	0.019	18.782	0.000	0.324	0.400

[a] Covariance structure: variance components; subject specification: id.
[b] Covariance structure: first-order autoregressive; subject specification: id.

TABLE 5.27 Alternative Model 2.3 (GEE Procedure): Parameter Estimates

Parameter	β	Std. error	95% Wald confidence interval		Hypothesis test			Exp(β)	95% Wald confidence interval for exp(β)	
			Lower	Upper	Wald χ²	df	Sig.		Lower	Upper
(Intercept)	0.937	0.0803	0.779	1.094	136.093	1	0.000	2.551	2.180	2.986
[female = 1]	0.447	0.0982	0.254	0.639	20.690	1	0.000	1.563	1.289	1.894
[female = 0]	0[a]							1		
[lowses = 1]	−0.732	0.0984	−0.925	−0.539	55.259	1	0.000	0.481	0.397	0.583
[lowses = 0]	0[a]							1		
time	−0.038	0.0292	−0.095	0.019	1.724	1	0.189	0.962	0.909	1.019
[female = 1]*time	0.030	0.0347	−0.038	0.098	0.768	1	0.381	1.031	0.963	1.103
[female = 0]*time	0[a]							1		
[lowses = 1]*time	−0.072	0.0349	−0.141	−0.004	4.326	1	0.038	0.930	0.869	0.996
[lowses = 0]*time	0[a]							1		
(Scale)	1									

Note: Dependent variable: readprof; model: (intercept), female, lowses, time, female*time, lowses*time.
[a] Set to 0 because this parameter is redundant.

the significant interaction, suggesting a lower probability of being proficient among low SES students compared with their average and high SES peers.

Building a Three-Level Model in GENLIN MIXED

The Beginning Model

We next focus our attention on building a three-level model to explain random variation in intercepts and the time slope across schools. In order to do this it helps to recode the student identifier (*id*) through the "ranking cases" procedure (*Rid*) to be numbered within each school unit *j* (*n* = 1, 2,..., *N*). We discuss how to do this in Chapter 2. We have found that this cuts down the required computer run time considerably for models with nested data. Then the new identifier is used at Level 2 for the nesting of students within schools. At Level 3, it is the school identifier (*nschcode*).

We will first develop a simplified three-level model, with random intercept at Levels 2 and 3, and we will fix the time variable at Level 2 (between students) because our primary interest

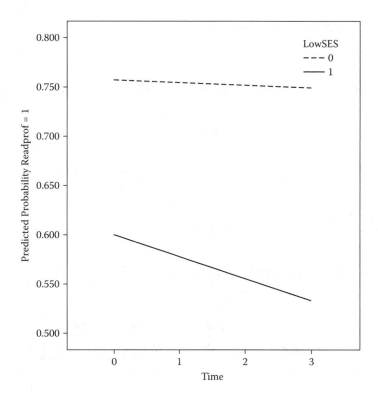

FIGURE 5.1 Observed proportions across time by SES background.

is the investigation of differences between schools in students' proficiency status. In this initial model, we will examine whether the time variable varies at Level 3 by specifying it as random. At Level 1, we have the following for student i in school j measured at time t:

$$\eta_{tij} = \log[\pi_{tij}/(1-\pi_{tij})] = \beta_{0ij} + \beta_{1ij}time_{tij} \qquad (5.17)$$

At Level 2, we have the following:

$$\beta_{0ij} = \gamma_{00j} + u_{0ij}$$

$$\beta_{1ij} = \gamma_{10j}. \qquad (5.18)$$

At Level 3, we will have the following model: We will use γ coefficients to define the variables also at Level 3, but will code the intercept as γ_{00j} for Level 2 and indicate the intercept at Level 3 as γ_{000}. We will use u_{00j} for the random effect at that level. At Level 3, then, we have the following:

$$\gamma_{00j} = \gamma_{000} + u_{00j} \qquad (5.19)$$

$$\gamma_{10j} = \gamma_{100} + u_{10j}. \qquad (5.20)$$

Through substitution, the combined single-equation model will be as follows:

$$\eta_{tij} = \gamma_{000} + \gamma_{100}time_{tij} + u_{10j}time_{tij} + u_{00j} + u_{0i} \qquad (5.21)$$

This suggests two fixed effects and three random effects in the model (as well as the two residual parameters related to the Level-1 covariance matrix for the repeated measures of *Y*). We continue to use an autoregressive covariance structure at Level 1. At Level 3, we will specify an unstructured covariance matrix, which will add one more random effect at Level 3 (for a total of four random effects). We also used robust SEs for these models.

Defining Model 3.1 With IBM SPSS Menu Commands

Note: If you are continuing after completing Model 2.3 in the prior section, we recommend clearing the default settings before proceeding.

Continue using the *ch5data1.sav* file.

1. Go to the tool-bar and select ANALYZE, MIXED MODELS, GENERALIZED LINEAR. This command opens the *Generalized Linear Mixed Models* (GENLINMIXED) main dialog box.

Note: If you are continuing this model immediately after completing Model 2.3, clear the default settings before proceeding by clicking the RESET button at the bottom of the display screen.

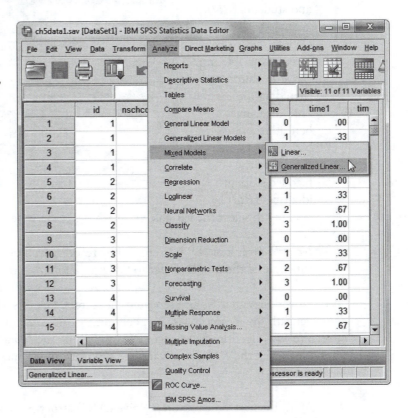

2a. We will begin con-
structing our model
by clicking to select
the *Data Structure*
command tab.

b. To have the model
reflect the hierarchi-
cal structure of the
data, we will assign
schools (*nschcode*)
and students (*Rid*)
as subjects. First,
click to select the
nschcode variable from
the left column and
then drag *nschcode*
onto the canvas area
beneath the *Subjects*
box.

c. Next, add the
recoded student
identifier (*Rid*) by
clicking to select the variable and then dragging it next to *nschcode*.

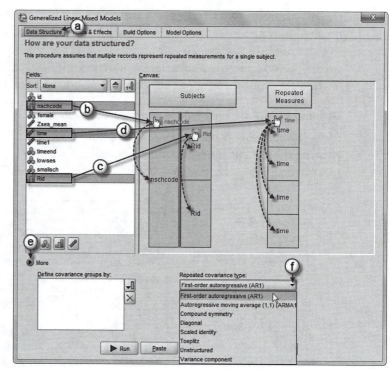

Note: Rid (rank of ID) was created by recoding individuals within schools to expedite the
run-time of the model. Refer to Chapter 2 for an example.

d. Now designate *time* as a repeated measure. Click to select *time* and then drag the variable
beneath the *Repeated Measures* column.

e. We will select the covariance type by clicking the *More* button to reveal additional options.

f. Now click the *Repeated* covariance type pull-down menu and select *First-order autoregressive (AR1)*.

Note: The first-order autoregressive (AR1) structure has homogeneous variances. The correlation between any two elements is equal to ρ for adjacent elements, ρ² for two elements
separated by a third, and so on. ρ is constrained to lie between −1 and 1.

3a. Click the *Fields & Effects* command tab to access settings for specifying the model's target (dependent variable), distribution, and relationship to model predictors through the link functions.

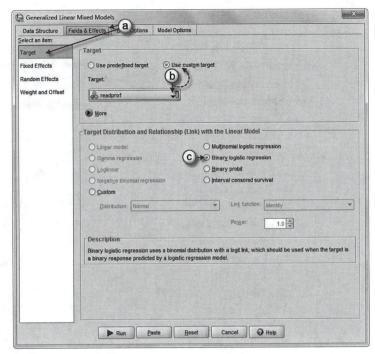

b. For this model we will continue using student reading proficency (*readprof*) as the dependent variable. Click the *Target* pull-down menu and select *readprof*.

Note: Selecting the target variable will automatically change the target setting to *Use custom input*.

c. Next, we will designate a binomial distribution and logit link function for the model. Click to select *Binary logistic regression*.

Note: The binary logistic regression uses a binomial distribution with a logit link and is used due to the binary aspect of the *readprof* dependent variable.

4a. Now click the *Fixed Effects* item to access the main dialog box; this enables introducing predictor variables into the model.

b. Note that the *Include intercept in model* option is preselected and displayed in the *Effect builder* as *intercept*.

c. We will begin to build the model by adding one predictor variable (*time*) at Level 1. Click to select *time* and then drag the variable into the *Main* (effect) column of the *Effect builder*.

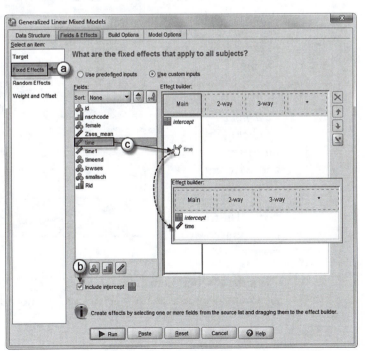

$$\eta_{tij} = \log[\pi_{tij}/(1 - \pi_{tji})] = \beta_{0ij} + \beta_{1ij} time_{tij} \quad (\text{Eq. 5.17})$$

To demonstrate using a different random effect covariance structure for the residuals, we will change the default VC setting.

5a. Click on the *Random Effects* item. The *Random Effects* command is used to specify which variables are to be treated as randomly varying across groups (random effects).

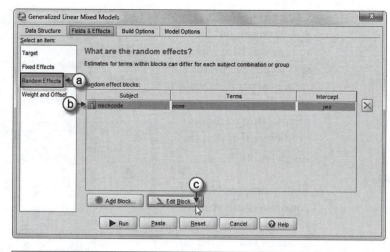

$$\gamma_{00j} = \gamma_{000} + u_{00j} \quad (Eq. 5.19)$$

$$\gamma_{10j} = \gamma_{100} + u_{10j} \quad (Eq. 5.20)$$

b. At Level 3, notice that the Level-1 random effect (subject is school with no effects, intercept only) has automatically been added to the model due to the variable's measurement setting and layout in the *Data Structure*. We will now add a second random effect (*time*) at the school level. Click to select and activate the *nschcode* random effect block.

c. Click the EDIT BLOCK button.

d. Add the second random effect (*time*) to the model by clicking to select *time* and then dragging it into the *Effect builder* section.

e. We will also change the *Random effect covariance type* by clicking the pull-down menu and selecting *Unstructured*.

Note: "Unstructured" is a general covariance matrix.

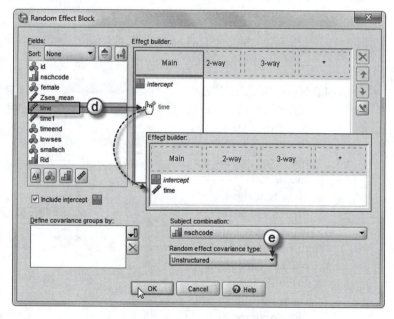

Click the OK button to close the *Random Effect Block* dialog box and return to the *Random Effects* main screen. The *Random effect block* now displays *time* as an added term.

f. The next step is to add a Level-2 random effect (intercept) to the model. Click on the *Random effects* block, which will highlight and activate the block to allow editing the content.

g. Now click the ADD BLOCK button to access the *Random Effect Block* main dialog box. We will add the Level-2 random intercept

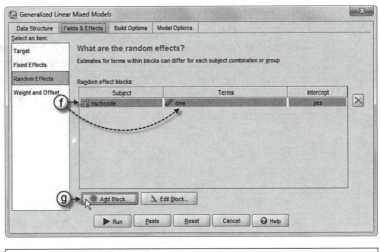

$$\beta_{0ij} = \gamma_{00j} + u_{0ij} \quad \text{(Eq. 5.18)}$$

(subject is *nschcode*Rid* with no effects, intercept only).

h. Click to select the *Include intercept* option; this changes the setting within the *Effect builder* to *intercept* (from *no intercept*). Because there is only one random effect (the intercept), we can use the default VC setting. This specifies a diagonal covariance matrix for the random effects; that is, it provides a separate variance estimate for each random effect,

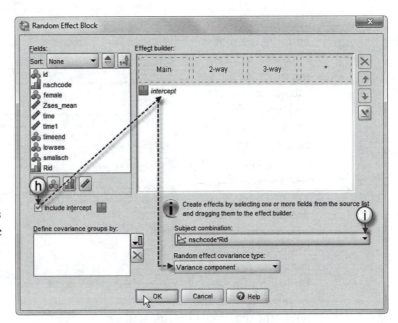

but not covariances between random effects (i.e., for one random effect, VC defaults to an identity covariance matrix).

i. Select the *Subject combination* by clicking the pull-down menu and selecting *nschcode*Rid*. Click the OK button to close the *Random Effect Block* dialog box and return to the *Random Effects* main screen.

j. In the main screen, notice that the Level-2 *Random effect blocks* displays a summary of the block just created: *nschcode*Rid* (Subject) and *intercept* (Yes).

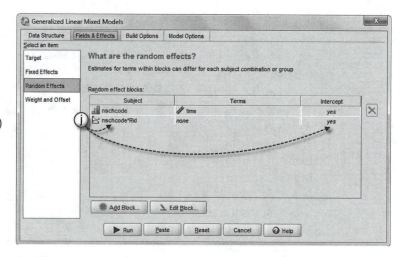

6a. Finally, click the *Build Options* command tab located at the top of the screen to access advanced criteria options for building the model.

b. Change the sorting order for both categorical targets and categorical predictors. Click the pull-down menu for each option and select *Descending*.

c. We will change the default to use robust covariances. Click to select *Use robust estimation to handle violations of model assumptions*. We will retain the remaining default settings.

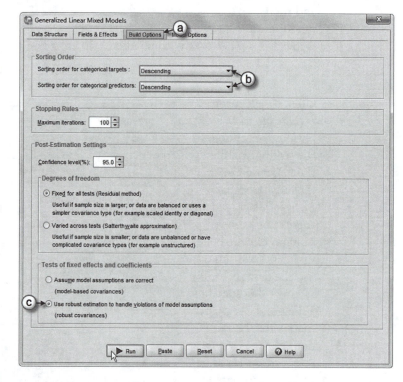

Click the RUN button to generate the model results.

Interpreting the Output of Model 3.1

We can confirm the specification of the model in Equation 5.21 in the model dimension table (Table 5.28). We can see that there are two fixed effects, four random effects, and two residual effects in the model. We note that the number of higher order units is now 154 schools instead of the number of individuals in the previous between-subjects model.

The results are presented in Table 5.29. Once again, the time variable is significant when there are no other covariates in the model (log odds = −0.090, $p < 0.02$).

In Table 5.30 we present the model's variance parameters. The estimates suggest that significant variance in intercepts exists at both Level 3 (school) and Level 2 (between subjects). Moreover, the time slope varies significantly across schools.

TABLE 5.28   Model 3.1: Covariance Parameters

Covariance parameters	
Residual effect	2
Random effects	4
Design matrix columns	
Fixed effects	2
Random effects	17
Common subjects	154

TABLE 5.29   Model 3.1: Fixed Effects

						95% Confidence interval for exp(coefficient)	
Model term	Coefficient	Std. error	t	Sig.	Exp(coefficient)	Lower	Upper
(Intercept)	1.414	0.114	12.410	0.000	4.113	3.290	5.143
time	−0.090	0.038	−2.369	0.018	0.914	0.848	0.985

Note: Probability distribution: binomial; link function: logit.

TABLE 5.30   Model 3.1: Variance Components

					95% Confidence interval	
Random and residual effects	Estimate	Std. error	z-Test	Sig.	Lower	Upper
Random block 1[a]						
Level-3 var(intercept)	1.304	0.243	5.373	0.000	0.905	1.878
Level-3 cov(slope)	−0.209	0.063	−3.328	0.001	−0.332	−0.086
Level-3 var(slope)	0.152	0.027	5.629	0.000	0.107	0.215
Random block 2[b]						
Level-2 var(intercept)	5.147	0.275	18.729	0.000	4.636	5.715
Residual[c]						
AR1 diagonal	0.444	0.013	33.028	0.000	0.418	0.471
AR1 rho	0.296	0.021	14.375	0.000	0.255	0.335

[a] Covariance structure: unstructured; subject specification: nschcode.
[b] Covariance structure: variance components; subject specification: nschcode*Rid.
[c] Covariance structure: first-order autoregressive (AR1); subject specification: nschcode*Rid.

We can calculate another estimate of the variance components that lie between schools and between subjects. The ICCs will be the following:

$$\frac{\sigma^2_{School}}{\sigma^2_{School} + \sigma^2_{Between} + \sigma^2_{Within}} = 1.304/9.741 = 0.134$$

$$\frac{\sigma^2_{Between}}{\sigma^2_{School} + \sigma^2_{Between} + \sigma^2_{Within}} = 5.147/9.741 = 0.528$$

This suggests that about 13.4% of the variance lies between schools and about 52.8% of the variance lies between individuals. A multilevel model might be formulated to explain outcomes at each level.

Adding Student and School Predictors

After specifying the basic three-level model, we will now add the predictors between subjects and at the school level. At Level 2, we add the two predictors (*female, lowses*) and their interactions. At the school level, these include a measure of student composition (*zses_mean*) and a dichotomous measure of school size (small school, less than 600, coded 1 vs. Else = 0). For this model, at Level 1, we will again specify the time-related variable as interval. The Level-1 model remains the same as Equation 5.17:

$$\eta_{tij} = \log[\pi_{tij}/(1-\pi_{tij})] = \beta_{0ij} + \beta_{1ij} time_{tij}$$

At Level 2, we have the following model to explain the random intercept:

$$\beta_{0ij} = \gamma_{00j} + \gamma_{01j} female_{ij} + \gamma_{02j} lowses_{ij} + u_{0ij}. \tag{5.22}$$

We fixed the time slope at Level 2 ($\beta_{1ij} = \gamma_{10j}$); however, we can model a Level-1 coefficient as nonrandomly varying among Level-2 units (Raudenbush, Bryk, Cheong, & Congdon, 2004). This allows us to account for cross-level interactions in the combined model specifying how between-individual background variables may alter the time effect over the repeated occasions of measurement during the study:

$$B_{1ij} = \gamma_{10j} + \gamma_{11} female_{ij} + \gamma_{12} lowses_{ij}. \tag{5.23}$$

As readers will note, there is no corresponding random coefficient (u_{1j}) in Equation 5.23.

At Level 3, then, we have the following model to explain the random intercept:

$$\gamma_{00j} = \gamma_{000} + \gamma_{001} zses\_mean_j + \gamma_{002} smallsch_j + u_{00j}. \tag{5.24}$$

We will allow the time-related slope to vary randomly across schools at Level 3:

$$\gamma_{10j} = \gamma_{100} + \gamma_{101} zses\_mean_j + \gamma_{102} smallsch_j + u_{10j}. \tag{5.25}$$

The combined model will then be as follows:

$$\begin{aligned}
\eta_{tij} = {} & \gamma_{000} + \gamma_{001} zses\_mean_j + \gamma_{002} smallsch_j + \gamma_{010j} female_{ij} + \gamma_{020j} lowses_{ij} \\
& + \gamma_{100} time_{tij} + \gamma_{101} zses\_mean_j * time_{tij} + \gamma_{102} smallsch_j * time_{tij} \\
& + \gamma_{110j} time_{tij} * female_{ij} + \gamma_{120j} time_{tij} * lowses_{ij} + u_{10j} time_{tij} + u_{00j} + u_{0ij},
\end{aligned} \tag{5.26}$$

where we can see the resulting cross-level interactions at levels 3 and 2.

Defining Model 3.2 With IBM SPSS Menu Commands

Note: The settings will default to those used in Model 3.1.

1. Go to the tool-bar and select ANALYZE, MIXED MODELS, GENERALIZED LINEAR. This command opens the *Generalized Linear Mixed Models* (GENLINMIXED) main dialog box.

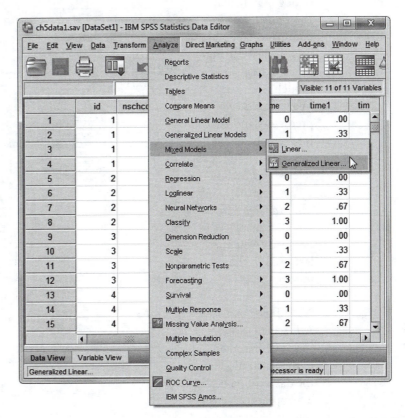

2a. The *Generalized Linear Mixed Models* dialog box displays the *Build Options* command tab, which was the last setting used in the prior model.

b. We will add two predictors and two interactions to the model. First, click to select the *Fields & Effects* command tab. The *Random Effects* screen is displayed because it was the last used in the prior model.

Note: The *Random Effects* item is preselected (represented in the illustration by the dashed-line box but appears on screen colored blue) because it was the last setting used from the prior model.

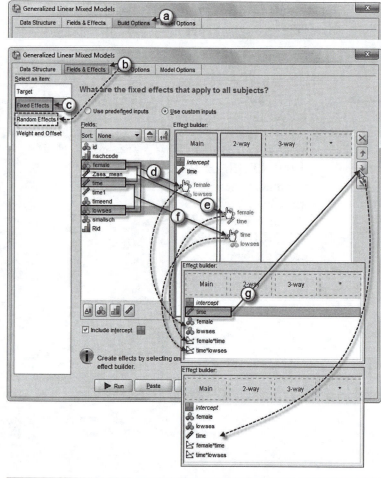

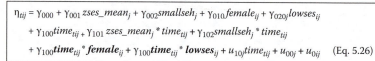

$$\eta_{tij} = \gamma_{000} + \gamma_{001} zses\_mean_j + \gamma_{002} smallseh_j + \gamma_{010} female_{ij} + \gamma_{020j} lowses_{ij}$$
$$+ \gamma_{100} time_{tij} + \gamma_{101} zses\_mean_j * time_{tij} + \gamma_{102} smallseh_j * time_{tij}$$
$$+ \gamma_{100} time_{tij} * female_{ij} + \gamma_{100} time_{tij} * lowses_{ij} + u_{10j} time_{tij} + u_{00j} + u_{0ij} \quad \text{(Eq. 5.26)}$$

c. Click to select the *Fixed Effects* item to access the main dialog box; this enables introducing predictor variables and interactions into the model.

d. We will begin by adding predictor variables into the model at Level 2. First, click to select two variables (*female*, *lowses*) and then drag them into the *Main* (effect) column of the *Effect builder*.

Adding Two Interactions to Model 3.2

e. Now we will add two interactions to the model. For the first interaction, click to select *time* and *female* and then drag them into the *2-way* column of the *Effect builder*.

f. Create the second interaction by clicking to select *time* and *lowses* and then drag them into the *2-way* column of the *Effect builder*.

g. To facilitate reading of the output tables, we will rearrange the variable order to place *time* at the bottom of the list. Click to select time above the first interaction (*time*lowses*). The variable order is now *female, lowses, time, time*female, lowses*time*. At Level 3, we will now add two more predictors and interactions to the model.

h. We will add two predictors (*Zses_mean*, *smallsch*) to the model. Click to select *Zses_mean* and *smallsch* and then drag both variables into the *Main* (effect) column of the *Effect builder*.

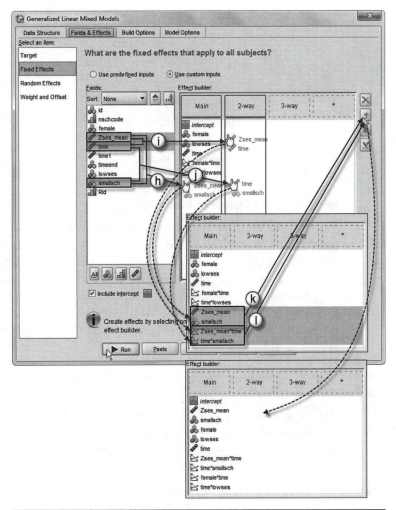

$$\eta_{tij} = \gamma_{000} + \gamma_{001}\, zses\_mean_j + \gamma_{002}smallseh_j + \gamma_{010}female_{ij} + \gamma_{020j}lowses_{ij}$$
$$+ \gamma_{100}time_{tij} + \gamma_{101}\, \textbf{\textit{zses\_mean}}_j * \textbf{\textit{time}}_{tij} + \gamma_{102}\textbf{\textit{smallseh}}_j * \textbf{\textit{time}}_{tij}$$
$$+ \gamma_{100}time_{tij} * female_{ij} + \gamma_{100}time_{tij} * lowses_{ij} + u_{10j}time_{tij} + u_{00j} + u_{0ij} \qquad \text{(Eq. 5.26)}$$

Adding Two More Interactions to Model 3.2

i. Next, we will add two interactions to the random-slope model. For the first interaction, click to select *Zses_mean* and *time* and then drag them into the *2-way* column of the *Effect builder*.

j. Create the second interaction by clicking to select *time* and *smallsch* and then drag them into the *2-way* column of the *Effect builder*.

k. To facilitate reading of the output tables, we will rearrange the variable order. Click to select *Zses_mean* and *smallsch* and then click the up-arrow button on the right to place them so that they are listed first and second, respectively.

l. We will also change the variable order for the two interactions (*Zses_mean*time*, *time*smallsch*). Click to select both interactions and then click the up-arrow button on the right to place them below *time*. The variable order is now *Zses_mean*, *smallsch*, *female*, *lowses*, *time*, *Zses_mean*time*, *time*smallsch*, *female*time*, *time*lowses*.

Click the RUN button to generate the model results.

Note: An alternative method would be to enter all the variables first and then rearrange the order prior to running the model. We have chosen to enter them in sequential order to facilitate illustrating the process.

TABLE 5.31   Model 3.2: Covariance Parameters

Covariance Parameters	
Residual effect	2
Random effects	4
Design Matrix Columns	
Fixed effects	16
Random effects	17
Common subjects	154

Interpreting the Output of Model 3.2

We can confirm this specification in the model dimension table (Table 5.31). We present the fixed effects in Table 5.32. Regarding intercepts, we can see that the average socioeconomic status of students is significantly related to likelihood to be proficient (OR = 0.623, $p < .01$). Regarding the random slope, we see a positive effect for small schools (OR = 1.208, $p < .05$). We might reasonably hypothesize that smaller schools are settings where students are more likely to change positively in terms of likelihood to be proficient in reading over time. The SES mean interaction effect is not significant, however (OR = 0.978, $p = .555$). The variance parameters appear in Table 5.33.

An Example Experimental Design

Before we leave the longitudinal approach developed in this chapter, we provide an example of a subset of the data illustrating a possible experimental design. Consider a study to examine students' likelihood to be proficient in reading over time and to assess whether their

TABLE 5.32   Model 3.2: Fixed Effects

Model term	Coefficient	Std. error	t	Sig.	Exp(coefficient)	95% Confidence interval for exp(coefficient)	
						Lower	Upper
(Intercept)	1.502	0.209	7.203	0.000	4.491	2.984	6.759
zses_mean	–0.474	0.101	–4.673	0.000	0.623	0.510	0.760
smallsch = 1	–0.221	0.220	–1.006	0.314	0.802	0.521	1.233
smallsch = 0	0[a]						
female = 1	0.730	0.173	4.219	0.000	2.075	1.478	2.914
female = 0	0[a]						
lowses = 1	–0.701	0.182	–3.845	0.000	0.496	0.347	0.709
lowses = 0	0[a]						
time	–0.176	0.081	–2.181	0.029	0.839	0.716	0.982
time*zses_mean	–0.023	0.038	–0.590	0.555	0.978	0.907	1.054
time*[smallsch = 1]	0.189	0.082	2.289	0.022	1.208	1.027	1.420
time*[smallsch = 0]	0[a]						
time*[female = 1]	0.084	0.074	1.132	0.258	1.087	0.941	1.257
time*[female = 0]	0[a]						
time*[lowses = 1]	–0.176	0.069	–2.546	0.011	0.839	0.733	0.960
time*[lowses = 0]	0[a]						

Note: Probability distribution: binomial; link function: logit.
[a] This coefficient is set to 0 because it is redundant.

TABLE 5.33   Model 3.1: Variance Components

Random and residual effects	Estimate	Std. error	z-Test	Sig.	95% Confidence interval Lower	Upper
Random block 1[a]						
Level-3 var(intercept)	0.959	0.203	4.730	0.000	0.633	1.451
Level-3 cov(slope)	–0.267	0.064	–4.204	0.000	–0.392	–0.143
Level-3 var(slope)	0.171	0.030	5.768	0.000	0.122	0.240
Random block 2[b]						
Level-2 var(intercept)	5.399	0.278	19.411	0.000	4.881	5.973
Residual[c]						
AR1 diagonal	0.430	0.012	35.894	0.000	0.407	0.454
AR1 rho	0.269	0.020	13.784	0.000	0.231	0.307

[a] Covariance structure: unstructured; subject specification: nschcode.
[b] Covariance structure: variance components; subject specification: nschcode*Rid.
[c] Covariance structure: first-order autoregressive (AR1); subject specification: nschcode*Rid.

participation in a treatment designed to target their reading problems helps increase their likelihood to become proficient over time. The data in this study consist of 219 elementary school students who were randomly assigned to a control ($N = 118$) group or to a treatment ($N = 101$) where they received individualized instruction with a classroom aid for 30 days. They were measured on four occasions (one pretest) regarding their proficiency in reading during one academic year.

The design is like the following:

$$R \quad O_1 \qquad O_2 \quad O_3 \quad O_4$$
$$R \quad O_1 \quad X \quad O_2 \quad O_3 \quad O_4$$

The probability of being proficient ($Y = 1$) is being modeled. Fixed effects include the treatment (coded 1) versus control (coded 0), time (coded 0–3 for the four occasions), and the treatment by time interaction. We expect no difference between the treatment and control groups initially. This is captured by the relationship between the treatment variable and the initial status at Time 1 (where time is coded 0). After implementing the treatment, we propose that, over time, the treatment group will increase in the probability of being proficient relative to the control group. This will be shown by a treatment by time interaction (i.e., different trends for the placebo and treatment).

We will compare the GEE (fixed-effects) model against a random-effects model. Recall that for the fixed-effects approach, there are no variance components at Level 2 because the nesting of observations within subjects is treated as a nuisance parameter. The GEE approach gives the average effect on all individuals (in the population), and the random-coefficients approach gives the effect on the individual subject. We can specify the model for individual i measured at time occasion t as follows:

$$\eta_{ti} = \left(\frac{\pi_{ti}}{1 - \pi_{ti}} \right) = \beta_0 + \beta_1 time + \beta_2 treatment + \beta_3 treatment * time. \qquad (5.27)$$

In contrast, we will specify the random-coefficients model at Level 1 as follows:

$$\eta_{ti} = \left(\frac{\pi_{ti}}{1 - \pi_{ti}} \right) = \beta_{0i} + \beta_{1i} time_t \qquad (5.28)$$

At Level 2, we assume that the intercept varies between subjects:

$$\beta_{0i} = \gamma_{00} + \gamma_{01} treatment_i + u_{0i}. \tag{5.29}$$

We will model the time slope at Level 2 as fixed (or nonrandomly varying):

$$\beta_{1i} = \gamma_{10} + \gamma_{11} treatment_i \tag{5.30}$$

The combined model will then be the following:

$$\eta_{ti} = \gamma_{00} + \gamma_{01} treatment_i + \gamma_{10} time_{ti} + \gamma_{11} time_{ti} {}^* treatment_i + u_{0j}. \tag{5.31}$$

We present a comparison of the fixed-effects and random-effects models in Table 5.34 with modeling instructions thereafter to construct the GENLIN MIXED portion of the model and note that the syntax to replicate the GEE portion is available in Appendix A. We note in passing that we investigated several different covariance structures for the repeated measures, but we decided to stay with a simple identity structure (which averages the variances due to occasions within subjects) because the within-subject covariance structure was really not the focus of the study.

The pattern of significant results across the two types of models in the table is quite consistent. We can see that individuals were declining in their probability of being proficient across time in both models. Similarly, in both models there was a significant treatment by time effect, suggesting that students who received the targeted intervention increased their probability of being proficient over time.

We can note one primary difference. The population average model (fixed-effects model) describes how the average response across subjects changes with the covariates. This is useful when one wants to compare the students who received the treatment versus the students who did not receive the treatment. Only the link function need be correctly specified to make consistent inferences about coefficients (Zeger et al., 1988). As before, the model estimates are a little smaller in the population-average model.

In contrast, the random-coefficient model uses the information contained in the population averaged response and also a distributional assumption about the existing heterogeneity among subjects to estimate subject-specific coefficients (Zeger et al., 1988). As we noted previously, this accounts for the slightly larger coefficients describing the variable effects. In this two-level situation, both the link function and the random effects distribution must be correctly specified for consistent inferences. More specifically, we found that the model with random intercept provided one plausible representation of the data (AIC corrected = 4170.846). It fit the data considerably better than a model with random intercept and random time*treatment slope (AIC corrected = 5036.400).

TABLE 5.34 Comparison of GEE and Random-Coefficients Approach

	Fixed-Effects Model (GEE)			Random-Effects Model 4.1 (GENLIN MIXED)		
Parameter	Estimates	Std. error	p	Estimates	Std. error	p
(Intercept)	−0.203	0.187	0.277	−0.267	0.272	0.327
Time	−0.204	0.084	0.016	−0.307	0.128	0.017
Treatment	0.238	0.282	0.397	0.286	0.421	0.497
Treatment by time	0.299	0.125	0.017	0.461	0.195	0.018
Random effect[a]:						
Var(intercept)				3.530	0.532	0.000

[a] Covariance structure: scaled identity; subject specification: id.

Defining Model 4.1 With IBM SPSS Menu Commands

Launch the IBM SPSS application program and select the *ch5data2.sav* data file.

1. Go to the toolbar, select ANALYZE, MIXED MODELS, GENERALIZED LINEAR. This command opens the *Generalized Linear Mixed Models* (GENLINMIXED) main dialog box.

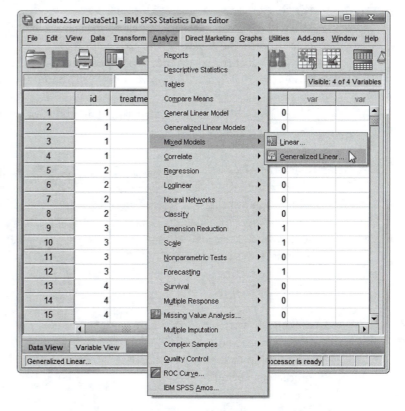

2. The *Generalized Linear Mixed Models* display screen shows four command tabs: *Data Structure, Fields & Effects, Build Options,* and *Model Options*.

a. The *Data Structure* command tab is the default screen when creating a model for the first time. The options enable specifying structural relationships between data set records when observations are correlated.

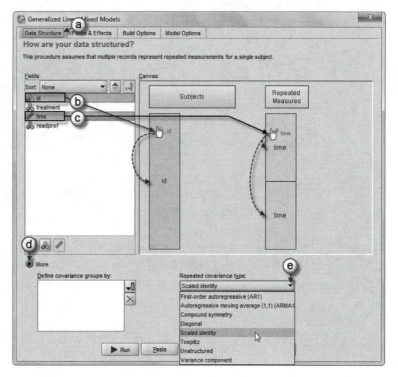

b. To build our model, begin by defining the subject variable. A "subject" is an observational unit that can be considered independent of other subjects. Click to select the *id* variable from the left column and then drag and drop *id* onto the canvas area beneath the *Subjects* box.

c. We will designate *time* as a repeated measure. Click to select *time* and then drag the variable beneath the *Repeated Measures* column.

d. We will select the covariance type by clicking the *More* button to reveal additional options.

e. Now click the *Repeated covariance type* pull-down menu and select *Scaled identity*.

Note: The scaled identity structure has constant variance and assumes no correlation between elements.

3a. Click the *Fields & Effects* command tab, which displays the default *Target* display screen when creating a model for the first time. The settings allow specifying a model's target (dependent variable), distribution, and relationship to model predictors through the link functions.

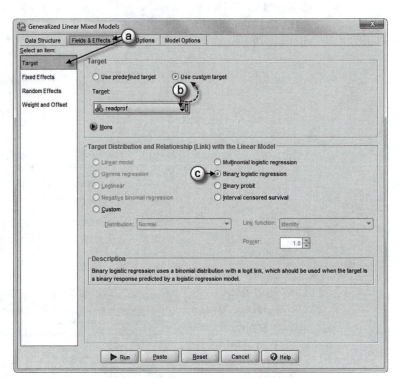

b. For this model we will use student reading proficiency (*readprof*) as the dependent variable. Click the *Target* pull-down menu and select *readprof*.

Note: Selecting the target variable will automatically change the target setting to *Use custom input*.

c. Next, we will designate a binomial distribution and logit link function for the model. Click to select *Binary logistic regression*.

Note: The binary logistic regression uses a binomial distribution with a logit link and is used due to the binary aspect of the *readprof* dependent variable.

4a. Now click the *Fixed Effects* item to access the main dialog box; this enables introducing predictor variables into the model.

b. Note that the *Include intercept in model* option is preselected and displayed in the *Effect builder* as *intercept*. This is because a Level-1 model is intercept only (default), so other model effects must be specified.

c. We will begin to build the model by adding two predictor variables (*time, treatment*). Time is a Level-1 predictor and treatment is a Level-2 predictor. Click to select both variables and then drag them into the *Main* (effect) column of the *Effect builder*.

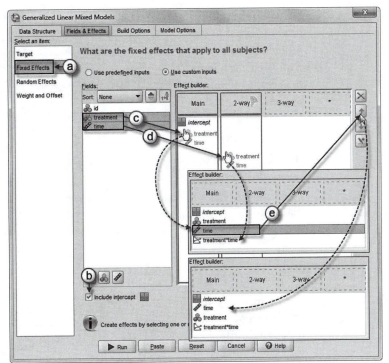

$$\eta_{ti} = \left(\frac{\pi_{ti}}{1 - \pi_{ti}}\right) = \beta_{0i} + \beta_1 \textit{time}_{ti} \quad \text{(Eq. 5.28)}$$

$$\beta_{0i} = \gamma_{00} + \gamma_{01}\textit{treatment}_i + u_{0i} \quad \text{(Eq. 5.29)}$$

$$\beta_{1i} = \gamma_{00} + \gamma_{11}\textit{treatment}_i \quad \text{(Eq. 5.30)}$$

d. We will also add the interaction to the model at Level 2. Click to select both variables and then drag them to the *2-way* column to create *treatment*time* (see Eq. 5.31).

e. To facilitate reading of the output table, we will rearrange the variable order. Click to select *time* and then click the up arrow on the upper right to place the variable at the top of the list. The variable sequence is now *time, treatment, treatment*time*.

5a. We will now add a random effect to the model. Click the *Random Effects* item to access the main screen.

b. Click the ADD BLOCK button to access the *Random Effect Block* main dialog box.

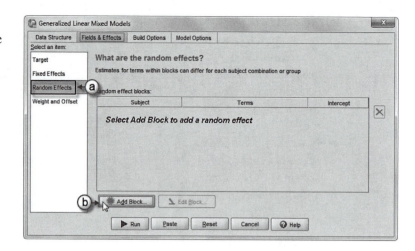

c. In this model, only the intercept is going to be randomly varying, so select the *Include intercept* option. Because there is only one random effect (intercept), we can use the default VC setting. This specifies an identity covariance matrix for the random intercept effect.

d. Select the *Subject combination* by clicking the pull-down menu and selecting

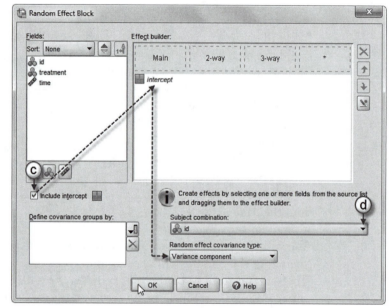

id. Click the OK button to close the *Random Effect Block* dialog box and return to the *Random Effects* main screen.

e. In the main screen notice that the *Random effect blocks* displays a summary of the block with *id* and *intercept* added to the model.

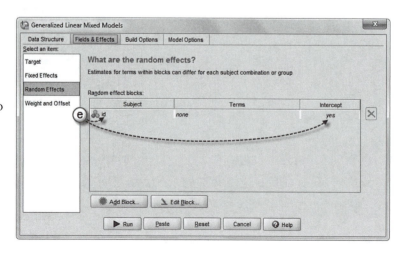

6a. Finally, click the *Build Options* command tab located at the top of the screen to access advanced criteria options for building the model.

b. Change the sorting order for both categorical targets and categorical predictors. Click the pull-down menu for each option and select *Descending*.

c. We will retain most of the default settings but change the *Tests of fixed effects and coefficients* to use robust covariances. Click to select *Use robust estimation to handle violations of model assumptions*.

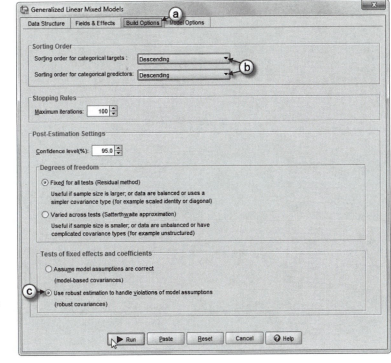

Click the RUN button to generate the model results.

Summary

In this chapter we provided examples of fixed-effect and random-effect longitudinal models for a dichotomous outcome. The GEE approach is often useful when one can assume that subjects are not nested in higher level groups and the concern is with an effect within a population. In contrast, the unit-specific results are useful in determining the effect of several predictors on a particular individual or a particular group. Both approaches can be extended to consider other types of outcomes consisting of nominal, ordinal, or count data. In the next chapter, we consider single-level and multilevel models in which the outcome is nominal or ordinal.

CHAPTER 6

Two-Level Models With Multinomial and Ordinal Outcomes

Introduction

In Chapter 4 we introduced the two-level model for dichotomous outcomes and extended this basic model to one with three levels. Next, in Chapter 5 we examined a longitudinal model where there were repeated dichotomous measures on individuals within groups. In this chapter we extend the basic multilevel model for dichotomous outcomes to models that include nominal outcomes (i.e., three or more unordered categories) and ordinal outcomes.

Multinomial logistic regression represents a logical extension of the binary case because it involves comparing successive categories to a reference category. A multinomial analysis of nominal outcomes produces one set of estimates less than the number of categories ($C - 1$) because the successive categories are each separately compared to the reference category. Conceptually, this type of model does not differ from the dichotomous case because the focus is on comparing the probability of being in one selected outcome category versus the reference category.

When there are more than two outcome categories, however, more possible combinations of outcome categories can be compared, so a bit more care must be taken to ensure that the comparisons being made are the ones intended. Multinomial logistic regression is a good option when the data are obviously not ordered, or in cases where the data may imply some type of "intended" order but do not meet the criteria for an ordinal regression analysis. More specifically, multinomial logistic regression can be used with either nominal or ordinal data, but the reverse is not true for ordinal logistic regression.

Methods designed to be used with ordinal variables cannot be used with nominal outcomes because the latter type of variable does not have ordered categories (Agresti, 2007). One typical type of ordered variable is a Likert scale, where individuals are asked to respond in terms of their relative agreement with a statement: for example, "I intend to leave my place of employment within the next 90 days" (1 = strongly disagree to 5 = strongly agree). Another example is when respondents provide their perception of the personal importance of a statement such as "How important is it for you to see the doctor rather than a nurse practitioner when you come to the clinic?" (1 = not important to 5 = very important). Provided that the assumption of ordered categories can be supported, the advantage of an ordinal model is that it is generally easier to interpret because the analysis produces only one set of estimates.

Sometimes, however, it is not clear whether the outcome categories actually can be conceived as ordered, and even if they can reasonably be conceived as ordinal, there may not be a clear empirical pattern of how the independent variables explain membership in the ordered categories. This is referred to as requiring a type of *parallel* structure; that is, the independent variables must have the same effect on the odds of being in each successive category (Hox, 2010). This is similar to the assumption in a linear model that an increase in X (e.g., previous learning) beginning at any level of X will produce a similar increase in Y (current learning). Therefore, it is often useful to determine beforehand whether the outcome data might be likely to support an ordinal analysis. Once this is determined, we emphasize that when using IBM SPSS for these two types of outcomes, it is important to specify the scale of measurement of the outcome variable appropriately as either nominal or ordinal so that the desired analysis is conducted in GENLIN MIXED.

Building a Model to Examine a Multinomial Outcome

In situations where there is an outcome with more than two possible events that are not ordered, or where the ordered nature of the outcome is not very clear or does not meet the necessary assumptions for ordinal analysis, multinomial logistic regression is an appropriate analytic procedure. The approach is similar to binary logistic regression, but the dependent variable is not restricted to two categories. As we noted in Chapter 3, because the outcome represents a probability between 0 and 1, linear regression would not be appropriate as it would result in predictions that fall outside the allowable range of the dependent variable. The multinomial probability distribution therefore represents an extension of the Bernoulli model for dichotomous outcomes; the difference is that the outcome consists of more than two unordered categories, which are separately compared to the selected reference category.

Research Questions

Our first example refers to the postsecondary plans of high school seniors. We may wish to consider whether individual-level (e.g., background) and school-level (e.g., academic press or focus, teacher quality) variables might also contribute to students' likelihood to have particular postsecondary plans. We might begin by simply investigating whether students' backgrounds influence their plans to attend postsecondary institutions or enter the workforce directly. We can then extend our investigation to consider the multilevel nature of our data. Our first research question might focus on whether students' likelihood to seek particular types of postsecondary education varies across schools. We can ask simply: Do students' postsecondary plans vary across high schools? We might then build a model to investigate how school-level variables help explain variability in likelihood to have particular plans.

Second, we might ask whether the effects of an observed relationship between a Level-1 slope (e.g., gender, student economic status [SES] or grade point average [GPA]) and likelihood to pursue certain types of postsecondary study vary across schools. More specifically, we ask: Does a specific Level-1 slope of theoretical interest vary across schools? If it does, we could develop a school-level model to explain this variability. Our research questions, therefore, provide an illustration of building a two-level model with a multinomial outcome to investigate (1) a randomly varying intercept (likelihood to be proficient) and, subsequently, (2) a randomly varying slope (i.e., background-reading proficiency relationship).

The Data

The data consist of 9,166 high school seniors who responded to a survey about their postsecondary plans. They were attending 43 high schools in a diverse geographic region (Table 6.1).

Defining the Multinomial Model

We will present our initial discussions of the multinomial and ordinal models in this chapter within the single-level context. As in previous chapters, for the multilevel setting we add the subscript j (for groups) after the subscript i for individuals. For a multinomial outcome, we will assume that the probability distribution is multinomial and we will use a logit link function. As

TABLE 6.1 Data Definition of *ch6data1.sav* (N = 9,166)

Variable	Level[a]	Description	Values	Measurement
schcode	School	School identifier (43 schools).	Integer	Ordinal
lowses	Individual	Dichotomous predictor variable representing low social economic status	0 = not low socioeconomic status 1 = low socioeconomic status	Nominal
male	Individual	Demographic predictor variable representing gender	0 = female 1 = male	Nominal
plans	Individual	Dependent variable with three categories classifying the postsecondary plans of high school seniors to attend college	0 = attend 4 year 1 = attend tech/voc/2 year 2 = no school	Nominal
gmgpa	Individual	Predictor interval variable (grand-mean centered) representing student grade point average	–2.47 to 1.53	Scale
gmacadpress	School	Predictor interval variable (grand-mean centered) representing the school's academic focus	–1.66 to 2.40	Scale
gmlicense	School	Predictor interval variable (grand-mean centered) representing percentage of fully certified teachers	–0.18 to 0.10	Scale
gmperlowses	School	Predictor interval variable (grand-mean centered) representing percentage of students with low socioeconomic status	–0.27 to 0.40	Scale

[a] Individual = Level 1; school = Level 2.

we noted in Chapter 3, for a nominal outcome, only the logit link function is available (referred to as a generalized logit link as opposed to cumulative logit link).

For a multinomial distribution, the total number of outcome categories can be denoted as C, and each individual category can be indexed by c such that the probability of being in the cth outcome category $P(Y = c)$ is π_c ($c = 1, 2,..., C$). The cumulative probabilities of each possible outcome ($\pi_1, \pi_2,..., \pi_c$) can be expressed such that their sum is 1.0. The last category is usually the reference, so there will actually be $C - 1$ equations to be estimated. The probability of membership in one of the other categories is compared against the probability of being in the reference category.

For individual i, then, the probability of being in category c ($c = 1, 2,..., C - 1$) versus the reference group (C) can be defined as follows:

$$\eta_c = \log\left(\frac{\pi_c}{\pi_C}\right). \tag{6.1}$$

In general, then, $c = 1, 2,..., C - 1$ logits for the categories can be defined:

$$\log(\pi_1/\pi_C), \log(\pi_2/\pi_C), \cdots, \log(\pi_{C-1}/\pi_C). \tag{6.2}$$

With only two categories, this will be the same as the dichotomous case with which we have worked in previous chapters. The outcome scores are transformed using the cumulative logit link function into an unobserved (or latent) continuous variable η_c that describes the log odds of being in a particular category (*c*) versus the reference category. For a nominal outcome, we can select any group as the reference group. The variance is then

$$\text{Var}(Y_c | \pi_c) = \pi_c(1 - \pi_c), \tag{6.3}$$

Once again, this suggests that the variance is tied to the expected value in the population.

In our postsecondary plans example, the outcome originally consisted of four categories referring to the postsecondary plans of high school seniors. The categories were

1. Planning to attend a 4-year postsecondary institution
2. Planning to attend a community college
3. Planning to attend a vocational/technical institution (e.g., for profit)
4. Not planning to attend a postsecondary institution

In this case, there is not a clearly ordered set of categories comprising the outcome. We ended up collapsing the second and third categories into one category. The data are coded such that planning to attend a 4-year college or university = 2 and planning to attend a community college/technical institute = 1. We will consider having no plan to attend an institution after high school as the reference group (coded 0). By using the default (ascending) coding, we will obtain log odds coefficients comparing the odds of attending a 4-year institution versus no additional education and the odds of attending a 2-year or technical institution versus no additional education.

By examining the frequencies in Table 6.2, we can see that most students report planning to go to a 4-year institution (47.7%). A smaller percentage plans to attend a community college or vocational/technical institution (37%), and about 15.4% do not plan any further education.

Defining a Preliminary Single-Level Model

First, we will build a single-level model using GENLIN MIXED for illustrative purposes.[1] We note that the multinomial distribution (with generalized logit link function) is currently not available in the single-level GENLIN routine. At present, only the multinomial model with cumulative logit link function for ordinal analyses is available in that program. In GENLIN MIXED, we can develop a single-level model by simply not declaring a random intercept at the school level. We will again use robust standard errors for our analyses.

Because the analysis is conducted at one level, we do not need the customary *j* for schools:

$$\eta_c = \log(\pi_c / \pi_C) = \beta_{0(c)} + \beta_{1c}gmgpa + \beta_{2c}lowses + \beta_{3c}male. \tag{6.4}$$

We will also eliminate the customary *i* for individuals because it is an individual-level analysis. The subscript *c* is needed, however, because it is likely that the model parameters would differ depending on the outcome category being modeled (Azen & Walker, 2011). We note that this is in contrast to an analysis of ordinal outcomes, where the primary assumption of the cumulative odds model is that the same slope coefficient explains successive membership in the ordered

TABLE 6.2 Frequency Distribution: Plans

		Frequency	Percent	Valid percent	Cumulative percent
Valid	4 Year	4368	47.7	47.7	47.7
	Tech/voc/2 year	3390	37.0	37.0	84.6
	No school	1408	15.4	15.4	100.0
	Total	9166	100.0	100.0	

TABLE 6.3 Model 1.1: Fixed Effects

Model term	Coefficient	Std. error	*t*	Sig.	Exp (coefficient)	95% Confidence interval for exp(coefficient) Lower	Upper
Multinomial category (1): 4 year							
Intercept	2.150	0.075	28.706	0.000	8.587	7.414	9.945
gmgpa	2.031	0.050	40.419	0.000	7.620	6.906	8.409
lowses	–0.983	0.075	–13.023	0.000	0.374	0.323	0.434
male	–0.020	0.075	–0.266	0.791	0.980	0.846	1.136
Multinomial category (2): tech/voc2 year							
Intercept	1.599	0.075	21.438	0.000	4.948	4.275	5.727
gmgpa	1.252	0.044	28.214	0.000	3.498	3.206	3.816
lowses	–0.118	0.070	–1.694	0.090	0.888	0.775	1.019
male	0.167	0.070	2.365	0.018	1.181	1.029	1.356

Note: Target: plans; reference category: no school; probability distribution: multinomial; link function: generalized logit.

categories. We can also express this multinomial model in terms of the probability $Y = \pi_c$ by use of the exponential function in the following form:

$$\pi_c = \frac{\exp(\beta_0 + x'\beta)}{1 + \exp(\beta_0 + x'\beta)}, \tag{6.5}$$

where $x'\beta$ represents a vector of predictors and their corresponding regression coefficients.

Although this form of the model may seem simplest because it results in a predicted probability for $Y = 1$, there is no simple interpretation for the logistic regression coefficients in this form. It can be shown that the meaning of the intercept is easily extended to the case where its value is interpreted as the predicted response when all the predictors are 0 (Azen & Walker, 2011).

In Table 6.3 we can see the fully efficient comparison that makes use of all three categories simultaneously. We can see that SES status is relevant in one situation (attending a 4-year versus not attending school). Gender is also relevant only in differentiating whether students plan to attend a 2-year institution versus not to continue their educations. Student GPA has a positive effect in both comparisons.

Defining Model 1.1 With IBM SPSS Menu Commands

Note: Refer to Chapter 3 for discussion concerning settings and modeling options available in the IBM SPSS GENLIN MIXED routine.

Launch the IBM SPSS application program and select the *ch6data1.sav* data file.

1. Go to the tool-
bar and select
ANALYZE,
MIXED
MODELS,
GENERALIZED
LINEAR. This
command allows
access to the
*Generalized Linear
Mixed Models*
(GENLINMIXED)
main dialog box.

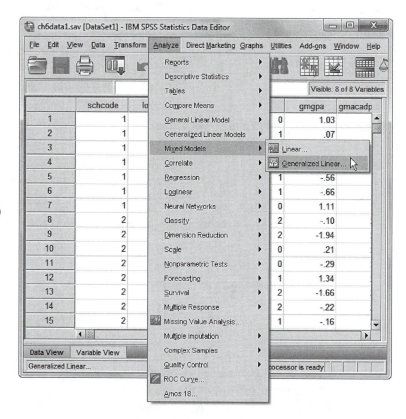

2a. The *Generalized
Linear Mixed Models*
display screen shows
four command tabs:
*Data Structure, Fields
& Effects, Build
Options,* and *Model
Options.* The *Data
Structure* command
tab is the default
screen display when
creating a model for
the first time. The
options enable speci-
fying structural rela-
tionships between
data set records
when observations
are correlated.

b. To build our model,
begin by defining

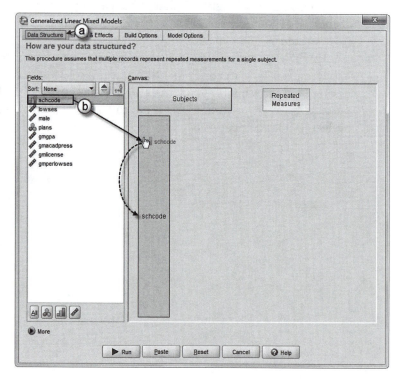

the subject variable. A "subject" is an observational unit that can be considered independent of other subjects. Click to select the *schcode* variable from the left column and then drag *schcode* onto the canvas area beneath the *Subjects* box.

3a. Click the *Fields & Effects* command tab, which displays the default *Target* display screen. The settings allow specifying a model's target (dependent variable), distribution, and relationship to model predictors through the link functions.

b. For this model we will use high school seniors' postsecondary plans (*plans*) as the dependent variable. Click the *Target* pull-down menu and select *plans*.

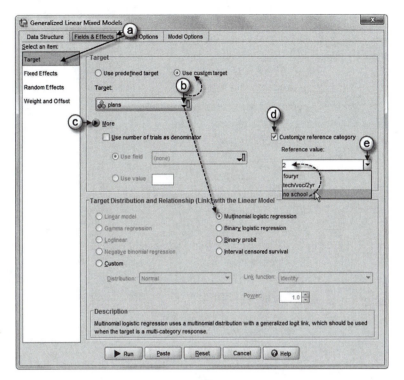

Note: Selecting the target variable will automatically change the setting to *Use custom input.* Also, the ordinal measurement setting of the dependent variable (*plans*) automatically designates the target distribution as *Multinomial logistic regression.*

c. Because *plans* consists of three categories, we will designate a reference category. First, click the *More* button to reveal additional options.

d. Click to select *Customize reference category.*

e. Click the *Reference value* pull-down menu and select the category: *no school.*

4a. Now click the *Fixed Effects* item to access the main dialog box.

b. Note that the *Include intercept in model* option is preselected and displayed in the *Effect builder* as *intercept*. This is because a Level-1 model is intercept only (default), so other model effects must be specified.

c. We will begin to build the model by adding three predictor variables. Click to select *lowses*, *male*, *gmgpa* and then drag the variables into the *Main* (effect) column of the *Effect builder*.

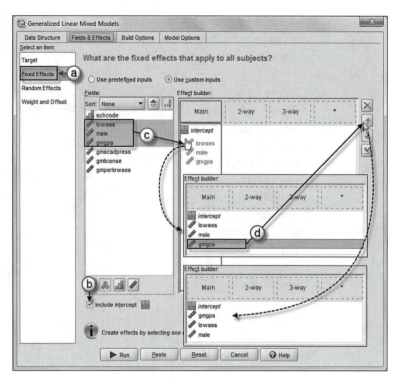

d. To facilitate reading of the output tables we will rearrange the variable order. Click to select *gmgpa* and then click the up-arrow button on the right to place *gmgpa* above *lowses*. The variable order is now *gmgpa*, *lowses*, *male*.

$$\eta_c = \log(\pi_c/\pi_C) = \beta_{0(c)} + \beta_{1c}\textit{gmgpa} + \beta_{2c}\textit{lowses} + \beta_{3c}\textit{male} \quad \text{(Eq. 6.4)}$$

Note: The order may also be controlled by entering the variables into the *Effect builder* beginning with *gmgpa*, followed by *lowses* and *male*.

5a. Finally, click the *Build Options* command tab located at the top of the screen to access advanced criteria for building the model. We will retain the default settings for *Stopping Rules, Post Estimation*, and *Degrees of Freedom*.

b. Change the *Tests of fixed effects and coefficients*. Click to select *Use robust estimation to handle violations of model assumptions*.

Click the RUN button to generate the model results.

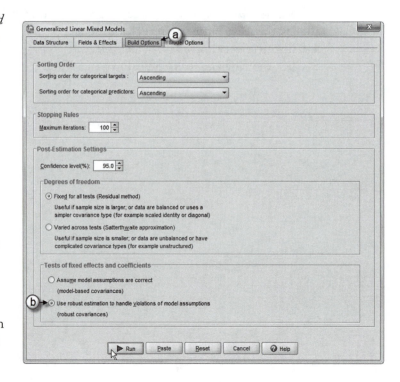

Interpreting the Output of Model 1.1

In Table 6.3, the estimated intercept ($\beta_0 = 2.150$) is the log odds in favor of planning to attend a 4-year institution versus no further postsecondary education on the underlying continuous logit scale (η_c). It describes the log of the odds that a student is planning to attend a 4-year institution versus not planning to attend school when the predictors in the model are equal to 0. In this case, this student would be female (coded 0) of average or high SES (coded 0) at the grand mean for GPA (0). The exponentiated intercept (e^β) can be interpreted as the predicted odds in favor of the category under consideration ($Y = 2$) compared against the reference category ($Y = 0$) when the other variables in the model are zero. The odds ratio (8.587) suggests that a female of average/high SES, who is at the average in GPA, is about 8.6 times more likely to plan on attending a 4-year institution than to plan on not continuing her education after high school.

Regarding SES, we note that the log of the odds of planning to attend a 4-year institution versus no further education is decreased by 0.983 units ($\beta = -0.983$) for low SES students compared with their more average and high SES peers. The odds ratio indicates that the predicted odds that a low SES individual is planning to attend a 4-year institution versus planning no further education are 0.374 the odds of her or his more average and high SES peers. In other words, the odds of planning to attend a 4-year institution versus planning to obtain no further education are reduced by about 62.6% (1 − 0.374 = 0.626) for low SES students compared to their higher SES peers, holding gender and GPA constant.

To illustrate the meaning of odds ratios a bit further, later, from Table 6.3, we can see that the odds of planning to attend a 4-year institution versus no further education for an average/high SES student (holding the other variables constant) would be e^{β_0}, which is the value of the odds ratio (8.587) in the table. The expected odds for a low-SES student would then be obtained from $e^{\beta_0}e^{\beta_2}$, which, through multiplying the odds ratios, is 3.212 (8.587*0.374 = 3.212). We could also obtain this result from adding the two log odds and then exponentiating ($e^{1.167} = 3.212$). The odds ratio is then defined as the difference between the predicted odds for low-SES versus high/average students (3.212/8.587) who are planning to attend a 4-year institution versus no further education. This results in the odds ratio for low SES displayed in the table (0.374). The interpretation for GPA would be that as GPA increases by 1 point, the expected odds of planning to attend a 4-year institution versus no further education are multiplied by 7.620. Gender is unrelated to the odds of planning to attend a 4-year institution versus not seeking further education ($p > .05$).

Turning to planning to attend a 2-year institution versus planning no further education, we can see that GPA has a similar positive effect on plans (OR = 3.498, $p < .001$). The odds of planning to attend a 2-year institution versus no further education are multiplied by 1.181 (or increased by 18.1%) for males versus females. Finally, student SES appears to have no effect on expected odds of attending a 2-year institution versus not obtaining any further education.

Developing a Multilevel Multinomial Model

We next extend the single-level model to consider the nesting of individuals within groups. At Level 1, then, the multinomial logistic regression model used to predict the odds of individual i in group j being in outcome category c relative to outcome C (reference category) using the set of q predictors will be

$$\eta_{cij} = \log(\pi_{cij}/\pi_{Cij}) = \beta_{0jc} + \beta_{1jc}X_{1ij} + \beta_{2jc}X_{2ij} + \ldots + \beta_{qjc}X_{qij} \quad (6.6)$$

Similarly to our previous models, at Level 1 there is no separate residual variance term because the variance is dependent upon the mean. The Level-1 variance is again set to a scale factor of

1.0. We can summarize this more general model to link the expected values of the outcome to the predicted values of η_{cij} as follows:

$$\eta_{cij} = \log\left(\frac{\pi_{cij}}{\pi_{Cij}}\right) = \beta_{0j(c)} + \sum_{q=1}^{Q} \beta_{qj(c)} X_{qij}. \tag{6.7}$$

Once again, at the lowest level, the model formulation suggests that the underlying continuous variate η_{cij} is a ratio of two odds (i.e., the probability of each category c versus the selected reference category C) that is explained by a set linear combination of (X) predictors ($q = 1...Q$). At Level 2, we have the following model:

$$\beta_{qj(c)} = \gamma_{q0(c)} + \sum_{s=1}^{S_q} \gamma_{qs(c)} W_{sj} + u_{qj(c)} \tag{6.8}$$

At Level 2, we can model one or more Level-1 intercepts or slopes as a function of a set of Level-2 predictors (W) and corresponding variance terms (u_{qj}). Keep in mind that for each intercept or slope modeled, there will be $C - 1$ equations. With a nominal outcome consisting of four or five categories, it would be easy to develop a model that gets too complicated, so we suggest exercising caution in keeping random slopes at Level 2 to a minimum.

Unconditional Two-Level Model

We can start by estimating an unconditional (no predictors) model to examine the extent of variability of the nominal outcome across schools. As we have developed, the unconditional model at Level 1 will have $C - 1$ estimates for individual i in school j as follows:

$$\eta_{1ij} = \log\left(\frac{\pi_{1ij}}{\pi_{Cij}}\right) = \beta_{0j(1)},$$

$$\eta_{2ij} = \log\left(\frac{\pi_{2ij}}{\pi_{Cij}}\right) + \beta_{0j(2)}. \tag{6.9}$$

At Level 2, the combined set of models suggests that the intercepts vary between schools:

$$\beta_{0j(1)} = \gamma_{00(1)} + u_{0j(1)},$$

$$\beta_{0j(2)} = \gamma_{00(2)} + u_{0j(2)}. \tag{6.10}$$

All together, there will be four parameters to estimate across the two models.

Defining Model 2.1 With IBM SPSS Menu Commands

Note: Settings default to those used for Model 1.1.

1. Go to the toolbar and select ANALYZE, MIXED MODELS, GENERALIZED LINEAR. This command opens the *Generalized Linear Mixed Models* (GENLINMIXED) main dialog box.

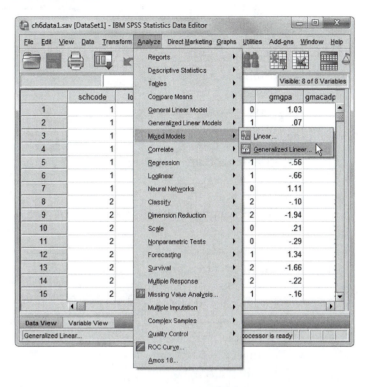

2a. The *Generalized Linear Mixed Models* dialog box displays the *Build Options* command tab, which was the last setting used in the prior model.

b. First, click to select the *Fields & Effects* command tab, which displays the *Fixed Effects* main screen.

c. We will remove the three predictor variables to create the unconditional (null) model. Click to select the three variables (*gmgpa, lowses, male*) and then click the red "X" button on the right to delete the variables.

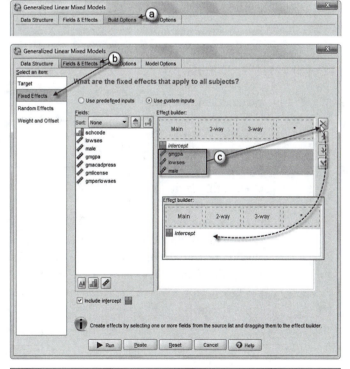

$$\eta_{1ij} = \log\left(\frac{\pi_{1ij}}{\pi_{Cij}}\right) + \beta_{0j(1)}$$

$$\eta_{2ij} = \log\left(\frac{\pi_{2ij}}{\pi_{Cij}}\right) + \beta_{0j(2)}$$

(Eq. 6.9)

3a. The next step is to add a random effect (intercept) to the model. Click on the *Random Effects* item. The *Random Effects* command is used to specify which variables are to be treated as randomly varying across groups (random effects).

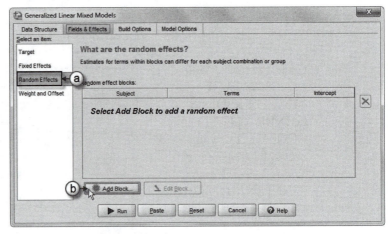

b. Click the ADD BLOCK button to access the *Random Effect Block* main dialog box.

c. In this model only the intercept is going to be randomly varying, so click to select the *Include intercept* option, which changes the setting within the *Effect builder* to *intercept* (from *no intercept*). Because there is only one random effect (the intercept), we can use the default variance component (VC) setting. This specifies a diagonal covariance matrix for the random effects; that is, it provides a separate variance estimate for

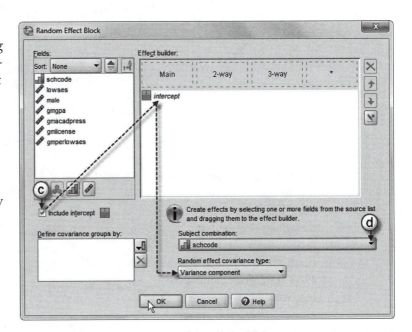

$$\beta_{0j(1)} = \Upsilon_{00(1)} + u_{0j(1)} \quad (Eq.\ 6.10)$$

$$\beta_{0j(1)} = \Upsilon_{00(2)} + u_{0j(2)}$$

each random effect, but not covariances between random effects (for one random effect, the VC setting defaults to an identity matrix).

d. Select the *Subject combination* by clicking the pull-down menu and selecting *schcode*. Click the OK button to close the *Random Effect Block* dialog box and return to the *Random Effects* main screen.

e. In the main screen notice that the *Random effect blocks* displays a summary of the block just created: *schcode* (Subject) and *intercept* (Yes).

Click the RUN button to generate the model results.

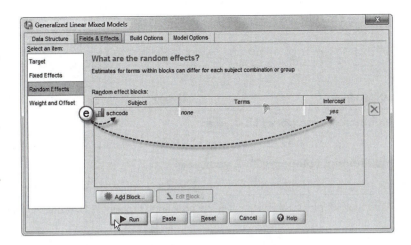

Interpreting the Output of Model 2.1

In Table 6.4 we summarize the measurement scale of the dependent variable, the relevant probability distribution, and the link function for the unconditional model. This table also provides information about the fit of the model (i.e., AIC, BIC indices). We note again that multilevel models should be compared cautiously because the models are estimated using quasimaximum likelihood techniques. The fixed-effect results of this unconditional model are summarized in Table 6.5.

The intercept for Category 1 suggests that for students in the average high school, the predicted log of the odds of planning to attend a 4-year college is greater than for planning on no further education (log odds = 1.193, $p < .001$). Similarly, for Category 2, the odds of planning to attend a vocational/technical or community college are also greater than for planning no additional schooling (log odds = 1.042, $p < .001$).

Computing Predicted Probabilities

When there are more than two categories for the dependent variable, computing probabilities is a little more complicated than it is for a dichotomous outcome. For a dependent variable with C categories, we must calculate $C - 1$ log odds equations. We will assume that the last category is the reference group (but we can use any category of choice). To calculate the probability for each

TABLE 6.4 Model 2.1: Model Fit Criteria

Target	Plans
Measurement level	Nominal
Probability distribution	Multinomial
Link function	Generalized logit
Information criteria	
Akaike corrected	69,642,652
Bayesian	69,656,897

Note: Information criteria are based on the –2 log pseudolikelihood (69,638,651) and are used to compare models. Models with smaller information criterion values fit better. When comparing models using pseudolikelihood values, caution should be used because different data transformations may be used across the models.

TABLE 6.5 Model 2.1: Fixed Effects

Model term	Coefficient	Std. error	*t*	Sig.	Exp (coefficient)	95% Confidence interval for exp(coefficient)	
						Lower	Upper
Multinomial category (1): 4 year							
Intercept	1.193	0.111	10.750	0.000	3.297	2.652	4.098
Multinomial category (2): 2 year/tech							
Intercept	1.042	0.107	9.743	0.000	2.835	2.999	3.496

Note: Target: plans; reference category: no school; probability distribution: multinomial; link function: generalized logit.

category, we take each one of the $C - 1$ log odds computed and then exponentiate it. Once the log odds are exponentiated, we simply divide each by the sum of the odds to obtain the probability for each category. For the $C - 1$ contrasts, beginning with $c = 1$ to $C - 1$, from Equation 6.5 we can represent the predicted probabilities as follows:

$$\pi_{cij} = \frac{\exp(\eta_{cij})}{1 + \sum_{c=1}^{C-1} \exp(\eta_{cij})}, \tag{6.11}$$

where η_{cij} is the value of the linear component for specific values of the predictors (if there are any in the model). Note that for the reference category the value of η_{cij} is 0 and $\exp(0) = 1$; the log odds are therefore 0 $[\log(1) = 0]$, and the odds ratio is 1 $(e^0 = 1)$. The probability of landing in the reference category is then

$$\pi_{Cij} = \frac{1}{1 + \sum_{c=1}^{C-1} \exp(\eta_{cij})}. \tag{6.12}$$

The estimated probability for each response category in our first model (with just the intercept), then, is as follows:

$$P(Y = 1) = \frac{\exp(\eta_{1ij})}{1 + \sum_{c=1}^{C-1} \exp(\eta_{cij})} = \frac{\exp(1.193)}{1 + \exp(1.193) + \exp(1.042)} = 0.462$$

$$P(Y = 2) = \frac{\exp(\eta_{2ij})}{1 + \sum_{c=1}^{C-1} \exp(\eta_{cij})} = \frac{\exp(1.042)}{1 + \exp(1.193) + \exp(1.042)} = 0.397$$

$$P(Y = 3) = \frac{1}{1 + \sum_{c=1}^{C-1} \exp(\eta_{cij})} = \frac{1}{1 + \exp(1.193) + \exp(1.042)} = 0.141$$

TABLE 6.6   Model 2.1: Variance Components

Residual effect	Coefficient	Std. error	z-Test	Significance	95% Confidence interval Lower	Upper
Multinomial category (1): 4 year						
Intercept	0.381	0.096	3.992	0.000	0.233	0.623
Multinomial category (2): 2 year/tech						
Intercept	0.349	0.092	3.802	0.000	0.208	0.584

Note: Covariance structure: variance components; subject specification: schcode.

For planning to attend a 4-year institution, therefore, the average school-level probability is 0.462 (which is similar to the student-level probability of 0.472 in Table 6.1). The probability of attending a 2-year college or technical school is 0.397 (2.835/7.132), and for the reference group it is 0.141 (1.0/7.132).

The variance components table (Table 6.6) suggests that both intercepts vary across schools ($\sigma^2_{u0j(1)} = 0.381$, SE = .096; $\sigma^2_{u0j(2)} = 0.349$, SE = .092). In both cases, the z tests suggest significant variability in postsecondary plans between schools.

From the Level-2 variance components, we can estimate the proportion of variance in postsecondary plans between schools. As we have noted previously, the logistic distribution has a variance of $\pi^2/3$, or about 3.29. The intraclass correlation (ρ) describing the proportion of variance that is between schools is then

$$\rho = \frac{\sigma^2_{Between}}{\sigma^2_{Between} + 3.29_{Within}}. \tag{6.13}$$

For plans to attend 4-year institutions versus no further education, the intraclass correlation (ICC) is 0.381/3.671 = 0.104, and for 2-year versus no further education, the ICC is 0.349/3.639 = 0.096.

Level-1 Model

We next propose a set of within-school predictors that may explain students' postsecondary plans. The two underlying outcomes are proposed to be explained by student SES background and gender. We will grand-mean center student GPA (mean = 2.471) for interpretability. The Level-1 model can be written as follows:

$$\eta_{1ij} = \log\left(\frac{\pi_{1ij}}{\pi_{Cij}}\right) = \beta_{0j(1)} + \beta_1 lowSES_{ij(1)} + \beta_2 male_{ij(1)} + \beta_3 gmgpa_{ij(1)}$$

$$\eta_{2ij} = \log\left(\frac{\pi_{2ij}}{\pi_{Cij}}\right) = \beta_{0j(2)} + \beta_1 lowSES_{ij(2)} + \beta_2 male_{ij(2)} + \beta_3 gmgpa_{ij(2)}. \tag{6.14}$$

As noted previously, η_{1ij} represents the log odds of planning to attend a 4-year college versus having no postsecondary plans, and η_{2ij} represents the log odds of attending a vocational/technical or 2-year college versus no postsecondary plans. We will treat the predictors ($\beta_1 - \beta_3$) as fixed at Level 2. The single-equation models will then be as follows:

$$\eta_{1ij} = \gamma_{00(1)} + \gamma_{10}lowSES_{ij(1)} + \gamma_{20}male_{ij(1)} + \gamma_{30}gmgpa_{ij(1)} + u_{0\,j(1)}$$

(6.15)

$$\eta_{2ij} = \gamma_{00(2)} + \gamma_{10}lowSES_{ij(2)} + \gamma_{20}male_{ij(2)} + \gamma_{30}gmgpa_{ij(2)} + u_{0\,j(2)}.$$

Defining Model 2.2 With IBM SPSS Menu Commands

Note: Settings default to those used for Model 2.1.

1. Go to the tool-bar and select ANALYZE, MIXED MODELS, GENERALIZED LINEAR. This command allows access to the *Generalized Linear Mixed Models* (GENLINMIXED) main dialog box.

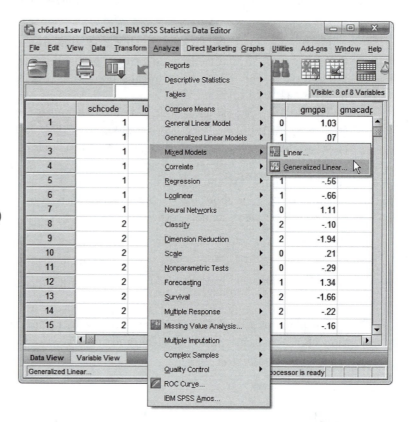

2a. The *Generalized Linear Mixed Models* dialog box displays the *Fields & Effects* command tab and the *Random Effects* screen, which was the last setting used in the prior model.

Note: The *Random Effects* item is preselected (represented in the illustration by the dashed-line box but would appear as grayed-out on the computer screen) as it was the last setting used from the prior model.

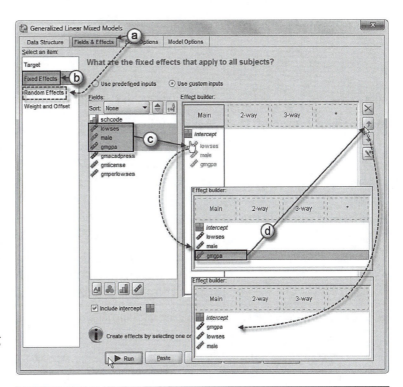

b. Click to select the *Fixed Effects* item to display the main screen, where we may add variables to the model.

$$\eta_{1ij} = \log\left(\frac{\pi_{1ij}}{\pi_{Cij}}\right) = \beta_{0j(1)} + \beta_1 lowSES_{ij(1)} + \beta_2 male_{ij(1)} + \beta_3 gmgpa_{ij(1)} \quad \text{(Eq. 6.14)}$$

$$\eta_{2ij} = \log\left(\frac{\pi_{2ij}}{\pi_{Cij}}\right) = \beta_{0j(2)} + \beta_1 lowSES_{ij(2)} + \beta_2 male_{ij(2)} + \beta_3 gmgpa_{ij(2)}$$

c. We will add three predictors to the model (*lowses, male, gmgpa*). Click to select the variables and then drag them into the *Main* (effect) column of the *Effect builder*.

d. To facilitate reading of the output tables, we will rearrange the variable order. Click to select *gmgpa* and then click the up-arrow button on the right to place *gmgpa* above *male*. The variable order is now *gmgpa, lowses, male*.

Note: The sequence may also be controlled by entering the variables into the *Effect builder* in the preferred order beginning with *gmgpa*, followed by the remainder (*lowses, male*).

Click the RUN button to generate the model results.

Interpreting the Output of Model 2.2

The results in Table 6.7 suggest that the importance of the predictors differs somewhat in explaining postsecondary plans across the categories. Again, because of how we have coded the predictors, the meaning of the intercept for Category 1 is now the log of the odds of planning to attend a 4-year institution versus planning no further schooling, where the predictors are equal to 0 (i.e., average/high SES = 0, female = 0, and grand-mean centered GPA = 0). Similarly to the single-level model, students planning on attending 4-year institutions versus no further schooling are less likely to have low SES standing (log odds = −0.815, $p < .001$) and to have higher GPAs (log odds = 2.135, $p < .01$). Students planning to attend vocational/technical or community institutions versus no further schooling are more likely to be male (log odds = 0.165, $p < .05$) and to have higher GPAs (log odds = 1.310, $p < .001$).

TABLE 6.7   Model 2.2: Fixed Effects

Model term	Coefficient	Std. error	t	Sig.	Exp (coefficient)	95% Confidence interval for exp(coefficient) Lower	Upper
Multinomial category (1): 4 year							
Intercept	2.112	0.188	11.326	0.000	8.269	5.720	11.954
gmgpa	2.135	0.077	27.763	0.000	8.458	7.275	9.834
lowses	−0.815	0.118	−6.906	0.000	0.443	0.351	0.558
male	−0.028	0.069	−0.401	0.689	0.973	0.849	1.114
Multinomial category (2): tech/voc/2 year							
Intercept	1.725	0.171	10.113	0.000	5.612	4.017	7.840
gmgpa	1.310	0.080	16.274	0.000	3.705	3.164	4.338
lowses	−0.100	0.909	−1.117	0.264	0.904	0.758	1.079
male	0.165	0.069	2.384	0.017	1.179	1.030	1.350

Note: Probability distribution: multinomial; link function: generalized logit; target: plans; reference category: no school.

We can use the results in the table to predict the probability of being in each category, given particular combinations of the predictors (as suggested in Equation 6.5). For example, if we want to predict the probability of being in each category for a female (coded 0) of average or high SES (coded 0) with a GPA one standard deviation (SD) above the grand mean, we would have the following log odds:

$$\eta_{1ij} = 2.112 - 0.815(0) - 0.208(0) + 2.135(1) = 4.247$$

$$\eta_{2ij} = 1.725 - 0.100(0) + 0.165(0) + 1.310(1) = 3.035$$

We then need to exponentiate again. Table 6.8 shows the probabilities for each category. The interpretation of these probabilities is that individuals who are female with average or high SES and a GPA of one standard deviation above the mean have a probability of 0.770 of being in Category 1, 0.229 of being in Category 2, and 0.011 of being in Category 3. Azen and Walker (2011) provide further discussion of predicting the probabilities associated with nominal outcome categories, given particular combinations of predictors.

We summarize the variance components in Table 6.9. Note that they are a bit different from the unconditional model (Table 6.7). The first one (Category 1) is a bit larger. As Hox (2010) notes, when variables are added to a model within continuous outcomes, we usually expect the variance components to be smaller relative to the intercept-only model. Each time Level-1 variables are added when the outcome is categorical, however, the underlying latent variable is rescaled (Hox, 2010). As a result, in addition to any actual changes due to the variables added

TABLE 6.8   Probabilities for Females of Average/High SES (Coded 1) and GPA 1 SD Above the Grand Mean

	Log odds	$e^{\eta_{ij}}$	Probability = e^{β}/sum
$P(Y = 1)$	4.247	69.895	0.770
$P(Y = 2)$	3.035	20.801	0.229
$P(Y = 3)$	0.000	1.000	0.011
		Sum 90.96	

TABLE 6.9   Model 2.2: Variance Components

Residual effect	Coefficient	Std. error	z-Test	Sig.	95% Confidence interval	
					Lower	Upper
Multinomial category (1): 4 year						
Intercept	0.415	0.107	3.882	0.000	0.251	0.688
Multinomial category (2): 2 year/tech						
Intercept	0.321	0.085	3.767	0.000	0.191	0.540

Note: Covariance structure: variance components; subject specification: schcode.

to the model, the values of the regression coefficients and second-level variances will also be rescaled. Therefore, we should not overly concern ourselves with how regression coefficients and variance components change between models.

This phenomenon is also present in single-level logit and probit analyses (Hox, 2010; Long, 1997; Tabachnick & Fidell, 2007). So-called pseudo r-square measures attempt to provide an estimate of explained variance by taking into consideration changes in the log likelihood [often expressed as the deviance or $-2*\log$ likelihood]. Pseudo r-square provides an indication of how much of the deviance is explained between two models being compared within the same data set (Hox, 2010; Tabachnick & Fidell, 2007). They are typically provided for single-level analyses. Several pseudo r-square measures are provided in IBM SPSS for the single-level logistic regression routine (but currently are not available in GENLIN MIXED). Others have attempted to examine the difference between predicting a continuous outcome using OLS regression and comparing the difference to a dichotomous version using pseudo r-square measures (Hox, 2010; McKelvey & Zavoina, 1975). Hox provides a detailed illustration of how the latent continuous outcome can be rescaled to the same underlying scale as the unconditional (no predictors) model and how to compare subsequent variance explained at Level 1 and Level 2.

Adding School-Level Predictors

We next add three school-level predictors that are proposed to explain differences in students' postsecondary educational plans. The variables represent student composition, academic press (i.e., perceived quality of educational opportunities and experiences), and proportion of the teachers who are fully certified to teach in their subject areas. The Level-1 model remains the same as Equation 6.14. At Level 2, we will have the following model for intercepts:

$$\beta_{0j(1)} = \gamma_{00(1)} + \gamma_{01(1)}gmacadpress_j + \gamma_{02(1)}gmlicense_j + \gamma_{03(1)}gmperlowses_j + u_{0j(1)}$$

$$\beta_{0j(2)} = \gamma_{00(2)} + \gamma_{01(2)}gmacadpress_j + \gamma_{02(2)}gmlicense_j + \gamma_{03(2)}gmperlowses_j + u_{0j(2)}.$$

(6.16)

The combined single-level equations will then be as follows:

$$\eta_{1ij} = \gamma_{00(1)} + +\gamma_{01(1)}gmacadpress_j + \gamma_{02(1)}gmlicense_j + \gamma_{03(1)}gmperlowses_j$$
$$+ \gamma_{10}lowSES_{ij(1)} + \gamma_{20}male_{ij(1)} + \gamma_{30}gmgpa_{ij(1)} + u_{0j(1)}$$

(6.17)

$$\eta_{2ij} = \gamma_{00(2)}\gamma_{01(2)}gmacadpress_j + \gamma_{02(2)}gmlicense_j + \gamma_{03(2)}gmperlowses_j$$
$$+ \gamma_{10}lowSES_{ij(2)} + \gamma_{20}male_{ij(2)} + \gamma_{30}gmgpa_{ij(2)} + u_{0j(2)}.$$

(6.18)

Defining Model 2.3 With IBM SPSS Menu Commands

Note: Settings default to those used for Model 2.2.

1. Go to the toolbar and select ANALYZE, MIXED MODELS, GENERALIZED LINEAR. This command allows access to the *Generalized Linear Mixed Models* (GENLINMIXED) main dialog box.

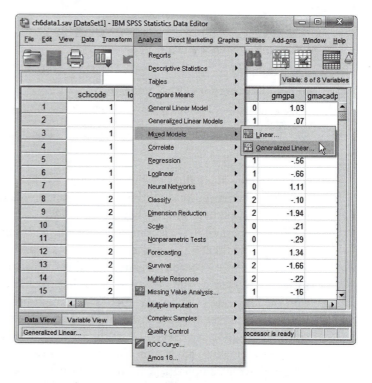

2a. The *Generalized Linear Mixed Models* dialog box displays the *Fields & Effects* command tab and the *Fixed Effects* screen, which was the last setting used in the prior model.

b. We will add three school-level predictor variables to the model. Click to select *gmacadpress, gmlicense,* and *gmperlowses* and then drag the variables into the *Main* (effect) column of the *Effect builder.*

c. To facilitate reading of the output tables, we will rearrange the variable order. Select the three school-level variables

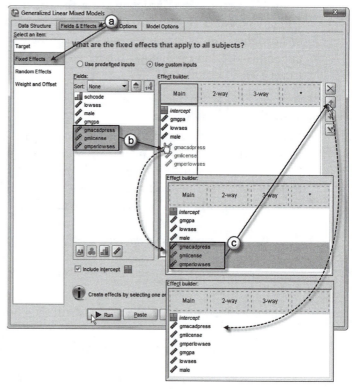

$$\beta_{0j(1)} = \gamma_{00(1)} + \gamma_{01(1)}gmacadpress_j + \gamma_{02(1)}gmlicense_j + \gamma_{03(1)}gmperlowses_j + u_{0j(1)}$$

$$\beta_{0j(2)} = \gamma_{00(2)} + \gamma_{01(2)}gmacadpress_j + \gamma_{02(2)}gmlicense_j + \gamma_{03(2)}gmperlowses_j + u_{0j(2)} \quad \text{(Eq. 6.16)}$$

(*gmacadpress, gmlicense, gmperlowses*) and then click the up-arrow button on the right to place them above *gmgpa*. The variable order is now *gmacadpress, gmlicense, gmperlowses, gmgpa, lowses, male*.

Click the RUN button to generate the model results.

Interpreting the Output of Model 2.3

We provide the fixed effects in Table 6.10. The individual-level predictors are similar to previous models, so we will not concern ourselves with discussing them further. At the school level, we can see that student composition (*gmperlowses*) is negatively related to plans to attend a 4-year institution versus no further schooling; that is, a 1-SD increase in students participating in the federal free/reduced lunch program decreases the log odds that a student will be proficient by 1.353 units ($p < .05$), holding the other variables in the model constant. The interpretation suggests that, for two students of similar standing on individual variables who attend schools that differ by one standard deviation in student composition, the predicted odds of planning to attend a 4-year institution versus no additional schooling would be decreased by a factor of 0.258 (or 74.2%), holding the other variables constant (including the school random parameter μ_{0j}). The other two school predictors (academic press and teacher certification) are unrelated to students' plans to attend 4-year institutions versus to pursue no further postsecondary education.

Holding the other variables constant at 0, we can calculate the expected probability of being in Category 1 and Category 2 versus the reference group for a 1-SD increase in student composition as follows:

$$\eta_{1ij} = 2.161 - 1.353(lowses\_m) = 0.808$$

$$\eta_{2ij} = 1.726 - 0.307(lowses\_m) = 1.419$$

We can again exponentiate the new log odds (using Equation 6.5). For planning to attend a 4-year institution versus no further schooling, we have $e^{0.808} = 2.243$. This suggests that the odds of planning to attend a 4-year institution versus planning no further schooling are about 2.24:1. That can be compared with about 8.7:1 for students at the grand mean of school SES. For 2-year/technical school versus no school, we have $e^{1.419} = 4.133$. So the odds of planning to attend a 2-year institution versus no additional schooling are about 4.1:1 (compared to 5.6:1 at the grand mean for student SES). Using Equation 6.11 and Equation 6.12, if we sum the probabilities ($2.243 + 4.133 + 1 = 7.376$), we obtain the expected probabilities for each category for individuals in schools 1-SD above the mean in student composition (all other variables held constant) as 0.304 (2.243/7.376) for Category 1, 0.560 (4.133/7.376) for Category 2, and 0.136 (1/7.376) for Category 3 (not tabled).

Turning to Category 2 (planning to attend a 2-year institution versus no further schooling), we can observe that student composition is unrelated to plans to attend a 2-year/technical school versus to pursue no further schooling (log odds = -0.307, $p > .05$). Similarly, teacher qualifications do not influence students' further postsecondary educational plans ($p > .05$). There was, however, cautious support for the view that school academic press may be a promising indicator of postsecondary plans (log odds = 0.239, $p < .10$), given the relatively small number of high schools in the sample ($N = 43$ schools). More specifically, students in high schools with higher academic press are more likely to report that they plan to attend a 2-year or technical school versus not to obtain any further education after high school.

TABLE 6.10   Model 2.3: Fixed Effects

Model term	Coefficient	Std. error	t	Sig.	Exp (coefficient)	95% Confidence interval for exp(coefficient)	
						Lower	Upper
Multinomial category (1): 4 year							
Intercept	2.161	0.169	12.806	0.000	8.681	6.236	12.085
gmacadpress	0.099	0.147	0.675	0.500	1.105	0.828	1.474
gmlicense	2.893	1.910	1.514	0.130	18.051	0.427	763.563
gmperlowses	−1.353	0.618	−2.190	0.029	0.258	0.077	0.868
gmgpa	2.135	0.078	27.396	0.000	8.460	7.262	9.857
lowses	−0.798	0.117	−6.844	0.000	0.450	0.358	0.566
male	−0.029	0.069	−0.417	0.676	0.971	0.848	1.113
Multinomial category (2): 2 year/tech							
Intercept	1.726	0.155	11.157	0.000	5.618	4.148	7.608
gmacadpress	0.239	0.142	1.684	0.092	1.270	0.962	1.678
gmlicense	−0.092	1.837	−0.050	0.960	0.912	0.025	33.413
gmperlowses	−0.307	0.448	−0.685	0.493	0.736	0.306	1.770
gmgpa	1.308	0.081	16.185	0.000	3.700	3.158	4.335
lowses	−0.091	0.090	−1.018	0.309	0.913	0.765	1.088
male	0.164	0.069	2.370	0.018	1.178	1.029	1.349

Note: Probability distribution: multinomial; link function: generalized logit; target: plans; reference category: no school.

The variance components are presented in Table 6.11. In each model, there is still significant variance to be explained between schools.

Investigating a Random Slope

We next investigate whether there might be a random slope that varies across the school level. We will assume that perhaps the relationship between student GPA and postsecondary plans varies across schools. We will simply include a random slope parameter in this model run. At Level 2, we will have the following:

$$\beta_{1j(1)} = \gamma_{10(1)} + u_{1j(1)}, \tag{6.19}$$

$$\beta_{1j(2)} = \gamma_{10(2)} + u_{1j(2)}.$$

The combined models, then, just add the random GPA slope to each:

TABLE 6.11   Model 2.3: Variance Components

Residual effect	Coefficient	Std. error	z-Test	Sig.	95% Confidence interval	
					Lower	Upper
Multinomial category (1): 4 year						
Intercept	0.334	0.093	3.599	0.000	0.194	0.576
Multinomial category (2): 2 year/tech						
Intercept	0.300	0.083	3.623	0.000	0.175	0.515

Note: Covariance structure: variance components; subject specification: schcode.

$$\eta_{1ij} = \gamma_{00(1)} + +\gamma_{01(1)}gmacadpress_j + \gamma_{02(1)}gmlicense_j + \gamma_{03(1)}gmperlowses_j$$
$$+ \gamma_{10}lowSES_{ij(1)} + \gamma_{20}male_{ij(1)} + \gamma_{30}gmgpa_{ij(1)} + u_{1j(1)}gmgpa_{ij(1)} + u_{0j(1)} \qquad (6.20)$$

$$\eta_{2ij} = \gamma_{00(2)}\gamma_{01(2)}gmacadpress_j + \gamma_{02(2)}gmlicense_j + \gamma_{03(2)}gmperlowses_j$$
$$+ \gamma_{10}lowSES_{ij(2)} + \gamma_{20}male_{ij(2)} + \gamma_{30}gmgpa_{ij(2)} + u_{1j(2)}gmgpa_{ij(2)} + u_{0j(2)}. \qquad (6.21)$$

Defining Model 2.4 With IBM SPSS Menu Commands

Note: Settings default to those used for Model 2.3.

1. Go to the tool-bar and select ANALYZE, MIXED MODELS, GENERALIZED LINEAR. This command allows access to the *Generalized Linear Mixed Models* (GENLINMIXED) main dialog box.

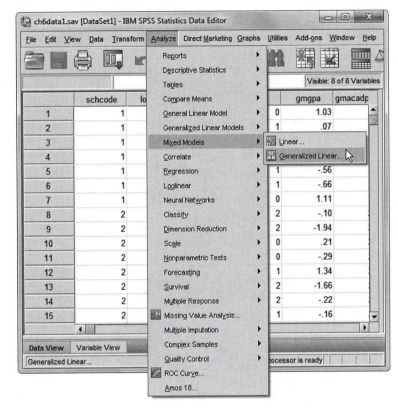

2a. The *Generalized Linear Mixed Models* dialog box displays the *Fields & Effects* command tab and the *Fixed Effects* screen, which was the last setting used in the prior model.

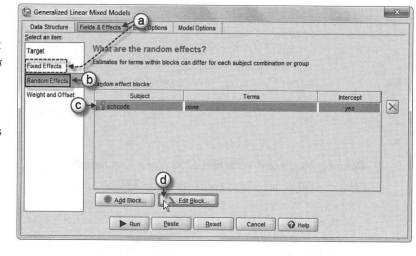

Note: The *Fixed Effects* item is preselected (represented in the illustration by the dashed-line box but would appear as grayed-out on the computer screen) as it was the last setting used from the prior model.

b. We will add a random slope to the model to examine whether the relationship between student grade point average and postsecondary plans varies across schools. To do so we will add a Level-2 randomly varying effect to the model. First, click to select the *Random Effects* item to display the main screen.

c. Now click on *schcode* to select and activate the *Random Effect Block* to enable further editing changes.

d. Click the EDIT BLOCK button.

e. We will add a random effect (*gmgpa*) to the model. Click to select *gmgpa* and then drag the variable to the *Main* section of the *Effect builder* box.

f. Note that we continue using the default settings for *Include intercept*, *schcode*, and *variance components*. Click the OK button to close the *Random Effect Block* dialog box and return to the *Random Effects* main screen.

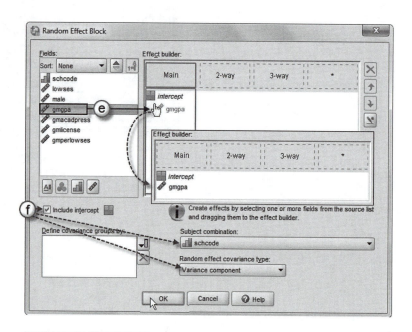

$$\eta_{1ij} = \gamma_{00(1)}\gamma_{01(1)}gmacadpress_j + \gamma_{02(1)}gmlicense_j + \gamma_{03(1)}gmperlowses_j +$$
$$\gamma_{10}lowSES_{ij(1)} + \gamma_{20}male_{ij(1)} + \gamma_{30}gmgpa_{ij(1)} + u_{1j(1)}gmgpa_{ij(1)} + u_{0j(1)} \quad \text{(Eq. 6.20)}$$

$$\eta_{2ij} = \gamma_{00(2)}\gamma_{01(2)}gmacadpress_j + \gamma_{02(2)}gmlicense_j + \gamma_{03(2)}gmperlowses_j +$$
$$\gamma_{10}lowSES_{ij(2)} + \gamma_{20}male_{ij(2)} + \gamma_{30}gmgpa_{ij(2)} + u_{1j(2)}gmgpa_{ij(2)} + u_{0j(2)^*} \quad \text{(Eq. 6.21)}$$

g. In the main screen notice that the *Random Effect Block* displays a summary of the block with *gmgpa* added to the model.

Click the RUN button to generate the model results.

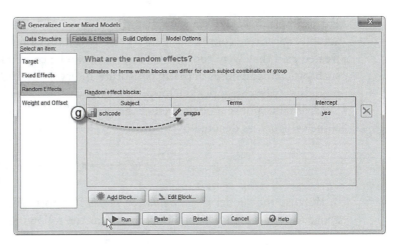

Interpreting the Output of Model 2.4 Model Results

The fixed-effect parameters are very similar, so we do not report them here. Instead, we will concentrate on examining the variance components for the random slope at Level 2. In Table 6.12, we summarize the variance components for 4-year institutions and 2-year/technical institutions. Once again, the intercepts vary across schools; however, for the Category 1 model (4-year institutions), there is not sufficient variance in GPA–plans slopes to build a model ($\sigma^2_{Slope} = .006$, $p > .10$). For 2-year institutions, we might conclude that there is some support for the idea that GPA–postsecondary plans slopes vary across schools, remembering that the significance levels for variance parameters of categorical outcomes are often too conservative (e.g., especially in small sample sizes) (Agresti, 2002; Hox, 2010).

For this demonstration, we would likely conclude that student composition appears to affect students' postsecondary plans after completing high school. In addition, we found some support for the academic process variable in differentiating plans of attending a 2-year institution versus obtaining no further postsecondary education. Regarding possible moderating effects of school features on within-school slopes such as GPA, it may be that the sample size is too small to detect this type of between-school variation.

Developing a Model With an Ordinal Outcome

Ordinal outcomes accommodate an ordered set of categories ($c = 1, 2, \ldots, C$) defining the dependent variable. The model is similar to the multinomial outcome except that, because the outcome categories are assumed to be ordered, the relationships can be captured in one set of estimates.

TABLE 6.12 Model 2.4: Variance Components

Residual effect	Coefficient	Std. error	z-Test	Sig.	95% Confidence interval Lower	Upper
Multinomial category (1): 4 year						
Intercept	0.341	0.095	3.585	0.000	0.198	0.590
gmgpa	0.006	0.013	0.439	0.661	0.000	0.510
Multinomial category (2): 2 year/tech						
Intercept	0.297	0.082	3.606	0.000	0.172	0.511
gmgpa	0.024	0.016	1.575	0.115	0.007	0.085

Note: Covariance structure: variance components; subject specification: schcode.

For example, a Likert scale implies the concept of "greater than" or "less than" but does not actually constitute continuous data. As we noted previously, ordinal response patterns are often treated as if they were continuous, but there can be consequences resulting from assuming that such ordered responses actually constitute interval data (Olsson, 1979). The predictors in the models may be categorical factors or continuous covariates.

Ordinal response models predict the probability of a response being at or below the cth outcome category, as denoted by

$$P(Y \leq c) = \pi_1 + \pi_2 + \cdots + \pi_c. \tag{6.22}$$

Because the model is a *cumulative probability* formulation, the complement is being at or above the cth category, which is expressed as the following:

$$P(Y > c) = 1 - P(Y \leq c). \tag{6.23}$$

In contrast to considering the probability of an individual event, such as in the dichotomous case $[\pi/(1 - \pi)]$, in the cumulative (or proportional) odds model, one considers the probability of the event of interest and all the events that *precede* it. As a result, we need to include only the first $C - 1$ outcome categories ($Y_o, \ldots, Y_{c-1}$) because the cumulative probability must always be 1 for the set of all possible outcomes (Azen & Walker, 2011).

For a response variable with three categories (1 = left school early, 2 = still in school, 3 = graduated) for individual i, the cumulative probabilities (which sum to 1) are then

$$P(Y_{1i} = 1) = P(c_i = 1) = \pi_{1i} \tag{6.24}$$

$$P(Y_{2i} = 1) = P(c_i = 1) + P(c_i = 2) = \pi_{2i}$$

$$P(Y_{3i} = 1) = P(c_i = 1) + P(c_i = 2) + P(c_i = 3) = 1.$$

The last category does not have a probability associated with it because the sum of the probabilities up to and including the last ordered category is 1.

Most often, ordinal outcomes specify a multinomial distribution and a cumulative logit or cumulative probit link function (see Chapter 3 for other link functions that can be used with the multinomial distribution and ordinal outcome). As with the other categorical models we have developed, what is predicted is not the observed values of the dependent variable but rather some transformation of it; that is, the ordered categories comprising the outcome are considered as comprising an underlying continuous predictor η_{ic}. The odds of the cumulative probability are defined as being at or below the cth outcome category relative to being above the cth category, and the log odds for that ratio are as follows:

$$\eta_{ic} = \log\left(\frac{P(Y \leq c)}{P(Y > c)}\right) = \log\left(\frac{\pi_c}{1 - \pi_c}\right), \tag{6.25}$$

where $c = 1, \ldots, C - 1$.

Similarly to the dichotomous case, the cumulative logit scale has a variance of $\pi^2/3$ (and scale factor of 1.0) and, once again, there is no separate residual error term. The cumulative probit function is defined as

$$\eta_{ic} = \Phi^{-1}(\pi_c), \tag{6.26}$$

where Φ^{-1} is used to designate the inverse of the normal cumulative density function and $c = 1,\ldots, C - 1$.

The probit model assumes that a continuous latent variable η_{ic} underlies the observed categorical outcomes. The model results in a standardized coefficient, which represents the change in the latent variable (rather than the observed outcome Y). Because the coefficients are standardized, the cumulative probit scale has a variance set of 1.0, so there is again no separate residual error term included in the model (see Chapter 3 for specification of the ordinal probit model).

It is important to keep in mind that, in cumulative odds formulations, the probability of an event occurring is redefined in terms of cumulative probabilities; that is, ordinal regression models the odds of *cumulative counts*, rather than the odds of individual levels of the dependent variable. The key feature of the model is that it assumes a common slope between each set of odds modeled; that is, the effects of the predictors are specified to be constant across the categories of the outcomes (this is also referred to as a parallel regression model). Thresholds (θ_c), or cut points, at particular values of the latent continuous variable η_{ic} determine which categorical response is observed; one less threshold than the number of ordered categories ($C-1$) is required to specify the estimated probabilities (Hox, 2010). Therefore, only the thresholds are different in modeling the cumulative logits (referred to as the proportional odds assumption).

The event of interest is observing a particular score or *less*, which may be summarized as follows for an ordinal response variable with three ordered categories:

$$Y_i = \begin{cases} 0,\ldots\ldots if\, \eta_i \leq \theta_1 \\ 1,\ldots\ldots if\, \theta_1 < \eta_i \leq \theta_2 \\ 2,\ldots\ldots if\, \theta_2 < \eta_i \end{cases}$$

This suggests that if the value of the continuous underlying predictor (η_i) is less than the first threshold, we observe the lowest category of the outcome. If η_i is greater than the first threshold, but less than or equal to the second threshold, we observe the second category. Finally, if η_i is larger than the second threshold, we observe the third category.

Because the proportional odds model with cumulative logit link function focuses on the likelihood of a variable falling into a particular category c or *lower* [$P(Y \leq c)$] versus falling above it [$P(Y > c)$], the model is often presented in the literature with the following structure:

$$\eta_{ic} = \log\left(\frac{\pi_{ic}}{1 - \pi_{ic}}\right) = \theta_c - \beta_q X_q, \qquad (6.27)$$

where θ_c are increasing model thresholds dividing the latent continuous η_{ic} of expected probabilities of Y_i.

Each logit has its own threshold term, but the slope coefficients (β_q) are the same across categories. Therefore, the coefficients do not have c subscripts. As we have noted, thresholds contain information about cut points between observed categories of the ordinal outcome, which are estimated for all but the highest level ($C-1$), similarly to the intercept in ordinary regression. Thresholds are the negative of intercepts, however. The thresholds are useful in determining predicted probabilities (as they are fixed cut points), but they are not of substantive interest in interpreting the study's results because they are not influenced by the levels of the X predictors for individual cases in the sample.

The manner in which the outcome categories are coded determines the direction of the odds ratios as well as the signs of the β coefficients (Tabachnick & Fidell, 2007). We draw

TABLE 6.13 Parameter Estimates

Parameter		β	Std. error	95% Wald confidence interval		Hypothesis test		
				Lower	Upper	Wald χ^2	df	Sig.
Threshold	[persist3 = 0]	–2.256	0.0428	–2.340	–2.172	2782.516	1	0.000
	[persist3 = 1]	–1.883	0.0393	–1.960	–1.806	2293.576	1	0.000
zses		–0.976	0.0375	–1.050	–0.903	679.1610	1	0.000
(Scale)		1[a]						

Note: Dependent variable: persist3; model: (threshold), zses.
[a] Fixed at the displayed value.

attention to the fact that in the proportional odds models, the β in Equation 6.27 represents the increase in log odds of falling into category *c* or lower (against being in the higher category) associated with a one-unit increase in *X*, holding the other *X* variables constant. Therefore, as Equation 6.27 indicates, a positive slope indicates a tendency for the response level to *decrease* as the level of *X decreases*. A negative slope indicates a tendency for the response level to *increase* as the level of *X decreases*. Hence, the regression coefficients in the cumulative probability model will have a sign that is the opposite of what we might expect in the corresponding multiple regression model (Hox, 2010). This can be confusing to interpret at first.

An example may help to illustrate this point. IBM SPSS assumes that the response variable is ordered from the lowest to the highest category. This type of coding is referred to as *ascending*, with the first category corresponding to the lowest value. If *persist* (coded 2) is the reference category, the focus is on the probability of being in category *c* or lower (i.e., still in school but behind = 1; dropping out = 0) against Category 2. In Table 6.13, we observe a coefficient of –0.976 for *zses*.

So, if for η_{1c}, we have the following expression from Equation 6.27:

$$\eta_{1c} = -2.256 - [-0.976(zses)],$$

we can see that when student socioeconomic status (*zses*) is high (e.g., 2.5), the estimate for η_{ic} will be 0.184 [–2.256 – (–0.976*2.5) = –2.256 + 2.44 = 0.184].

This coefficient is considerably above the first (–2.256) and second (–1.883) thresholds. A negative coefficient suggests that as the values of the predictor decrease, the likelihood of being in a lower category increases (because η_{ic} gets smaller); hence, there will be larger cumulative probabilities for lower categories. More specifically, in Figure 6.1, we see that as *zses* decreases (e.g., falls from 2.5 to 0), the value of η_{ic} will get smaller (decreasing from 0.184 to –2.256), so the probability of being in category *c* or lower *increases*. That is, as student SES decreases, the probability of being in a category *c* or less [$P(c \leq 2)$] increases; therefore, we conclude that students will be more likely to fall below Category 2.

If the ordinal outcome is ordered from highest to lowest, one can reverse the response variable's ordering (referred to as *descending*) or may have to recode manually to achieve this result. The *descending* option tells the program to reverse the order of the outcome so that the highest category is considered the "first" category and lowest category becomes the "last" category, or reference group. We note that for factors (i.e., dichotomous or categorical predictors), it depends on how they are defined. In the IBM SPSS syntax language, if a dichotomous variable is defined as *nominal* and specified as a "factor" (using the *By* syntax command), the reference category will be the last category. Therefore, for a dichotomous predictor, the coefficient will indicate the log odds for individuals coded 0 on that variable.

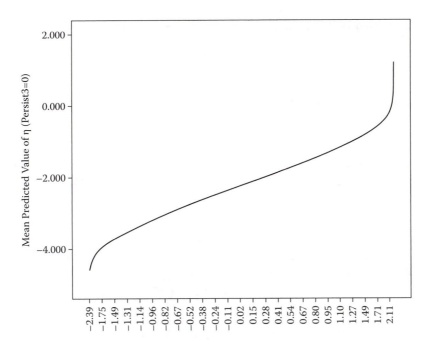

FIGURE 6.1   Predicted values of η related to student SES.

Suppose that we have the dichotomous variable *male* with females coded 0 and males coded 1. A positive coefficient will indicate that higher scores (i.e., persisting) are more likely for females versus males. We recommend coding all dichotomous variables as 0 and 1 to achieve this result.

Because of the confusion that can result from the *inverse* relationship between the ordered outcome categories and the direction of the predictors in the linear model to predict η_{ic}, some software programs (e.g., IBM SPSS, Mplus) simply multiply the linear predictors $\beta_q X_q$ by −1 to restore the direction of the regression coefficients such that positive coefficients increase the likelihood of being in the highest category and negative coefficients decrease it (Hox, 2010). After this change, the proportional odds model then becomes more familiar looking:

$$\eta_{ic} = \theta_c + B_1 X_{1i} + \cdots + \beta_q X_q, \tag{6.28}$$

where $c = 1, 2, \ldots, C - 1$ (i.e., suggesting one fewer thresholds than the number of ordered outcome categories).

In Table 6.14, we illustrate this difference in coding for three often used multilevel modeling programs (HLM, IBM SPSS, and Mplus) for the data set that we use in our next example. HLM uses the model underlying Equation 6.27; IBM SPSS and Mplus use Equation 6.28. Even though we can see that the direction of the coefficients is reversed in some output compared with other output, this does not change the substantive interpretation of the results. For example, if we add the second threshold reported in HLM (0.395) to the first threshold, the second threshold reported in IBM SPSS is obtained (−2.043). As shown in the table, the IBM SPSS and Mplus software retain the same thresholds but multiply the set of predictors by −1 to restore the signs to the direction they would have in a typical multiple regression model (Hox, 2010). It is important, however, to keep in mind what the particular software program is providing in terms of output so that the estimates are presented in the direction one intends to facilitate their interpretation. This emphasizes the point that results should make intuitive sense.

TABLE 6.14   Comparing HLM, IBM SPSS, Mplus: Parameter Estimates

Parameter	Coefficient	Std. error	Est./SE	Sig.	Exp(coefficient)	95% Confidence interval for exp(coefficient)	
						Lower	Upper
HLM: $t - (\beta X)$							
Intercept	−2.438	0.053	−45.965	0.000	0.087	0.080	0.096
threshold2	0.395	0.026	15.192	0.000	1.485	1.421	1.553
hiabsent	1.664	0.114	14.556	0.000	5.281	4.335	6.435
zses	−0.925	0.041	−22.763	0.000	0.397	0.367	0.428
IBM SPSS: $t - (-1)\beta X$							
Intercept	−2.438	0.046	−52.853	0.000	0.087	0.080	0.096
threshold2	−2.043	0.042	−48.506	0.000	0.130	0.119	0.141
hiabsent	−1.664	0.104	−16.020	0.000	0.189	0.154	0.232
zses	0.925	0.039	24.013	0.000	2.522	2.339	2.720
Mplus: $t - (-1)\beta X$							
Intercept	−2.438	0.047	−51.664	0.000	0.087	0.080	0.096
threshold2	−2.043	0.043	−47.919	0.000	0.130	0.119	0.141
hiabsent	−1.664	0.101	−16.428	0.000	0.189	0.154	0.232
zses	0.925	0.039	23.679	0.000	2.522	2.339	2.720

The Data

This example examines students' likelihood to persist through high school (Table 6.15). The data set (*ch6data2.sav*) consists of 6,983 students in 934 high schools (approximately 10 students per school).

The frequency distribution shown in Table 6.16 suggests that about 83% of the students graduated, a little over 4% were still in school, and nearly 13% dropped out before graduating. In our example, we will examine student persistence (Y_i), which is defined by three categories (coded 0 = dropped out, 1 = still in school but behind peers, 2 = persisted).

We can see that the cumulative percentages in the table (i.e., the proportions they represent) match up to the cumulative odds model; that is, the first percentage represents the likelihood of being in the first category (0.127) versus the higher categories; the second cumulative percentage represents the likelihood of being in the first or second category (0.169) versus the higher category. Being at or below the last category is redundant because all respondents are at or below it (1.00).

Developing a Single-Level Model

We will first build a single-level model. The first model, the unconditional model, simply consists of the two thresholds:

$$\eta_{ci} = \log\left(\frac{\pi_{ic}}{1 - \pi_{ic}}\right) = \theta_1 + \theta_2. \tag{6.29}$$

Keep in mind that, for Y, Category 0 is students who dropped out, and Category 1 is students who are still in school but behind their peers. The last category, students who persisted, is the reference group (coded 2). It is instructive to determine how the ordered categories are modeled in a cumulative odds model. From Equation 6.25, we can express the log odds as the ratio of the probability of

TABLE 6.15 Data Definition of *ch6data2.sav* (*N* = 6,983)

Variable	Level[a]	Description	Values	Measurement
schcode	School	School identifier (934 schools)	Integer	Ordinal
persist3	Individual	Dependent variable with three categories classifying students' completion of high school	0 = dropped out 1 = still enrolled 2 = graduated	Ordinal
hiabsent	Individual	Dichotomous predictor variable representing high school absenteeism	0 = average or low absenteeism 1 = high absenteeism	Scale
zses	School	Predicator continuous variable (converted into a factor score) representing student socioeconomic status	–3.55 to 2.39	Scale
zgpa	Individual	Predictor interval variable converted to a standardized factor score representing student grade point average	–3.54 to 1.86	Scale
zqual	School	Predictor interval variable converted to standardized factor score[b] representing school expectations and educational processes	–4.69 to 3.77	Scale
zabsent	School	Predictor interval variable converted to a standardized factor score[b] representing percentage of students with high absences	–0.57 to 9.20	Scale
zenroll	School	Predictor interval variable converted to a standardized factor score representing student enrollment, which indicates school size	–4.68 to 2.17	Scale
zcomp	School	Predictor interval variable converted to a standardized factor score[b] representing student socioeconomic status and language background	–0.35 to 2.89	Scale
persistnom	Individual	Three-category dependent variable representing nominal measurement of *persist3* (ordinal) classifying student completion of high school	0 = left school without graduating 1 = still enrolled 2 = graduated	Nominal
persistcat1	Individual	Recoded *persist3* (three categories) into a dichotomous variable representing students' completion of high school	0 = still enrolled/graduated 1 = dropped out	Nominal
persistcat1or2	Individual	Recoded *persist3* (three ategories) into a dichotomous variable representing students' completion of high school	0 = graduated 1 = still enrolled/dropped out	Nominal
enroll3	Individual	Three-category dependent variable representing ordinal measurement of school size based on the number of students	2 = schools with >1,100 students 1 = schools with 600–1,099 students 0 = schools with <600 students	Ordinal

[a] Individual = Level 1; school = Level 2.
[b] Standardized scores (mean = 0, SD = 1).

TABLE 6.16   Frequencies and Cumulative Outcome Category Percentages (*persist3*)

		Frequency	Percent	Valid percent	Cumulative percent
Valid	Dropped out	885	12.7	12.7	12.7
	Still enrolled	292	4.2	4.2	16.9
	Graduated	5806	83.1	83.1	100.0
	Total	6983.0	100.0	100.0	

being in the lowest category (dropped out = 0) versus the probability of being in a category greater than 0 $[\eta_{1i} = \log(\pi_{1i}/(1 - \pi_{1i}))]$. For the first threshold, from Table 6.17, we have the following:

$$\eta_{1i} = \log\left(\frac{\pi_{1i}}{1 - \pi_{1i}}\right) = \log(0.1455) = -1.930.$$

More specifically, if the proportion who drop out (π_{1i}) is equal to 0.127, then the probability $(1 - \pi_{1i})$ will be 0.873 (1 − 0.127 = 0.873) for the higher two categories. We can express this ratio as 0.127/0.873, which is equal to 0.1455. Taking the natural log, we arrive at the log odds of −1.930. This is the value of the first threshold (θ_1), or cut point, summarized in Table 6.17. Observed values of η_{ic} less than the first threshold will predict membership in Category 0.

The other threshold (θ_2) is the cut point greater than −1.596, which represents the probability of having an observed response of 0 or 1 versus the probability of having a score of 2. This will be the ratio of the sum of the first two probabilities divided by π_{3i} or $[\pi_{1i} + \pi_{2i}/(1 - (\pi_{1i} + \pi_{2i}))]$. From Table 6.18, we have the following:

$$\eta_{2i} = \log\left(\frac{\pi_{1i} + \pi_{2i}}{1 - (\pi_{1i} + \pi_{2i})}\right) = \log(0.203) = -1.595.$$

The sum of the first two probabilities (0.127 + 0.042) is 0.169. Subtracting this result from 1.0 yields 0.831 (π_{3i}). The natural log of this ratio (0.169/0.831 = 0.203) is approximately −1.595. Hence, if η_{ic} is larger than −1.930 but less than or equal to −1.595, we would predict a score of $Y = 1$. If η_{ic} is larger than the second threshold, we would predict a score of $Y = 2$.

From the estimated cumulative probabilities, we can then calculate the estimated probability for those individuals who have a score of 1 (still in school but behind peers) by subtraction. As Azen and Walker (2011) note, the definition of the cumulative probability model

TABLE 6.17   Parameter Estimates

Parameter		β	Std. error	95% Wald confidence interval		Hypothesis test			Exp(β)	95% Wald confidence interval for exp(β)	
				Lower	Upper	Wald χ²	df	Sig.		Lower	Upper
Threshold	[persist3 = 0]	−1.930	0.0360	−2.001	−1.860	2879.129	1	0.000	0.145	0.135	0.156
	[persist3 = 1]	−1.596	0.0320	−1.659	−1.533	2492.501	1	0.000	0.203	0.190	0.216
(Scale)		1[a]									

Note: Dependent variable: persist3; model: (null).
[a] Fixed at the displayed value.

TABLE 6.18 Model 3.1: Comparing Two Separate Dichotomous Models[a]

| Parameter | β | Std. error | 95% Profile likelihood confidence interval | | Hypothesis test | | | Exp(β) | 95% Profile likelihood confidence interval for exp(β) | |
			Lower	Upper	Wald χ^2	df	Sig.		Lower	Upper
Model: dropped out (persistcat1)										
(Intercept)	−2.483	0.050	−2.581	−2.385	2455.169	1	0.000	0.083	0.076	0.092
zses	−0.892	0.046	−0.982	−0.801	373.337	1	0.000	0.410	0.374	0.449
hiabsent	1.626	0.124	1.383	1.869	171.752	1	0.000	5.082	3.985	6.480
zabsent	0.035	0.040	−0.043	0.114	0.775	1	0.379	1.036	0.958	1.120
zqual	−0.183	0.042	−0.265	−0.100	18.723	1	0.000	0.833	0.767	0.905
zenroll	0.074	0.044	−0.011	0.159	2.899	1	0.089	1.077	0.989	1.173
zcomp	0.090	0.035	0.021	0.158	6.542	1	0.011	1.094	1.021	1.172
Model: enrolled/ graduated (persistcat1 or persistcat2)										
(Intercept)	−2.062	0.043	−2.147	−1.978	2276.899	1	0.000	0.127	0.117	0.138
zses	−0.869	0.042	−0.951	−0.787	431.740	1	0.000	0.419	0.386	0.455
hiabsent	1.546	0.119	1.313	1.779	168.887	1	0.000	4.691	3.716	5.923
zabsent	0.053	0.036	−0.017	0.123	2.181	1	0.140	1.054	0.983	1.130
zqual	−0.150	0.039	−0.227	−0.073	14.551	1	0.000	0.861	0.797	0.930
zenroll	0.119	0.039	0.042	0.196	9.165	1	0.002	1.126	1.043	1.217
zcomp	0.102	0.032	0.040	0.164	10.246	1	0.001	1.107	1.040	1.179

[a] Dependent variable: persistcat1 (model: dropped out); persistcat1 or persistcat2 (model: enrolled/graduated); (intercept), zses, hiabsent, zabsent, zqual, zenroll, zcomp.

(Equation 6.24) allows us to compute the probabilities for a specific category as the difference between two adjacent probabilities:

$$P(Y = c) = P(Y \le c) - P(Y < c). \tag{6.30}$$

Hence, we have the probability that Y is 0 or 1 (0.169) minus the probability that Y is 1 (0.127), which equals 0.042.

For three ordered outcome categories, the predicted cumulative log odds can also be converted back to predicted probabilities as follows:

$$P(y = 2 \mid x) = \frac{1}{1 + e^{-(\eta_{2i})}}$$

$$P(y = 1 \text{ or } 2 \mid x) = \frac{1}{1 + e^{-(\eta_{1i})}}, \tag{6.31}$$

where η_{2i} and η_{1i} represent the value of the cumulative logit equations, given the specific values for a set of predictors.

This provides a means of calculating the probability of observing the highest category versus the middle and low categories for a given value of X and for calculating the probability of observing the middle or higher category versus the lowest category for a given value of X. Because there are only thresholds in the current model, for the probability that Y falls into the lowest category (0), we have $1/(1 + e^{-(-1.93)})$ or 0.127 $[1/(1 + 6.89) = 0.127]$.

Alternatively, as we have noted, we can also use the inverse transformation $[\exp(\beta)/(1 + \exp(\beta))]$ to calculate the probability of being in each particular ordered category of the outcome. To illustrate, we can exponentiate the log odds for Category 0 (dropped out) ($e^{\eta_{1i}} = 0.145$). We next apply the inverse transformation (0.145/1.145), which equals 0.127. The odds ratio for Category 1 (still in school) is 0.203 ($e^{\eta_{2i}} = 0.203$). The odds ratio for the reference group is 1.0 (because the log odds = 0). We then obtain the following probabilities (where θ_c are the threshold odds ratios):

$P(Y=0) = \pi_1 = \theta_1$ 0.145/1.145 = 0.127, or 12.7% dropped out

$P(Y=1) = \pi_2 = \theta_2 - \theta_1$ 0.203/1.203 = 0.169 and 0.169 − 0.127 = 0.042, or 4.2% still in school

$P(Y=2) = \pi_3 = 1 - (\theta_2 - \theta_1)$ 1 − 0.169 = 0.831, or 83.1% persisted

These probabilities match the percentages given in Table 6.16.

Preliminary Analyses

We note that the cumulative odds assumption of equal slopes across the ordered categories of the outcome may not always hold empirically. This is why we suggest conducting a series of preliminary analyses (as we did for our first example) to determine whether a cumulative odds model seems appropriate in a given instance. Various alternative formulations are well known for single-level analyses (e.g., multinomial or modeling the thresholds separately), but they have not been widely implemented in multilevel software (e.g., see Azen & Walker, 2011; Hedeker & Mermelstein, 1998; Hox, 2010). Before we build our ordinal models, therefore, it is important to establish whether this outcome might be considered ordinal for purposes of our proposed analyses or whether it would be better to consider it as multinomial. This can be accomplished by first creating two dummy variables for the two lower categories of cumulative odds (i.e., students who left school early = 1, Else = 0) or who dropped out early or were still in school (coded 1, Else 0). The reference group is students who graduated.

One can then use the predictors to estimate a model for each of the $C-1$ dummy variable outcomes and note the pattern of significant predictors. In this case, we will use two within-school predictors (*zses, hiabsent*) and four between-school predictors (*zqual, zabsent, zenroll, zcomp*) that we are planning to use subsequently in building our ordinal model. In this preliminary investigation of whether the predictors will support an ordinal model, we next provide the menu commands for building the dichotomous models summarized in Table 6.18. For convenience, we will use the single-level GLM routine (GENLIN) for this purpose.

Defining Model 3.1 With IBM SPSS Menu Commands

Note: Refer to Chapter 3 for discussion concerning settings and modeling options available in the IBM SPSS GENLIN routine.

Launch the IBM SPSS application program and select the *ch6data2.sav* data file.

1. Go to the toolbar and select ANALYZE, GENERALIZED LINEAR MODELS, GENERALIZED LINEAR MODELS. This command allows access to the *Generalized Linear Models* (GENLIN) main dialog box.

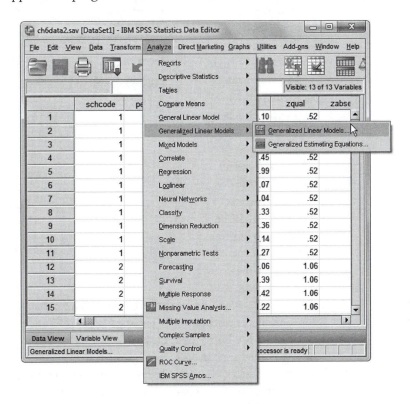

2a. The *Generalized Linear Models* display screen displays nine command tabs: *Type of Model, Response, Predictors, Model, Estimation, Statistics, EM Means, Save, Export.* The default command tab when creating a model for the first time is *Type of Model.* The command tab when creating a model for the first time is *Type of Model.* The command tab enables specifying

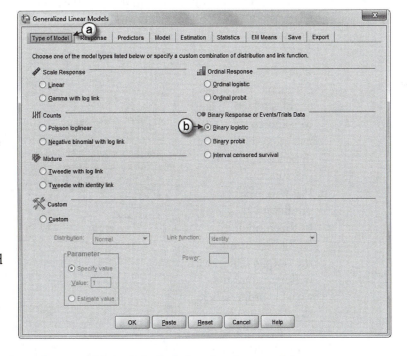

the distribution of the dependent variable and the link function for a model.

b. For this first model we will designate a binary logistic distribution and log link function. Click the radio button to select *Binary logistic*.

3a. Click the *Response* command tab, which enables specifying a dependent variable for the model.

b. For this model we will use students' completion of high school (*persistcat1*) as the dependent variable. (This is a dichotomous variable coded 0, 1.) Click to select *persistcat1* from the *Variables* list and then click the right-arrow button to move the variable into the *Dependent Variable* box.

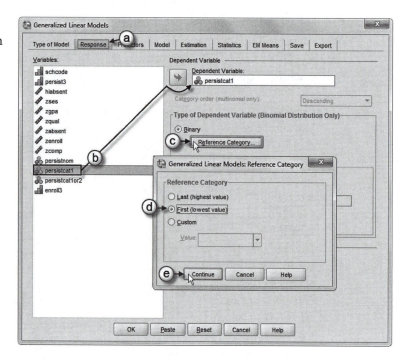

c. When the dependent variable has only two values, we can specify a reference category for parameter estimation. To specify a reference category, click the REFERENCE CATEGORY button, which will activate the *Generalized Linear Models Reference Category* dialog box to appear on screen.

d. We will change the default reference category by clicking to select *First (lowest value)*. Interpretation of the resulting parameter estimates will now be relative to the likelihood of the value "0" (not completing high school) category.

e. Click the CONTINUE button to close the *Generalized Linear Models Reference Category* dialog box.

4a. Click the *Predictors* command tab, which enables selecting factors and covariates to use in predicting the dependent variable (*persistcat1*).

b. We will use six predictor variables in this model. To facilitate reading of the output tables, we will rearrange the variable order. Click to select *hiabsent*, *zses*, *zqual*, *zabsent*, *zenroll*, and *zcomp* and then click the right-arrow button to move the variables into the *Covariates* box.

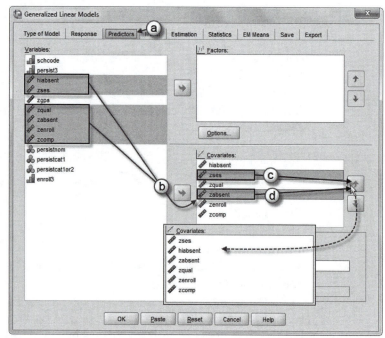

c. To facilitate reading of the output tables, we will rearrange the variable order. Click to select *zses* and then click the up-arrow button on the right to place *zses* above *hiabsent*.

d. Now click to select *zabsent* and then click the up-arrow button on the right to place *zabsent* above *zqual*. The variable order is now *zses, hiabsent, zabsent, zqual, zenroll, zcomp*.

Note: An alternative method for managing the variable order is to enter them in sequence into the *Covariates* box.

5a. Click the *Model* command tab, which enables specifying model effects based upon the covariate variables (*zses, hiabsent, zabsent, zqual, zenroll, zcomp*).

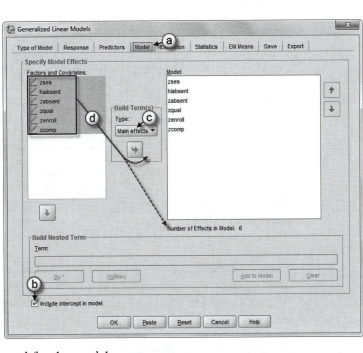

b. Note that the *Include intercept in model* option is preselected. This is because a Level-1 model is intercept only (default), so other model effects must be specified.

c. We will use the default *Main Effects*, which creates a main-effects term for each variable selected for the model.

d. Click to select the six variables (*zses, hiabsent, zabsent, zqual, zenroll, zcomp*) and then click the right-arrow button to move them into the *Model* box. Note that the six effects are now identified in the model (*Number of Effects in Model: 6*).

6a. Click the *Estimation* command tab, which enables specifying estimation methods and providing initial values for the parameter estimates.

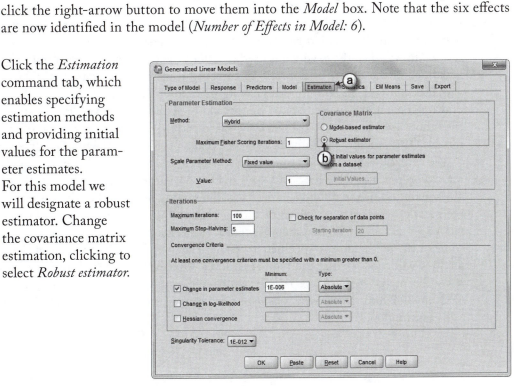

b. For this model we will designate a robust estimator. Change the covariance matrix estimation, clicking to select *Robust estimator*.

7a. Click the *Statistics* command tab, which enables specifying model effects (analysis type and confidence intervals) and print output.

b. We will retain the default *Model Effects* settings but add an additional print option to obtain exponential parameter estimates. Click to select *Include exponential parameter estimates.*

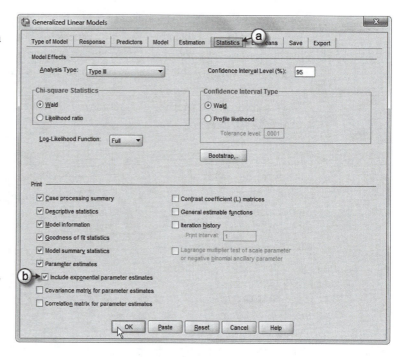

Click the OK button to generate the model results.

Note: To replicate Model B results shown later in Table 6.20, change the model's dependent variable to *persistcat1for2*. Perform Steps 1 and 3 to replace the dependent variable and then click the OK button to generate the model results

Interpreting the Output of Model 3.1

The results in Table 6.18 indicate similar directions for the individual- and school-level predictors in both models. For example, holding the other variables in the model constant, increasing student SES by one standard deviation decreases the log of the odds of dropping out or still being in school, but behind the peer cohort versus persisting, by 0.892 or 0.869 units, respectively. The Wald statistic, which is used to determine the significance of individual parameters, is defined as the square of the ratio of the coefficient to its standard error. Holding the other variables constant, in both models, having high absenteeism also increases the log of the odds of dropping out or being in school, but behind the cohort group versus persisting. We note that the school-level predictors also have the same pattern of effects on the log of the odds in both models. This provides preliminary evidence that an ordinal model might suffice for the analysis.

We can also examine the test of parallel slopes, which is available in the single-level ordinal regression routine (REGRESSION > ORDINAL). The results in Table 6.19, which include a chi-square test of the difference in –2 log likelihoods, suggest that the slopes of the predictors are parallel across the categories of the outcomes. The significance level of the chi-square test (above $p = .05$) indicates that an ordinal model is consistent with the data, keeping in mind that some of the predictors are specified at school level but that, for purposes of the initial test, the analysis was conducted at the individual level. Our experience with this test, however, is that it can be pretty stringent, especially with large sample sizes and several predictors (see also O'Connell, 2006). This is why we suggest using a variety of criteria to examine the proportional odds assumption.

TABLE 6.19   Test of Parallel Lines[a]

Model	−2 Log likelihood	χ^2	df	Sig.
Null hypothesis	6566.330			
General	6558.925	7.404	6	0.285

Note: The null hypothesis states that the location parameters (slope coefficients) are the same across response categories.

[a] Link function: logit.

TABLE 6.20   Comparing Variable Means Across Outcome Categories

persist3	hiabsent	zses	zqual	zabsent	zenroll	zcomp
Left school without graduating	0.25	−0.7782	−0.2961	0.4174	0.1588	0.2857
Still enrolled	0.12	−0.5360	−0.1696	0.2096	0.2140	0.2516
Graduated	0.04	0.1456	0.0537	−0.0742	−0.0350	−0.0562
Total	0.07	0.0000	0.0000	0.0000	0.0000	0.0000

Finally, it is also sometimes useful to examine differences in the means of the predictors across categories of the persist outcome. Table 6.20 suggests that the predictors either increase or decrease across the categories of the outcome. The one exception is *zenroll*, where the size of the school does not quite decrease systematically as the categories of the outcome increase from leaving school (0.159), still being in school (0.214), and graduating (−0.035). Given these preliminary results, we have established the case for treating the persist outcome as ordinal, as opposed to multinomial.

Adding Student Background Predictors

We can build a single-level model with an ordinal outcome in GENLIN. We will add two student predictors to our threshold-only model in Equation 6.29. These variables are student socioeconomic status (which is a standardized score with mean = 0, SD = 1) and high absenteeism (coded 1 vs. Else = 0). The model will be as follows:

$$\eta_{ic} = \log\left(\frac{\pi_{ic}}{1 - \pi_{ic}}\right) = \theta_c + \beta_1 zses + \beta_2 hiabsent. \tag{6.32}$$

Defining Model 3.2 With IBM SPSS Menu Commands

Note: Settings default to those used for Model 3.1.

1. Go to the tool-bar and select ANALYZE, GENERALIZED LINEAR MODELS, GENERALIZED LINEAR MODELS. This command allows access to the *Generalized Linear Models* (GENLIN) main dialog box.

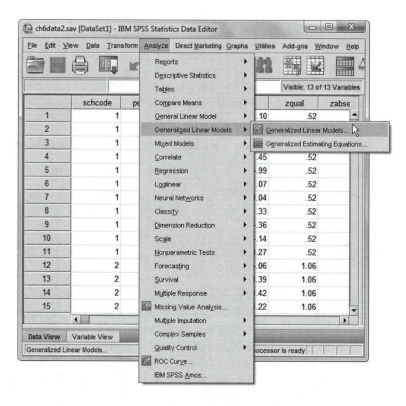

2a. The *Generalized Linear Mixed Models* dialog box displays the *Statistics* command tab, which was the last setting used in the prior model.

b. We will begin by changing the model settings. First, click the *Type of Model* command tab to access the dialog box.

c. The prior model (Model 3.1) used a binary logistic model for the dichotomous outcome variable. We now plan to use a three-category outcome variable (*persist3*) for the model, so click to select *Ordinal logistic*.

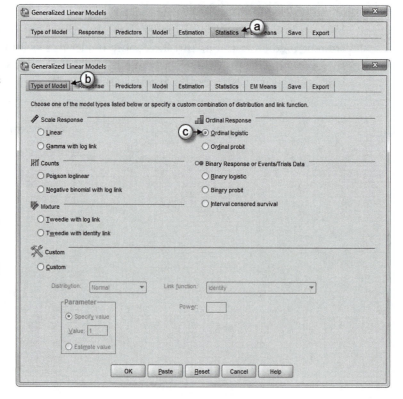

Note: The *Ordinal logistic* specifies a multinomial (ordinal) as the distribution and cumulative probit as the link function.

3a. Click the *Response* command tab, which enables specifying a dependent variable for the model.

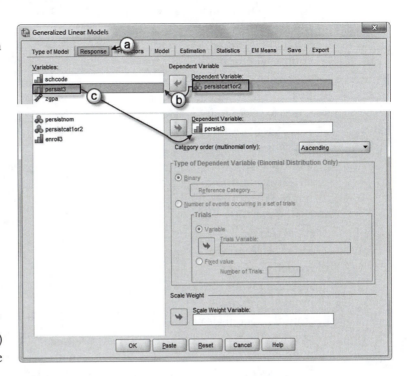

b. First, remove the variable (*persist1or2*) specified from the prior model as the dependent variable. Click to select the variable and then click the left-arrow button to remove it from the *Dependent Variable* box.

c. We will use students' completion of high school (*persist3*) as measured by three categories (coded 0, 1, 2) as the dependent variable. Click to select *persist3* from the *Variables* list and then click the right-arrow button to move the variable into the *Dependent Variable* box.

4a. We will change the model by removing four variables, leaving two predictors. Click the *Predictors* command tab to access the dialog box.

b. Click to select four variables (*zabsent, zqual, zenroll, zcomp*) and then click the left-arrow button to remove them from the model. This will leave two predictor variables in the model: *zses, hiabsent*. Removing the predictor variables from the model also removes them as main effects from

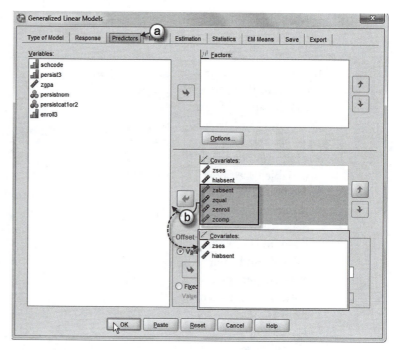

$$\eta_{ic} = \log\left(\frac{\pi_{ic}}{1 - \pi_{ic}}\right) = \theta_c + \beta_1 zses + \beta_2 hiabsent \quad \text{(Eq. 6.32)}$$

the *Model* dialog box, so no further changes are necessary.

Click the OK button to generate the model results.

Interpreting the Output of Model 3.2

The results are presented in Table 6.21. Keeping in mind that IBM SPSS reverses the regression coefficients to improve interpretability, logistic coefficients for ordinal outcomes can be interpreted in the usual manner, though as in logistic regression they give the effects of a unit increase of the predictor on the ordered log odds of the underlying dependent variable η_i (rather than on the dependent variable Y itself).

TABLE 6.21   Model 3.2: Parameter Estimates

Parameter		β	Std. error	95% Wald confidence interval Lower	Upper	Wald χ²	df	Sig.	Exp (β)	95% Wald confidence interval for exp(β) Lower	Upper
Threshold	[persist3 = 0]	–2.438	0.0461	–2.529	–2.348	2793.506	1	0.000	0.087	0.080	0.096
	[persist3 = 1]	–2.043	0.0421	–2.125	–1.960	2352.843	1	0.000	0.130	0.119	0.141
zses		0.925	0.0385	0.850	1.001	576.636	1	0.000	2.522	2.339	2.720
hiabsent		–1.664	0.1039	–1.868	–1.461	256.633	1	0.000	0.189	0.154	0.232
(Scale)		1ᵃ									

Note: Dependent variable: persist3; model: (threshold), zses, hiabsent.
ᵃ Fixed at the displayed value.

We can say that student SES is positively associated with the tendency to persist. An increase of one standard deviation in *zses* increases the expected log odds of persisting by 0.925 units as one moves to the next higher category of persistence (holding other variables constant). We note that this increase is the same regardless of the persistence category under scrutiny. It is less confusing to say that the positive value for *zses* suggests that higher levels of student SES increase the probability that the individual is in highest or second highest persistence categories (i.e., persisted or still in school but behind one's cohort) rather than left school early (Category 0).

For ordinal outcomes, the odds ratio for the logit model represents the odds of the higher category as compared to all lower categories combined. In other words, it is a cumulative odds ratio representing the increased predicted odds to be the next highest category relative to the lower category for each unit increase in the predictor. In terms of odds ratios, because *zses* is a standardized score, we would say that for a 1-SD increase in *zses*, the predicted odds of persisting versus the combined middle and low categories are increased by a factor of 2.52 ($e^{0.925}$), or are 2.52 times greater, holding any other variables in the model constant. Similarly, the predicted odds of being in the combined middle and high category versus low category (left school early) are also 2.52 times greater, holding the other variables constant.

As we have noted elsewhere (Heck, Thomas, & Tabata, 2010), standardizing or grand-mean centering (in the multilevel context) the predictor each produces a mean of 0, which can aid in providing a meaningful interpretation of the intercept. Standardizing also changes the standard deviation (making it 1.0); grand-mean centering retains the original standard deviation. For multilevel models, grand-mean centering the predictor is often preferable to standardizing it because the latter approach can affect the interpretation of the regression slopes (e.g., the metric of standardized and unstandardized estimates is changed) as well as the residual variances (Hox, 2010).

In contrast, the negative value for absenteeism (*hiabsent* = –1.664) shows that having high absences decreases the log of the odds of being in higher versus combined middle and lower categories. Note that we coded this such that the reference group is 0 (students with low absences). Therefore, the negative estimate is in terms of the category coded 1 (students with high absences). In terms of odds ratios, increasing absenteeism from 0 to 1 (high absenteeism) suggests that the odds of being in the high category versus combined middle and low categories are multiplied by 0.189 (or are only about one fifth as likely). Alternatively, we could also say the odds are reduced by 81.1% (i.e., we convert the proportions to percentages, 100% – 18.9% = 81.1%). As we have noted previously, it is sometimes convenient to interpret odds ratios less than 1.0 as the probability of the event coded 0. In this case, for a one-unit increase in absenteeism (from 0 to 1), the odds of being in the low category versus combined middle and high categories are 5.29 times greater (1/0.189).

Testing an Interaction

We may decide to examine whether there is an interaction between student SES and absenteeism. We will propose an interaction between *hiabsent* (a dichotomous variable) and *zses* (a continuous variable). The model will now look like the following:

$$\eta_{ic} = \log\left(\frac{\pi_{ic}}{1 - \pi_{ic}}\right) = \theta_c + \beta_1 zses + \beta_2 hiabsent + \beta_3 zses * hiabsent. \tag{6.33}$$

When we propose an interaction, we should take care to ensure that the interaction has the intended meaning. For interactions, the slope coefficient is the expected value of one of the variables when the other is zero (Hox, 2010). Sometimes there is no meaningful zero in the data. For example, if we are measuring family income, zero may not be a meaningful value (if we would not expect family income to be zero). Grand-mean centering family income on the sample average (e.g., $50,000) will create a new mean (0) that represents the average in the sample. For two continuous variables, grand-grand-mean centering will provide an interaction that is more easily interpreted. We can also standardize the variable; however, if that approach is taken, the interaction is most easily interpreted if it is not standardized (Hox, 2010).

In Equation 6.33, therefore, a *zses* value of 0 refers to the sample mean; for high absenteeism, the 0 refers to students who have average or low absences. If the interaction between a categorical and continuous variable is significant, it can be interpreted as the effect of the continuous variable on the outcome is moderated by (dependent on) the level of the factor. The interaction then will be the added effect of having high absenteeism (coded 1) versus low absenteeism on the relationship between student SES and likelihood to persist.

For a dichotomous predictor (coded 0, 1), the slopes for variables interacting with the dichotomous variable will refer to the reference category (which we will code as 0). The interaction indicates the joint effects that the two predictors exert on the outcome in addition to their individual main effects (Azen & Walker, 2011). In this case, the intercept would be interpreted as the expected log of the odds for the category of each variable that is coded 0. If we then specified an interaction between gender and high absenteeism, a significant interaction would indicate that the effect of gender on the persistence outcome differs by levels of absenteeism.

Defining Model 3.3 With IBM SPSS Menu Commands

Note: Settings default to those used for Model 3.2.

1. Go to the toolbar and select ANALYZE, GENERALIZED LINEAR MODELS, GENERALIZED LINEAR MODELS. This command allows access to the *Generalized Linear Models* (GENLIN) main dialog box.

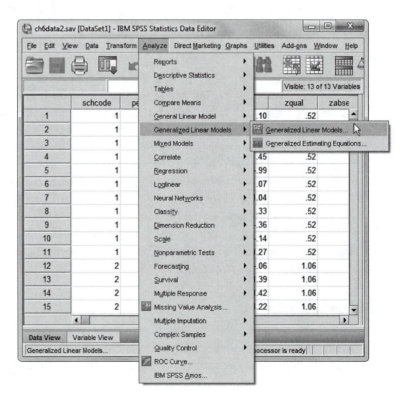

2a. The *Generalized Linear Mixed Models* dialog box displays the *Predictors* command tab, which was the last setting used in the prior model.

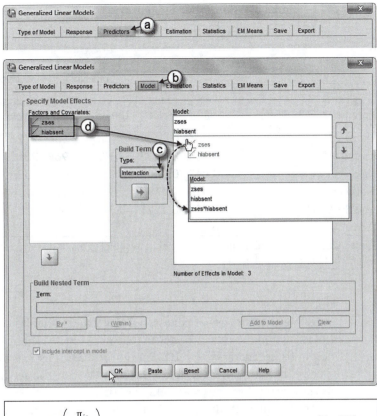

$$\eta_{ic} = \log\left(\frac{\pi_{ic}}{1 - \pi_{ic}}\right) = \theta_c + \beta_1 zses + \beta_2 hiabsent + \beta_3 zses * hiabsent \qquad \text{(Eq. 6.33)}$$

Adding Interactions to Model 3.3

    b. We will add an interaction to the model. First, click the *Model* command tab to access the dialog box.

    c. Next, select the *Build Term* "type" command. Click the *Type* pull-down menu to select *Interaction*.

    d. Click to select the two variables for the interaction (*zses, hiabsent*) and then click the right-arrow button to move them into the *Model* box. This creates the interaction *zses*hiabsent*. Note that the three effects are now identified in the model (*Number of Effects in Model: 3*).

    Click the OK button to generate the model results.

Interpreting the Output of Model 3.3

The results are presented in Table 6.22. As the table suggests, the interaction is not significant ($p = .201$). The direction suggests that having high absenteeism reduces the effect of *zses* (0.943) on the predicted log of the odds of persisting by 0.157 units. Because the effect is not significant, however, we could remove it from the model. We note that when the interaction is significant, both predictors should be retained, even if one or both are not significant.

Following Up With a Smaller Random Sample

To illustrate how a significant interaction might look, we provide an example in Table 6.23 with a smaller random sample of the data, where we obtained a significant interaction.

We will first investigate the meaning of the main effects for *high absenteeism* and *zses*. The main effect for high absenteeism suggests that as absenteeism increases (i.e., from 0 to 1), the expected log odds for persisting are reduced by –2.128 units ($p < .001$) as one moves to the

TABLE 6.22 Model 3.3: Parameter Estimates

Parameter		β	Std. error	95% Wald confidence interval		Hypothesis test			Exp(β)	95% Wald confidence interval for exp(β)	
				Lower	Upper	Wald χ²	df	Sig.		Lower	Upper
Threshold	[persist3 = 0]	−2.448	0.0469	−2.540	−2.356	2726.261	1	0.000	0.086	0.079	0.095
	[persist3 = 1]	−2.053	0.0432	−2.137	−1.968	2262.680	1	0.000	0.128	0.118	0.140
zses		0.943	0.0408	0.863	1.023	535.396	1	0.000	2.568	2.371	2.782
hiabsent		−1.758	0.1197	−1.993	−1.524	215.916	1	0.000	0.172	0.136	0.218
zses*hiabsent		−0.157	0.1225	−0.397	0.083	1.635	1	0.201	0.855	0.673	1.087
(Scale)		1[a]									

Note: Dependent variable: persist3; model: (threshold), zses, hiabsent, zses*hiabsent.
[a] Fixed at the displayed value.

next highest category of persist, holding other variables constant. Similarly, the main effect for socioeconomic status indicates that as *zses* is increased by one standard deviation, the predicted log odds are increased by 1.103 units ($p < .001$) as one moves to the next highest category of persist, holding other variables constant. We can also see in Table 6.23 that the interaction term is significant (log odds $= -0.548$, $p < .05$).

If the interaction is significant, it often becomes the focus of interpreting the results. However, one should also pay attention to the other parameters. One way to facilitate the interpretation of the interaction is to examine the effect visually as in Figure 6.2. We first saved the predicted log odds values of the underlying predictor η_{ic}. Along the Y axis, the results indicate that positive log odds of η_{ic} increase the predicted odds of being in the lowest category (left school early) versus the combined higher categories. We can observe that this is also related to negative *zses* values (plotted on the X axis). We can also see the contingent effect of absenteeism on the SES–persistence relationship. When student SES is low, there is little accompanying effect of absenteeism on the odds of being in the lowest category versus

TABLE 6.23 Parameter Estimates

Parameter		β	Std. error	95% Wald confidence interval		Hypothesis test			Exp(β)	95% Wald confidence interval for exp(β)	
				Lower	Upper	Wald χ²	df	Sig.		Lower	Upper
Threshold	[persist3 = 0]	−2.517	0.1177	−2.748	−2.287	457.319	1	0.000	0.081	0.064	0.102
	[persist3 = 1]	−2.119	0.1081	−2.331	−1.907	384.154	1	0.000	0.120	0.097	0.148
zses		1.103	0.1031	0.901	1.305	114.458	1	0.000	3.013	2.462	3.688
hiabsent		−2.128	0.2811	−2.679	−1.577	57.330	1	0.000	0.119	0.069	0.207
zses*hiabsent		−0.548	0.2702	−1.078	−0.019	4.115	1	0.043	0.578	0.340	0.982
(Scale)		1[a]									

Note: Dependent variable: persist3; model: (threshold), zses, hiabsent, zses*hiabsent.
[a] Fixed at the displayed value.

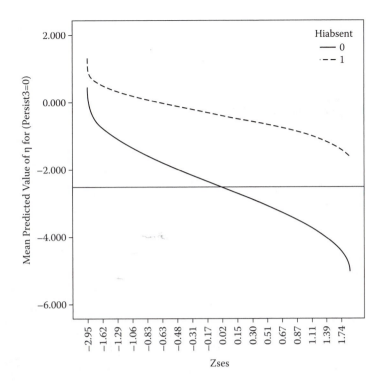

FIGURE 6.2 Relationship between student SES and persistence for two levels of absenteeism.

the combined higher categories. As SES increases, so does size of the interaction effect of high absenteeism.

We can also plot some points to illustrate the impact of the interaction. For the odds that $\eta_{ic} < \theta_1$ (Category 0), the threshold is –2.517, which is represented by the horizontal line in the figure. This represents the log of the odds of persisting for an individual who is 0 on *hiabsent* (i.e., low/average absenteeism) and at the sample average for *zses* (mean = 0). This is where the horizontal line and solid line representing *hiabsent* = 0 intersect. Readers should keep in mind that in the IBM SPSS cumulative odds formulation, for *ascending* coding of the ordinal outcome (the default), larger values of η_{ic} are associated with greater likelihood of being in the lowest category (persist = 0), as opposed to the two higher categories. We can see at this point that the individual who has high absenteeism (*hiabsent* = 1) has considerably higher predicted log odds of being in the lowest category [log odds = –2.517 – (–2.128) = –0.389] than the individual with low or average absenteeism [log odds = –2.517 – (0) = –2.517].

Because of the significant negative interaction (log odds = –0.548, $p < .05$), in Figure 6.2 we can see that the slope of the effect of student SES on the log odds favoring persistence also depends on absenteeism; that is, the SES effect on persistence is contingent on levels of student absenteeism. More specifically, the positive effect of SES on the log odds favoring persistence is reduced considerably by high absenteeism, at least under some conditions. To illustrate, at the sample mean (i.e., when *zses* is 0), there is no additional interaction effect [(0)(–0.548) = 0]. When SES is 1.0, however, the combined effect of high absenteeism and *zses* is 0.555 units [1.103 + (1)(–0.548) = 0.555]. When SES is 2.0, the combined effect will be 1.11 units (2.206 – 1.096 = 1.11). In contrast, when *zses* is –1.0, the combined effect will be –0.555 units [(1.103)(–1) + (–0.548)(–1)].

This suggests that, at low levels of student SES, the combined effects of *zses* and its interaction with absenteeism are less consequential in decreasing the log of the odds that an individual is in category *c* or lower [$P(Y \leq c)$] versus above category *c* [$P(Y > c)$]. This is shown visually in Figure 6.2 in that the two lines are closer together at low levels of student SES. Readers should keep in mind that when estimating differences in effects using odds ratios (as opposed to log odds), the effects are multiplicative (i.e., increasing SES from 1.0 to 2.0 does not simply double the odds ratio).

Developing a Multilevel Ordinal Model

Level-1 Model

We can next extend this single-level model to a multilevel ordinal model using GENLIN MIXED. At Level 1, for the log odds (logit), each of the ordinal responses of individual i in group j can be specified as a linear expression assuming cumulative odds:

$$\eta_{cij} = \log\left(\frac{\pi_{ijc}}{1 - \pi_{cij}}\right) = \beta_{0j} + \sum_{q=1}^{Q}\beta_{qj}X_{qij} + \sum_{c=2}^{C}\theta \tag{6.34}$$

Equation 6.34 indicates that a series of thresholds (θ_c), beginning with the second threshold ($c = 2...C$), separates the ordered categories of the outcome. As we have noted previously, we actually only need $C - 1$ cut points, including the intercept, because the cumulative logit for the last ordered category is redundant. As Equation 6.34 implies, the lowest threshold (i.e., for η_{1ij}) is redefined as the intercept β_{0j} (so θ_1 is assumed to be zero). In the multilevel formulation, we do this so that the intercept parameter can be allowed to vary randomly across groups (as specified with the subscript j). However, the second threshold (θ_2) and any subsequent thresholds in Equation 6.34 are fixed to be invariant across groups (i.e., they have no j subscript) to maintain measurement invariance across the groups (Hox, 2010). The second cut point will be the sum of the first cut point, or intercept (β_0), and the second threshold, and so forth. We note that IBM SPSS output provides the cut points in the continuous indicator (η_{ij}) and not the actual threshold value (which is the same for the second and subsequent cut points defining observed ordinal categories).

Unconditional Model

We can look at the unconditional model (no predictors) first. For individual i in school j, we have the following:

$$\eta_{cij} = \log\left(\frac{\pi_{cij}}{1 - \pi_{cij}}\right) = \beta_{0j} + \theta_2. \tag{6.35}$$

Remember that the intercept (first threshold) can vary across schools but the second threshold is treated as a fixed parameter. At Level 2, then, the random intercept will be as follows:

$$\beta_{0j} = \gamma_{00} + u_{0j}. \tag{6.36}$$

Using substitution, the single equation version of Equations 6.35 and 6.36 will be the following:

$$\eta_{cij} = \gamma_{00} + \theta_2 + u_{0j}. \tag{6.37}$$

This suggests three parameters to estimate (two fixed effects and one random effect).

Defining Model 4.1 With IBM SPSS Menu Commands

Note: If you completed Model 3.3 before starting this model, we recommend in Step 2 to click the RESET button to clear the default settings.

Continue using the *ch6data2.sav* data file.

1. Go to the tool-
bar and select
ANALYZE,
MIXED
MODELS,
GENERALIZED
LINEAR. This
command allows
access to the
*Generalized Linear
Mixed Models*
(GENLINMIXED)
main dialog box.

Note: If you are con-
tinuing this model
immediately after
completing Model
3.3, clear the default
settings before pro-
ceeding by clicking
the RESET button
at the bottom of the
display screen.

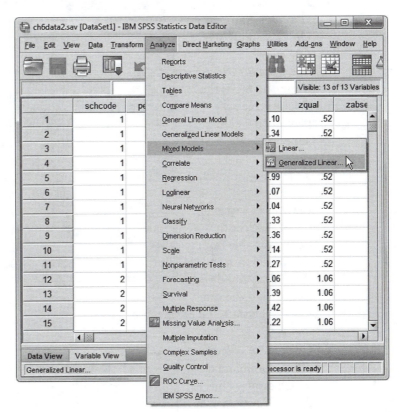

2a. The *Generalized Linear Mixed Models* display screen shows four command tabs: *Data Structure, Fields & Effects, Build Options,* and *Model Options.*

b. The *Data Structure* command tab is the default screen display when creating a model for the first time.

c. To build our model, begin by defining the subject variable. Click to select the *schcode* variable from the left column and then "drag and drop" *schcode* onto the canvas area beneath the *Subjects* box.

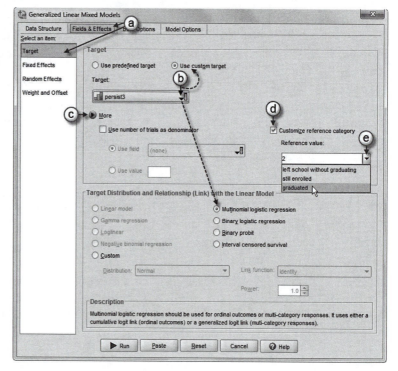

3a. Click the *Fields & Effects* command tab, which displays the default *Target* display screen.

b. For this model we will use students' completion of high school (*persist3*) as the dependent variable. Click the *Target* pull-down menu and select *persist3*.

Note: Selecting the target variable will automatically change the setting to *Use custom input.* Also, the ordinal measurement setting of the dependent variable (*plans*) automatically designates the target distribution as *Multinomial logistic regression.*

c. Because *plans* consists of three categories, we will designate a reference category. First, click the *More* button to reveal additional options.

d. Click to select *Customize reference category.*
e. Click the *Reference value* pull-down menu and select the category *enrolled.*

4a. Click to select the *Fixed Effects* item to access the options for adding "fixed effects" such as predictor variables and interactions.

b. This first model is an unconditional (null) model that is intercept only with no predictors. The default setting for a Level-1 model includes the intercept so no further changes will be made to the model.

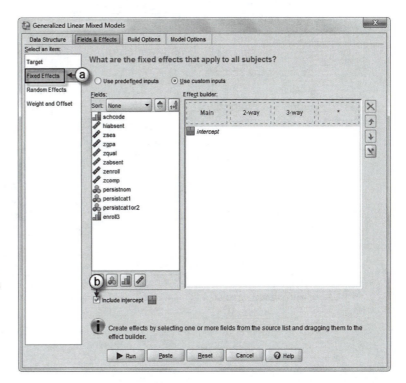

$$\eta_{cij} = \log\left(\frac{\pi_{cij}}{1 - \pi_{cij}}\right) = \beta_{0j} + \theta_2 \quad \text{(Eq. 6.35)}$$

5a. Next, we will add a random effect for the model. Click the *Random Effects* command tab.

b. Click the ADD BLOCK button to access the *Random Effect Block* main dialog box.

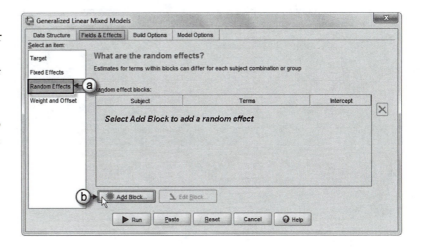

c. In this model only the intercept is going to be randomly varying, so click to select the *Include intercept* option, which changes the setting within the *Effect builder* to *intercept* (from *no intercept*). Because there is only one random effect (the intercept), we can use the default variance component (VC) setting.

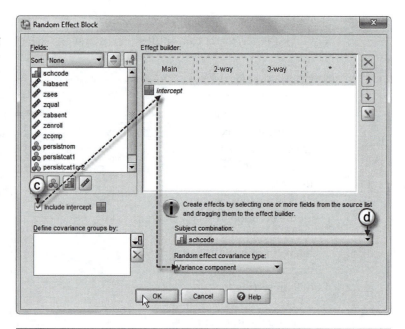

d. Select the *Subject combination* by clicking the pull-down menu and selecting *schcode*.

$$\beta_{0j} = \gamma_{00} + u_{0j} \quad \text{(Eq. 6.36)}$$

Click the OK button to close the *Random Effect Block* dialog box and return to the *Random Effects* main screen.

e. In the main screen notice that the *Random Effect Block* displays a summary of the block just created: *schcode* (Subject) and *intercept* (Yes).

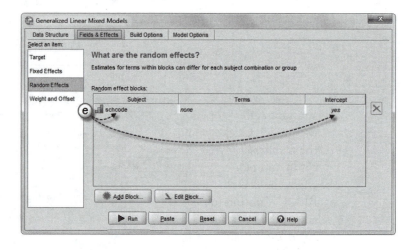

6a. Finally, click the *Build Options* command tab located at the top of the screen to access advanced criteria for building the model. We will retain the default settings for *Stopping Rules, Post Estimation,* and *Degrees of Freedom.*

b. Change the *Tests of fixed effects and coefficients.* Click to select *Use robust estimation to handle violations of model assumptions.*

Click the RUN button to generate the model results.

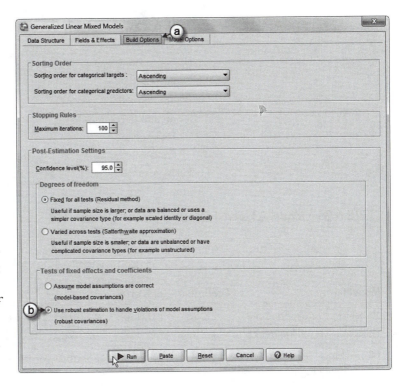

Interpreting the Output of Model 4.1

The model description is summarized in Table 6.24. As shown, the outcome is ordinal, the probability distribution is multinomial, and the link is the cumulative logit. If we use the –2 log psuedolikelihood as a baseline against which to compare successive models, we should keep in mind that such comparisons are approximate only.

We can next confirm in Table 6.25 that we have three effects to estimate (two fixed effects and one random intercept). Table 6.26 provides the two thresholds (i.e., the reference category is persist = 2).

In the two-level setting, the intercept represents the unit-level intercept conditional on the random effects. The unit-specific intercept (β_{0j}) can be interpreted as the expected log odds of

TABLE 6.24 Model 4.1: Model Fit Criteria

Target	persist3
Measurement level	Ordinal
Probability distribution	Multinomial
Link function	Cumulative logit
Information criteria	
Akaike corrected	46,971.726
Bayesian	46,978.576

Note: Information criteria are based on the –2 log pseudolikelihood (46,969.726) and are used to compare models. Models with smaller information criterion values fit better. When comparing models using pseudolikelihood values, caution should be used because different data transformations may be used across the models.

TABLE 6.25   Model 4.1: Covariance Parameters

Covariance parameters	
Residual effect	0
Random effects	1
Design matrix columns	
Fixed effects	2
Random effects	1
Common subjects	934

TABLE 6.26   Model 4.1: Fixed Effects

Model term	Coefficient	Std. error	*t*	Sig.	Exp(coefficient)	95% Confidence interval for exp(coefficient)	
						Lower	Upper
Threshold (0): dropped out	−2.086	0.050	−41.305	0.000	0.124	0.113	0.137
Threshold (1): still enrolled	−1.709	0.047	−35.983	0.000	0.181	0.165	0.199

Note: Target: persist3; reference category: graduated; probability distribution: multinomial; link function: cumulative logit.

persisting within a school where the other predictors are 0 and the random effect is 0 (i.e., average in terms of variability in log odds at Level 2 in the sample). Using Equation 6.5, we can transform the odds ratios e^β into probabilities for each school-level category:

$$\exp(\beta)/(1+\exp(\beta)) = 0.124/1.124 = 0.110$$

$$\exp(\beta)/(1+\exp(\beta)) = 0.181/1.181 = 0.153$$

and $0.153 - 0.110 = 0.042$. Therefore, we have 0.110 (11%) as the "left school early" mean at the school level (Category 0). To determine the adjacent category (Category 1) we can subtract 0.110 from $0.153 = 0.042$ (or about 4.2%).

Let us look at the variance components. Table 6.27 suggests that the intercept varies significantly across schools (z-test = 9.850, $p < .001$).

Within-School Predictor

We next add two within-school predictors: student background (*zses*) and high absenteeism. The model is as follows:

$$\eta_{cij} = \log\left(\frac{\pi_{cij}}{1 - \pi_{cij}}\right) = \beta_{0j} + \beta_{1j}zses_{ij} + \beta_{2j}hiabsent_{ij} + \theta_2. \tag{6.38}$$

TABLE 6.27   Model 4.1: Variance Components

Residual effect	Coefficient	Std. error	z-Test	Sig.	95% Confidence interval	
					Lower	Upper
Var(Intercept)	0.880	0.089	9.850	0.000	0.721	1.074

Note: Covariance structure: variance components; subject specification: schcode.

At the school level, we allow the intercept to vary randomly, but we hold the slopes of the predictors to be fixed across schools:

$$\beta_{0j} = \gamma_{00} + u_{0j} \tag{6.39}$$

$$\beta_{1j} = \gamma_{10}$$
$$\beta_{2j} = \gamma_{20}. \tag{6.40}$$

The combined model will be as follows:

$$\eta_{cij} = \gamma_{00} + \theta_2 + \gamma_{10} zses_{ij} + \gamma_{20} hiabsent_{ij} + \mu_{0j}. \tag{6.41}$$

This suggests five total effects to estimate (four fixed effects and one random Level-2 effect).

Defining Model 4.2 With IBM SPSS Menu Commands

Note: Settings default to those used for Model 4.1.

1. Go to the tool-
   bar and select
   ANALYZE,
   MIXED
   MODELS,
   GENERALIZED
   LINEAR. This
   command allows
   access to the
   *Generalized Linear
   Mixed Models*
   (GENLINMIXED)
   main dialog box.

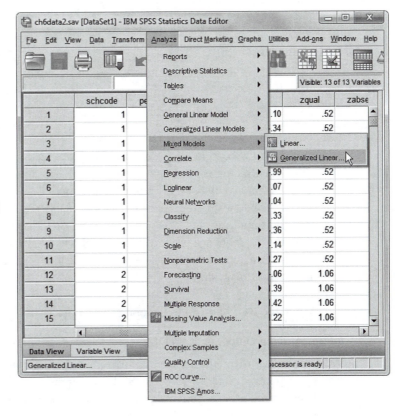

2a. The *Generalized Linear Mixed Models* dialog box displays the *Build Options* command tab, which was the last setting used in the prior model.

b. We will add predictors to our model, so click the *Fields & Effects* command tab.

Note: The *Random Effects* item is preselected (represented in the illustration by the dashed-line box but would appear as grayed-out on the computer screen) as it was the last setting used from the prior model.

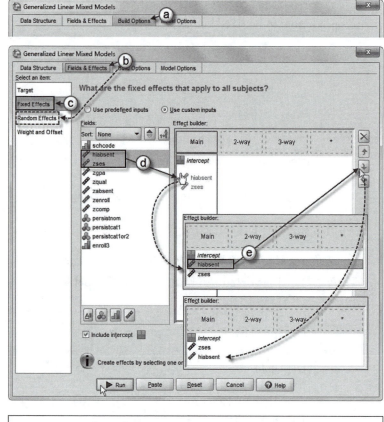

c. Now click to select the *Fixed Effects* item to display the main screen where we may add variables to the model.

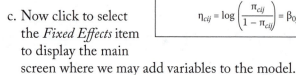

$$\eta_{cij} = \log\left(\frac{\pi_{cij}}{1 - \pi_{cij}}\right) = \beta_{0j} + \beta_{1j}zses_{ij} + \beta_{2j}hiabsent_{ij} + \theta_2 \quad \text{(Eq. 6.38)}$$

d. We will add two predictors to the model (*hiabsent, zses*). Click to select both *hiabsent* and *zses* and then drag the variables into the *Main* (effect) column of the *Effect builder*.

e. Change the sequence order of the variables by clicking to select *hiabsent* and then clicking the down-arrow button to place the variable below *zses*. The sequence order of the variables is now *zses, hiabsent*.

Click the RUN button to generate the model results.

Interpreting the Output of Model 4.2

We can first confirm the model parameters to be estimated in Table 6.28. The fixed-effects results are summarized in Table 6.29. They suggest, similarly to the single-level model, that SES background affects students' predicted log odds of persisting. More specifically, for a 1-SD increase in zses, the predicted odds of persisting versus being still enrolled or leaving school early are increased by 2.427 times ($p < .001$), holding absenteeism constant. Similarly, the expected odds of persisting versus the combined middle and lower categories are decreased by a factor of 0.180 (or reduced by 82%) for a one-unit increase in absenteeism, holding student SES constant.

In Table 6.30, we next provide a summary of the Level-2 intercept variance, which is significant.

Adding the School-Level Predictors

At the school level we will add four predictors. We have converted the school-level continuous variables to standardized scores (mean = 0, SD = 1). As we have noted previously, this is similar to grand-mean centering them and helps facilitate the meaning of the estimates for the

TABLE 6.28 Model 4.2: Covariance Parameters

Covariance parameters	
Residual effect	0
Random effects	1
Design matrix columns	
Fixed effects	4
Random effects	1
Common subjects	934

TABLE 6.29 Model 4.2: Fixed Effects

Model term	Coefficient	Std. error	*t*	Sig.	Exp(coefficient)	95% Confidence interval for exp(coefficient)	
						Lower	Upper
Threshold (0): dropped out	−2.510	0.053	−47.017	0.000	0.081	0.073	0.090
Threshold (1): still enrolled	−2.086	0.051	−41.023	0.000	0.124	0.112	0.137
zses	0.887	0.042	21.165	0.000	2.427	2.236	2.635
hiabsent	−1.712	0.115	−14.868	0.000	0.180	0.144	0.226

Note: Target: persist3; reference category: graduated; probability distribution: multinomial; link function: cumulative logit.

TABLE 6.30 Model 4.2: Variance Components

Residual effect	Coefficient	Std. error	z-Test	Sig.	95% Confidence interval	
					Lower	Upper
Var(Intercept)	0.481	0.072	6.655	0.000	0.358	0.645

Note: Covariance structure: variance components; subject specification: schcode.

school-level model. School quality (*zqual*) is a factor score summarizing school expectations and educational processes. School attendance (*zabsent*) provides the percentage of students with high absences. Student enrollment (*zenroll*) indicates the size of the school. Finally, student composition (*zcomp*) is a factor score summarizing the student socioeconomic status and language background. The school-level model is as follows:

$$\beta_{0j} = \gamma_{00} + \gamma_{01}zqual_j + \gamma_{02}zabsent_j + \gamma_{03}zenroll_j + \gamma_{04}zcomp_j + u_{0j}. \tag{6.42}$$

The combined model will be as follows:

$$\eta_{cij} = \gamma_{00} + \theta_2 + \gamma_{01}zqual_j + \gamma_{02}zabsent_j + \gamma_{03}zenroll_j + \gamma_{04}zcomp_j \\ + \gamma_{10}zses_{ij} + \gamma_{20}hiabsent_{ij} + u_{0j}. \tag{6.43}$$

As this suggests, there will now be eight fixed effects and one random effect to be estimated.

Defining Model 4.3 With IBM SPSS Menu Commands

Note: Settings default to those used for Model 4.2.

1. Go to the toolbar and select ANALYZE, MIXED MODELS, GENERALIZED LINEAR. This command allows access to the *Generalized Linear Mixed Models* (GENLINMIXED) main dialog box.

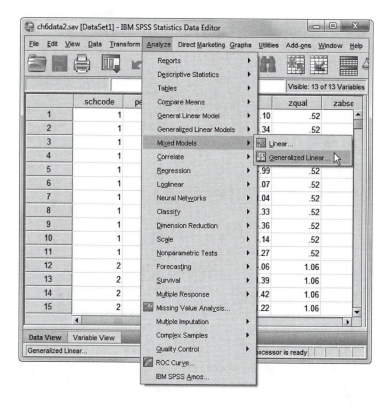

2a. The *Generalized Linear Mixed Models* dialog box displays the *Fields & Effects* command tab and *Fixed Effects* screen, which was the last setting used in the prior model.

b. We will add four predictors to the model (*zqual, zabsent, zenroll, zcomp*). Click to select the four variables and then drag them into the *Main* (effect) column of the *Effect builder.*

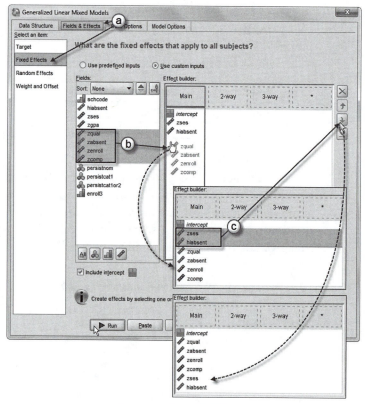

$$\beta_{0j} = \gamma_{00} + \gamma_{01}zqual_j + \gamma_{02}zabsent_j + \gamma_{03}zenroll_j + \gamma_{04}zcomp_j + u_{0j} \quad \text{(Eq. 6.42)}$$

TABLE 6.31 Model 4.3: Covariance Parameters

Covariance parameters	
Residual effect	0
Random effects	1
Design matrix columns	
Fixed effects	8
Random effects	1
Common subjects	934

c. We will change the sequence order of the variables in the model by clicking to select *zses* and *hiabsent* and then clicking the down-arrow button to place them in the last two positions. The variable sequence is now *zqual, zabsent, zenroll, zcomp, zses, hiabsent*.

Click the RUN button to generate the model results.

Interpreting the Output of Model 4.3

We can confirm the model effects in Equation 6.43 to be estimated in Table 6.31. The fixed effects are summarized in Table 6.32. As the within-school variables are consistent with previous models, we will concentrate on the effects of the school-level predictors. First, the results suggest that school quality affects students' odds of persisting (OR = 1.173). More specifically, for a 1-SD increase in the quality of the school's educational processes, students' expected odds of persisting versus the combined middle and lower categories would be expected to increase by a factor of 1.173 (or 17.3%), holding the other predictors (and random-variable component) in the model constant. Second, student composition is negatively related to odds of persisting. A 1-SD increase in percentages of students on the federal free/reduced lunch program and requiring English language services would be expected to decrease students' odds of persisting versus the combined middle and lower categories by a factor of 0.891 (or reduce them by 10.9%), holding other variables constant. For enrollment size, smaller schools were settings where students were more likely to persist. A 1-SD increase in enrollment size would decrease the expected odds of persisting versus the combined lower categories by a factor of 0.894 (or 10.6%), again holding the other variables constant. Average student absenteeism did not affect students' odds of persisting.

TABLE 6.32 Model 4.3: Fixed Effects

Model term	Coefficient	Std. error	t	Sig.	Exp(coefficient)	95% Confidence interval for exp(coefficient)	
						Lower	Upper
Threshold (0): dropped out	−2.544	0.054	−47.148	0.000	0.079	0.071	0.087
Threshold (1): still in school	−2.118	0.051	−41.360	0.000	0.120	0.109	0.133
zqual	0.160	0.048	3.325	0.001	1.173	1.068	1.289
zabsent	−0.048	0.044	−1.087	0.277	0.953	0.874	1.039
zenroll	−0.112	0.054	−2.054	0.040	0.894	0.804	0.995
zcomp	−0.116	0.040	−2.878	0.004	0.891	0.823	0.964
zses	0.839	0.044	19.142	0.000	2.313	2.123	2.520
hiabsent	−1.648	0.127	−12.996	0.000	0.192	0.150	0.247

Note: Target: persist3; reference category: graduated; probability distribution: multinomial; link function: cumulative logit.

TABLE 6.33   Model 4.3: Variance Components

Residual effect	Coefficient	Std. error	z-Test	Sig.	95% Confidence interval Lower	Upper
Var(Intercept)	0.458	0.071	6.494	0.000	0.339	0.620

Note: Covariance structure: variance components; subject specification: schcode.

The variance components for the intercept are in Table 6.33. The table suggests that odds of student persistence differ across schools ($z = 6.494$, $p < .001$).

Using Complementary Log–Log Link

We note that we could also use the complementary log–log link to estimate an ordinal model. Complementary log–log models are frequently used when the probability of an event is very small or very large. In this case, it might have some advantage because our categories are not equally distributed (i.e., most individuals are in the persist category). Unlike the logit link function (which is the canonical link function for the multinomial distribution), however, the complementary log–log function is asymmetrical, which suggests that it does not have some of the desirable mathematical properties of the specific canonical link function for each distribution. In this case, an odds ratio cannot be directly estimated. A rate ratio, however, can be calculated from the regression coefficients of complementary log–log models (i.e., by raising the unstandardized β complementary log–log to e^β). The rate ratio and the odds ratio calculated (using the logit link function) are similar for small values of π (i.e., below 0.2). For larger values of π, however, the two functions become more different. Therefore, the output does not provide rate ratios (e^β) for the complementary log–log link function (i.e., corresponding confidence intervals cannot be accurately calculated because the function is asymmetrical).

We can see in Table 6.34 that when we use this link function, the model estimates are slightly different but lead to the same substantive conclusion. In this case, we might be satisfied to stay with the logit link because we do not observe any real substantive differences.

Interpreting a Categorical Predictor

For purposes of demonstration, we also present results for a categorical predictor. We provide the model syntax in the appendix. In this case, we will assume that *enroll3* is an ordinal variable

TABLE 6.34   Alternative Model 4.3: Fixed Effects With Complementary Log–Log

Model term	Coefficient	Std. error	t	Sig.	95% Confidence interval for exp(coefficient) Lower	Upper
Threshold (0): dropped out	–2.510	0.049	–51.210	0.000	–2.606	–2.414
Threshold (1): still in school	–2.147	0.046	–46.339	0.000	–2.238	–2.057
zqual	0.145	0.041	3.556	0.000	0.065	0.226
zabsent	–0.030	0.036	–0.847	0.397	–0.101	0.040
zenroll	–0.107	0.047	–2.259	0.024	–0.199	–0.014
zcomp	–0.086	0.034	–2.518	0.012	–0.152	–0.019
zses	0.718	0.037	19.180	0.000	0.645	0.792
hiabsent	–1.336	0.097	–13.716	0.000	–1.527	–1.145

Note: Probability distribution: multinomial; link function: complementary log–log; target: persist3; reference category: graduated.

TABLE 6.35 Categorical Predictor Model: Covariance Parameters

Covariance parameters	
Residual effect	0
Random effects	1
Design matrix columns	
Fixed effects	10
Random effects	1
Common subjects	934

describing three categories of school size. The enrollment categories are arranged from low to high. Categorical indicators will add $C - 1$ dummy-coded variables to the model. We will assume that Category 0 is schools smaller than 600 students, Category 1 is schools between 600 and 1,099 students, and the last category is schools with over 1,100 students. We used small schools (enrollments less than 600 students) as the reference group (Category 0). In this case, then, both Category 1 and Category 2 schools are predictors of the predicted odds of student persistence. The first category does not have a coefficient because it is the reference group. In this formulation, there will now be 10 fixed effects (reflecting that enrollment now has three categories) and one random effect to estimate.

We can confirm the model effects to be estimated in Table 6.35. The fixed-effect estimates are provided in Table 6.36. We can see in this formulation that enrollment size affects students' expected odds of persisting versus being still in school or leaving school early ($p < .05$). More specifically, in schools with between 600 and 1,099 students (Category 1), the odds of persisting versus the combined middle and low categories are reduced by a factor of 0.793 (i.e., reduced by 20.7%) given that all of the other variables in the model are held constant. In the largest category of schools (Category 2), the odds of persisting versus the combined lower categories are reduced by a factor of 0.796 (20.4%), holding other variables constant.

TABLE 6.36 Categorical Predictor Model: Fixed Effects

Model term	Coefficient	Std. error	*t*	Sig.	Exp(coefficient)	95% Confidence interval for exp(coefficient)	
						Lower	Upper
School level							
Threshold (0): dropped out	−2.681	0.086	−31.046	0.000	0.068	0.058	0.081
Threshold (1): still in school	−2.255	0.083	−27.079	0.000	0.105	0.089	0.123
zqual	0.162	0.048	3.388	0.001	1.176	1.071	1.292
zabsent	−0.054	0.044	−1.224	0.221	0.947	0.868	1.033
zcomp	−0.114	0.041	−2.813	0.005	0.892	0.824	0.966
enroll3 = 2	−0.228	0.111	−2.056	0.040	0.796	0.641	0.989
enroll3 = 1	−0.232	0.109	−2.131	0.033	0.793	0.640	0.982
enroll3 = 0	0[a]						
Individual level							
zses	0.838	0.044	19.111	0.000	2.312	2.122	2.620
hiabsent	−1.650	0.127	−12.968	0.000	0.192	0.150	0.246

Note: Syntax for this model is available in Appendix A. Probability distribution: multinomial; link function: cumulative logit; target: persist3; reference category: graduated.

[a] This coefficient is set to 0 because it is redundant.

The random intercept variance component at Level 2 was similar to the last model, so we do not reproduce it here.

Other Possible Analyses

We could, of course, investigate a random Level-1 slope across schools. We might wonder whether student SES has a larger effect on odds of persisting in some schools than in others. As we have conducted this type of analysis in other places, however, we will not reproduce it here.

We note that readers might wish to compare the previous results with a model that has "persist" considered as multinomial outcome rather than ordinal. We created a new variable (*persist-nom*), which can be specified as nominal instead of ordinal. This is important so that GENLIN MIXED knows that the outcome is to be considered as nominal rather than ordinal. In the multinomial situation, we estimate coefficients that capture differences between all possible pairs of groups. In this instance, this decision will result in two sets of estimates instead of one set. Keep in mind, however, that the multinomial model ignores the ordering of the categories. For the unconditional model, we have estimates comparing "left school early" (dropped out) versus "persisted" and "in school but behind" versus persisted. Once again, the reference group is "persisted." In this case the measurement of the outcome is nominal, so the generalized logit link function is used (instead of the cumulative logit link function). The results are generally consistent with the ordinal model; however, when the assumptions of the proportional odds model hold (as they did in our example), it is the favored way to proceed.

Examining a Mediating Effect at Level 1

In this last section, we will demonstrate how users might test for a mediating effect at Level 1. More specifically, in some situations we may be interested in examining whether another variable, referred to as a mediating (M) or intervening variable, may exist in the causal chain between a predictor X and the outcome Y. This situation is often found in experimental designs (where X represents a treatment) when a particular treatment may produce an effect on a mediating variable (such as an attitude) that, in turn, influences a subsequent change in behavior. We can, however, apply the mediation model more generally within social science research. For example, we might test whether a mediating process exists between a student background variable (e.g., socioeconomic status) and student persistence that influences students' odds of persisting. Such mediating process variables often provide hints about how to make a change in students' school situations that might positively influence an outcome of interest.

Readers may recognize these types of relationships if they have previously encountered path models or structural equation models (SEMs). A mediating process implies the possibility of direct, indirect, and total effects. They suggest that a total effect (e.g., $X \rightarrow Y$) may be decomposed into a direct and an indirect effect. In this case, we will advance the argument that higher student SES may bring with it particular advantages (e.g., increased parent participation, assignment to higher curricular tracks, stronger academic outcomes) over time that increase students' odds of persisting. We could argue that perhaps the effect of student SES is entirely mediated by students' previous success in school (in this case, represented by their high school GPA). In Figure 6.3, the effect of the independent variable on the outcome is entirely indirect—that is, through its relationship to students' GPA and the relationship between GPA and likelihood to persist.

It is probably more likely, however, that the effect of SES background on persistence is partially mediated by students' cumulative GPA, as shown in Figure 6.4. This latter hypothesis seems logical, given the nature of most social science research, which typically identifies many causes of behavior (MacKinnon, 2008).

In advancing this argument, we must assume that X (student SES) comes before M (GPA), which represents a proxy for students' accumulated success in high school, and that

FIGURE 6.3 Examining a fully mediated effect.

student GPA is measured prior to student persistence (Y). At a minimum, when we test for a mediated effect, we expect the relationship between student SES and GPA (a) to be significant as well as the relationship between student GPA and persistence (b). It is often the case that the direct effect of X on Y alone is significant (with the direct effect estimate referred to as c), but it is not, strictly speaking, a requirement for examining mediation (MacKinnon, 2008).

A practical test of the partial-mediation hypothesis is whether or not the adjusted direct effect of X on Y (c' in Figure 6.4) is significant in the presence of the mediator variable (MacKinnon, 2008). As MacKinnon notes, one way to test the presence of a mediated effect is to examine the difference between the initial direct effect (c) and the adjusted direct effect (c'). This approach, however, is more difficult when there is a categorical outcome because, as we have noted, the variance in Y is rescaled when predictor variables are added to the model. It is easier simply to multiply the two paths that comprise the indirect effect (ab) between X and Y (see MacKinnon, 2008, for further discussion).

In the proposed model, student SES and GPA are continuous variables and normally distributed; however, persistence is ordinal and therefore not normally distributed. For purposes of our demonstration, we will assume that these are the only variables in which we are interested but, clearly, we could also introduce multiple mediators and other covariates into the model. We note in passing that if the proposed model becomes too complex (e.g., with several mediated effects), it is likely easier to estimate it with a software program that supports multilevel structural equation modeling because the proposed model may imply several direct and indirect effects, as well as covariances between variables (Heck & Thomas, 2009).

To test the proposed single-mediator model, we make use of several equations. First, we specify the direct effect of student SES on the log odds of persisting. This provides an estimate of the possible direct effect between X and Y (c) before testing for the mediated effect (MacKinnon, 2008):

$$\eta_{cij} = \beta_{0j} + \beta_{1j} zses_{ij} + \theta_2, \tag{6.44}$$

where $\eta_{cij} = \log(\pi_{cij}/(1 - \pi_{cij}))$.

We will allow the intercept to be random across schools (i.e., $\beta_{0j} = \gamma_{00} + u_{0j}$). In addition, we specify two additional equations. First, we add the moderator ($zgpa$) to the Level-1 model:

$$\eta_{cij} = \beta_{0j} + \beta_{1j} zses_{ij} + \beta_{2j} zgpa_{ij} + \theta_2. \tag{6.45}$$

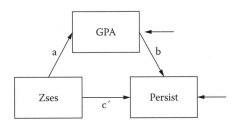

FIGURE 6.4 Examining a partially mediated effect.

This will provide path (*b*) in Figure 6.4. We will also treat its slope as a fixed parameter between schools (because we are examining a Level-1 effect):

$$\beta_{2j} = \gamma_{20}.\tag{6.46}$$

Defining Model 4.4 With IBM SPSS Menu Commands

Note: Settings default to those used for Model 4.3.

1. Go to the tool-bar and select ANALYZE, MIXED MODELS, GENERALIZED LINEAR. This command allows access to the *Generalized Linear Mixed Models* (GENLINMIXED) main dialog box.

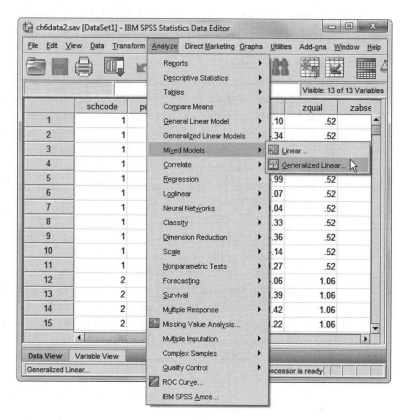

2a. The *Generalized Linear Mixed Models* dialog box displays the *Fields & Effects* command tab and *Fixed Effects* screen, which was the last setting used in the prior model

b. We will first remove five variables from the model (*zqual, zabsent, zenroll, zcomp, hiabsent*). Click to select the five variables then click the red "X" button to remove them from the model. This will leave only *zses* in the model.

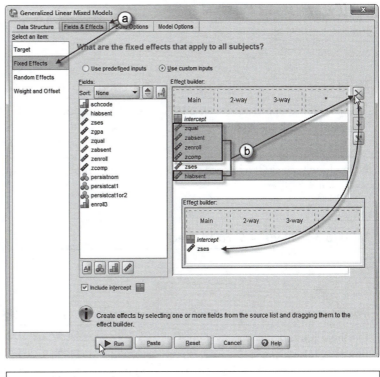

$$\eta_{cij} = \beta_{0j} + \beta_{1j}zses_{ij} + \theta_2 \quad \text{(Eq. 6.44)}$$

Click the RUN button to generate the model results.

We also need to obtain a separate estimate describing the relationship between student SES and the mediator, which is student GPA (i.e., path *a* in Figure 6.4). This model can be defined as follows:

$$Y_{ij} = \beta_0 + \beta_1 zses_{ij} + \varepsilon_{ij}. \tag{6.47}$$

Because both of these variables are continuous, we can estimate their relationship in GENLIN MIXED by specifying GPA as the dependent variable (Y_{ij}) and, because it is continuous, we change the link function to *identity*. Note that we have an error term because the outcome is presumed to be from a normal distribution. GPA is a Level-1 variable, so we will treat its intercept as fixed ($\beta_{0j} = \gamma_{00}$) at the school level. We provide the syntax for running this model in Appendix A. We note in passing that we could also estimate the relationship using multiple regression (or using MIXED).

Interpreting the Output of Model 4.4

The fixed effects for Equation 6.44 are presented in Table 6.37. We can see from the table that there is a direct relationship between student SES and odds of persisting (OR = 2.565, $p < .001$). We note that observing a significant direct effect between the independent variable and the outcome is not required, however, for a mediating effect model (MacKinnon, 2008).

Next, the fixed effects for Equation 6.45 are presented in Table 6.38. We can see in the table that student GPA is significantly related to odds of persisting. A 1-SD increase in *zgpa* would result in multiplying the odds of persisting versus still being in school or dropped out by about 2.5 times, holding *zses* constant. We note that *zses* is still significantly related to the odds of persisting. The size of the odds ratio has been reduced somewhat by the presence of *zgpa* in the model (from about 2.6 to about 2.2). However, keep in mind that assessing the strength of mediation based on the difference in these two coefficients is incorrect, unless the model

TABLE 6.37   Model 4.4: Fixed Effects

Model term	Coefficient	Std. error	t	Sig.	Exp(coefficient)	95% Confidence interval for exp(coefficient)	
						Lower	Upper
Threshold (0): dropped out	–2.319	0.051	–45.088	0.000	0.098	0.089	0.109
Threshold (1): still enrolled	–1.918	0.049	–39.451	0.000	0.147	0.134	0.162
zses	0.942	0.041	22.975	0.000	2.565	2.367	2.779

Note: Target: persist3; reference category: graduated; probability distribution: multinomial; link function: cumulative logit.

TABLE 6.38   Model 4.5: Fixed Effects

Model term	Coefficient	Std. error	t	Sig.	Exp(coefficient)	95% Confidence interval for exp(coefficient)	
						Lower	Upper
Threshold (0): dropped out	–2.620	0.059	–44.175	0.000	0.073	0.065	0.082
Threshold (1): still enrolled	–2.177	0.054	–40.318	0.000	0.113	0.102	0.126
zses	0.791	0.045	17.437	0.000	2.205	2.018	2.410
zgpa	0.917	0.039	23.449	0.000	2.502	2.317	2.701

Note: Probability distribution: multinomial; link function: cumulative logit; target: persist3; reference category: graduated.

parameters are standardized (MacKinnon, 2008), because of the rescaling that takes place when variables are added to the model. Overall, however, the results in Table 6.38 provide support for the view that the partial mediation model (Figure 6.4) should be accepted over the full mediation model because *zses* still significantly affects persistence.

Finally, in Table 6.39 we provide the estimate of the effect of *zses* on student GPA from Equation 6.47. We can interpret this as a 1-SD increase in standardized student SES results in an approximate 0.325-SD increase in students' standardized GPA ($p < .001$).

Estimating the Mediated Effect

We can now estimate the mediated effect between student SES and GPA (*a*) and GPA and odds of persisting (*b*). The indirect effect is found by multiplying path *a* and path *b* (0.325*0.917 = 0.298). We next need to calculate the standard error for the indirect effect and then determine whether it is significant. The most common way to calculate the standard error of the estimated mediation effect is based on first derivatives using the multivariate delta method (Folmer, 1981; MacKinnon, 2008; Sobel, 1982). We note that correctly calculating the standard errors is more complex when there are multiple mediators in the model, especially if using the $c - c'$ approach

TABLE 6.39   Estimates of Fixed Effects (Y = GPA)

Parameter	Estimate	Std. error	t	Sig.	95% Confidence interval	
					Lower	Upper
Intercept	0.000	0.011	0.000	1.000	–0.022	0.022
zses	0.325	0.011	29.440	0.000	0.303	0.346

Note: Probability distribution: normal; link function: identity.

because of the rescaling that takes place when successive variables are added and because covariances between the model coefficients may need to be taken into consideration (MacKinnon, 2008). For a single mediator model, we can make use of the associated t-values for the estimated a and b paths in calculating the standard error estimates for mediated effects (MacKinnon, 2008):

$$S_{First} = \frac{ab\sqrt{t_a^2 + t_b^2}}{t_a t_b}.$$
(6.38)

If we plug the values from the t-tests into the equation, we obtain the following coefficients:

$$0.016 = \frac{(0.325)(0.917)\sqrt{29.440^2 + 23.449^2}}{(29.440)(23.449)}.$$

The mediated effect (0.298) can then be divided by its standard error (0.016) and we obtain a Z-ratio of 18.625 (which exceeds the required z-score of 1.96 at $p = 0.05$). We can also place a 95% confidence interval around the estimate (i.e., $0.267 - 0.329$).

MacKinnon (2008) describes a number of possible multilevel mediated-effects models that can be specified. In one situation, for example, the independent variable might be at the group level, with the mediator and outcome at Level 1. The independent variable at Level 2 might take on a different value in each group j. For example, we might consider that the negative effect of a particular school context variable on persistence might be fully (or partially) mediated by students' accumulated GPA, which, in turn, increases their log odds of persisting. As MacKinnon notes, however, when there are multiple dependent categorical variables, the models are probably best estimated with SEM software that can take into consideration more complex model formulations with multiple direct and indirect effects.

Summary

In this chapter, we provided extended examples comparing the multinomial approach for nominal outcomes and the ordinal approach for outcomes with ordered categories. We noted that if the assumptions regarding ordered categories can be supported, it is generally preferable to conduct an ordinal analysis because it maximizes information about relationships between the categories comprising the outcome. If the assumptions do not hold, a multinomial analysis will suffice, provided there are not so many outcome categories that the relationships comprising the set of nominal categories are hard to interpret. We also illustrated how we might test for an interaction as well as explore a mediated-effect relationship. In the next chapter, we examine single-level and multilevel models where the outcomes consist of count data.

Note

1. We could also develop this model using regression (multinomial logit). Results will be virtually identical.

CHAPTER 7

Two-Level Models With Count Data

Introduction

In this chapter we develop two extended examples with count data. Count data involve zeros and positive integers. They are similar to ordinal data in the sense that they typically have a restricted range. For example, if there are considerable zeros in the data set, the distribution of a count outcome may be positively skewed—that is, with more individuals below the mean, which is at the low end of the count. If the counts are relatively rare events, they may follow a Poisson distribution. More frequent counts can be analyzed using a negative binomial model (Hox, 2010). The Poisson regression model assumes that the length of the observation period is fixed in advance, the events occur at a constant rate, and the number of events at disjoint intervals is statistically independent.

We first examine a Poisson distribution with *constant exposure;* that is, we examine the probability that students failed a course during an academic year in high school (their ninth grade year). The interval is the same for each student (equal exposure), which means the individuals in the study have an *equal* opportunity to fail a course during their first year of high school. For this model, we use a log link function. We compare the results using a Poisson regression model and a negative binomial regression model. We then examine a Poisson regression model with a varying exposure rate t. An example developed in the SPSS user's guide concerns an analysis of the number of traffic accidents a set of individuals has, with the varied exposure being their years of experience driving. In our second example, if students have been pursuing an undergraduate degree over differing lengths of time, the distribution of course failures would likely again be Poisson, and the exposure rate might be the number of semesters each student has been enrolled as an undergraduate.

A Poisson Regression Model With Constant Exposure

Our first example focuses on the number of core courses that students fail during their ninth grade year of high school. This is a critical year in students' educational careers because research has shown that, if they fall behind their peers early in terms of completing course requirements, they have a difficult time catching up and graduating with their cohort. The typical student takes four core courses (which we will define as English, math, social studies, and science) during this first year of high school.

The Data

The data were collected from 9,956 ninth grade students who were monitored over time as they transitioned from high school to postsecondary educational settings. They were in 44 high schools. We summarize the variables used in the example in Table 7.1. We note that the count

TABLE 7.1 Data Definition of *ch7data1.sav* (N = 9,956)

Variable	Level[a]	Description	Values	Measurement
nschcode	School	School identifier (44 schools)	Integer	Ordinal
lowses	Individual	Predictor dichotomous variable measuring student's social economic status (SES)	0 = average or high socioeconomic status 1 = low socioeconomic status	Scale
male	Individual	Demographic predictor variable representing student's gender	0 = female 1 = male	Scale
age	Individual	Predictor variable representing the number of years old the student was at the time he or she started the ninth grade	12.08 to 16.58	Scale
gmage	Individual	Predictor variable (grand-mean centered) representing the number of years old the student was at the time he or she started the ninth grade	−1.50 to 3.00	Scale
math	Individual	State math test score to describe student's achievement during Grade 8	100 to 500	Scale
gmmath	Individual	State math test score (grand-mean centered) to describe student's achievement during Grade 8	−144.66 to 255.34	Scale
fail	Individual	The number of courses that students fail during their ninth grade year of high school	0 to 4	Scale
gmAvgYearsExp	School	Predictor variable (grand-mean centered) representing average years of teaching experience of the staff	−6.78 to 6.52	Scale
gmlicensedper	School	Predictor variable (grand-mean centered) summarizing the percentage of teachers in the school who are fully licensed according to state standards for what they are teaching	−0.18 to 11	Scale

[a] Individual = Level 1; school = Level 2.

outcome (fail) must be defined as "scale" for the Poisson distribution. We can see in Table 7.2 that 69.5% of the students did not fail any core course during this first year. In contrast, 30.5% do fail at least one course, and 4.4% fail all of their core courses. We provide a histogram to show the skewness present in the data. This is summarized in Figure 7.1.

In our example, all students take the four core courses, so neither the number of courses taken nor the time can be thought of as varied exposure. We assume that any given student has some probability of failing any given course. As Figure 7.1 suggests, however, over two thirds of the students did not fail any courses. We therefore will assume a Poisson distribution (because failing a course is a relatively rare event) with constant exposure; that is, the interval during which the failures can accumulate is 1 year, and a student can fail a maximum of four core courses. Hence, this interval is the same for each student, which means that each has an equal opportunity to accumulate failures (Raudenbush & Bryk, 2002).

TABLE 7.2 **Frequency Distribution of "Fail" Variable**

		Frequency	Percent	Valid percent	Cumulative percent
Valid	0	6,916	69.5	69.5	69.5
	1	1,426	14.3	14.3	83.8
	2	712	7.1	7.1	90.9
	3	466	4.7	4.7	95.6
	4	436	4.4	4.4	100.0
	Total	9,956	100.0	100.0	

The Poisson distribution has only one parameter: the event rate. For the Poisson distribution, the mean and the variance are both expressed by $\mu = \sigma^2 = \lambda$. This suggests that the mean and variance are equal to the event rate (Hox, 2010). In real data, however, if the event rate increases, the frequency of the higher counts increases and thus the variance increases. This can introduce overdispersion into the model. This can happen, for example, if there are outliers or very small group sizes (Hox, 2010). Using the robust standard error option is recommended to adjust for mild violation of the distribution assumption (Hox, 2010).

Preliminary Single-Level Models

We begin with several single-level models to illustrate some of the different possibilities for analyzing these data. We will use four predictors. Student socioeconomic status (SES) is coded 0 = average or high SES and 1 = low SES. For gender, male is coded 1 and female is coded 0. Age is the number of years old the student was at the time she or he started ninth grade. We use a standardized state math test to describe students' achievement during Grade 8. The descriptive

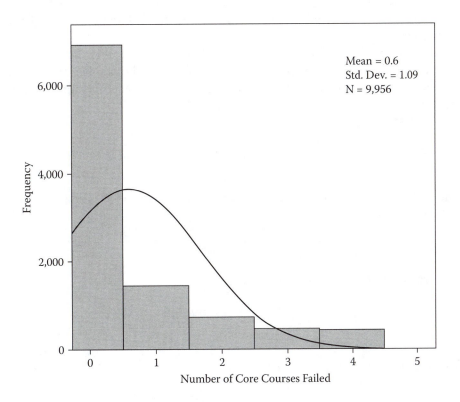

FIGURE 7.1 Distribution of course failures.

TABLE 7.3 Model 1.1: Continuous Variable Information

		N	Minimum	Maximum	Mean	Std. deviation
Dependent variable	fail	9956	0	4	0.6000	1.09000
Covariate	lowses	9956	0	1	0.4200	.49400
	male	9956	0	1	0.5100	.50000
	math	9956	100.00	500.00	244.6600	62.36500
	age	9956	12.08	16.58	13.5814	.43097

statistics are provided in Table 7.3. The table indicates that the mean event rate is 0.60 (with a variance of 1.19, if we square the standard deviation [SD]). This suggests some possible presence of overdispersion.

Our first model will simply describe the log of the counts:

$$\eta_i = \log(\lambda) = \beta_0, \tag{7.1}$$

where η_i is equal to the log of the expected event count λ_i or $\eta_i = \log(\lambda_i)$.

We note that in a Poisson distribution there is also no error term at the lowest level. We must assume that the distribution of course failures follows the Poisson distribution. The intercept with no predictors in the model can be interpreted as the estimated natural log of the expected count of failing a course during freshman year in the population (–0.508). If we exponentiate the logged count ($e^{-0.508}$), we obtain the expected event count ($\lambda = 0.60$) in Table 7.3 because the natural log is the canonical link function for the Poisson distribution (Azen & Walker, 2011). The inverse of the canonical link function returns the same value. For the Poisson distribution, the inverse function (or mean function) is then $\mu = e^{\beta}$. What we refer to as a count can also be called a rate (i.e., the number of events per a specific time interval) and is also referred to as an incident rate.

Once we know the expected incident rate (λ), we can determine the probability of failing a given number of courses in the population. We substitute the number of expected rate (1) into the following equation to obtain the predicted probability. The probability of a count of c events is defined as follows:

$$P(Y = c) = \frac{e^{-\lambda}\lambda^c}{c!}. \tag{7.2}$$

The probability of not failing a course can then be calculated as follows:

$$P(Y = 0) = \frac{e^{-\lambda}\lambda^c}{c!} = \frac{e^{-(0.6)}0.6^0}{0!} = \frac{(0.549)(1)}{1} = 0.549$$

The probability of failing one course is

$$P(Y = 1) = \frac{(0.549)(0.6)^1}{1} = 0.330.$$

The probability of failing two courses is

$$P(Y = 2) = \frac{(0.549)(0.6)^2}{(2)(1)} = 0.099.$$

The probability of failing three courses is

$$P(Y = 3) = \frac{(0.549)(0.6)^3}{(3)(2)(1)} = 0.020.$$

The probability of failing four courses is

$$P(Y = 4) = \frac{(0.549)(0.6)^4}{(4)(3)(2)(1)} = 0.003.$$

We return to the idea that distribution functions describe the behavior of means, variances, standard deviations, and proportions. Assuming the Poisson distribution and the variables in the model, the probability of failing one course is 0.33. We can see that the probability of failing each additional course gets progressively smaller. From Table 7.2, we can determine that 69.5% of students do not fail a course, and we can calculate that 30.5% fail at least one course. In the dichotomous case (failed a course = 1, did not fail = 0), the expected value (μ) would be 0.305 (i.e., instead of the event rate of 0.6 in Table 7.3). If we estimated the model using a binomial distribution (and logit link), the log odds are estimated as -0.822 (not tabled), and the corresponding odds ratio ($e^{-0.822}$) is then 0.44. We would calculate the probability of failing a course expressed in terms of odds $\left(\pi = \frac{e^{0.44}}{1+e^{0.44}} \right)$ as 0.305. The Poisson distribution therefore takes into consideration the extra information present in the original count outcome.

Our primary interest, however, is in the investigation of possible academic and background factors that might explain the distribution of course failure counts. Our proposed model can be described as follows:

$$\eta_i = \beta_0 + \beta_1 lowSES + \beta_2 male + \beta_3 math + \beta_4 age, \tag{7.3}$$

where $\eta_i = \log(\lambda)$. The model output confirms that we have a Poisson probability link with a log link function (see Table 7.4).

TABLE 7.4 Model 1.1: Model Information

Variable	Fail
Probability distribution	Poisson
Link function	Log

Defining Model 1.1 With IBM SPSS Menu Commands

Note: Refer to Chapter 3 for discussion concerning settings and modeling options available in the IBM SPSS GENLIN routine.

Launch the IBM SPSS application program and select the *ch7data1.sav* data file.

1. Go to the tool-bar and select ANALYZE, GENERALIZED LINEAR MODELS, GENERALIZED LINEAR MODELS. This command opens the *Generalized Linear Models* (GENLIN) main dialog box.

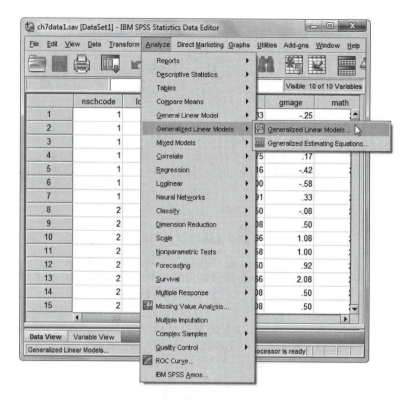

2. *The Generalized Linear Models* display screen displays nine command tabs: *Type of Model, Response, Predictors, Model, Estimation, Statistics, EM Means, Save, Export.* The default command tab when creating a model for the first-time is *Type of Model.*

a. The *Type of Model* command tab allows specifying the model's distribution and link function.

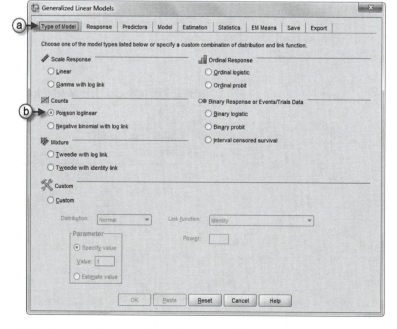

b. For this first model we will designate a Poisson distribution and log link function. Beneath the *Counts* section, click the radio button to select *Poisson loglinear.*

3a. Click the *Response* command tab, which enables specifying a dependent variable for the model.

b. For this model we will use the number of courses that students fail during their ninth grade year of high school. Click to select *fail* from the *Variables* list and then click the right-arrow button to move the variable into the *Dependent Variable* box.

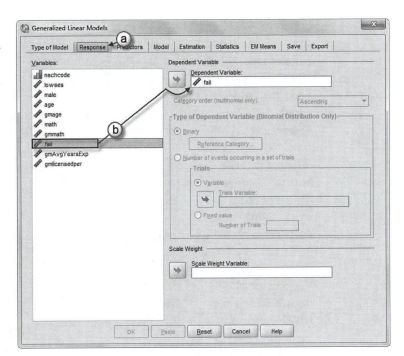

4a. Click the *Predictors* command tab, which enables selecting factors and covariates to use in predicting the dependent variable.

b. We will use four predictor variables in this model. To facilitate reading of the output tables we will rearrange the variable order. Click to select *lowses, male, age,* and *math* and then click the right-arrow button to move the variables into the *Covariates* box.

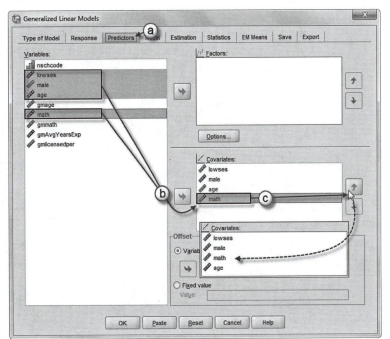

c. To facilitate reading of the output tables we will rearrange the variable order. Click to select *math* and then click the up-arrow button on the right to place *math* above *age*. The variable order is now *lowses, male, math, age*.

Note: An alternative method for managing the variable order is to enter them individually into the *Covariates* box.

5a. Click the *Model* command tab, which enables specifying model effects based upon the covariate variables (*lowses, male, math, age*).

b. Note that the *Include intercept in model* option is preselected. This is because a model is intercept only (default), so other model effects must be specified.

c. We will use the default *Main effects*, which creates a main-effects term for each variable selected for the model.

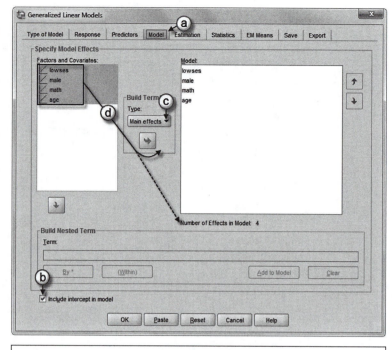

$$\eta_i = \beta_0 + \beta_1 lowSES + \beta_2 female + \beta_3 math + \beta_4 age \quad \text{(Eq. 7.3)}$$

d. Click to select the four variables (*lowses, male, math, age*) and then click the right-arrow button to move them into the *Model* box. Note that the four effects are now identified in the model (*Number of Effects in Model: 4*).

6a. Click the *Estimation* command tab, which enables specifying estimation methods and providing initial values for the parameter estimates.

b. Change the *Covariance Matrix* by clicking the *Robust estimator.*

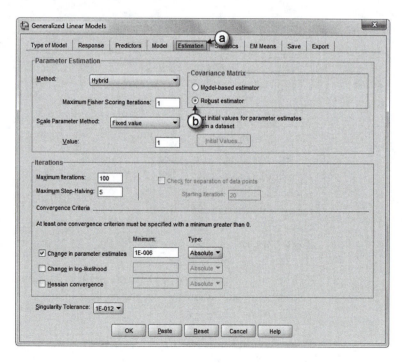

7a. Click the *Statistics* command tab, which enables specifying model effects (analysis type and confidence intervals) and print output.

b. We will retain the default *Model Effects* settings but add an additional print option to obtain exponential parameter estimates. Click to select *Include exponential parameter estimates*.

Click the OK button to generate the model results.

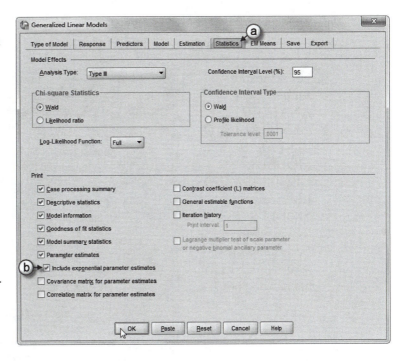

Interpreting the Output Results of Model 1.1

The fixed effects are summarized in Table 7.5. The intercept (−2.473) describes the log of expected counts (or incident rates) of a student failing a course when the value is 0 on all the predictors. As we have noted previously, this preliminary model may not produce the estimates that make the most sense for examining these data because in this case no one in the data set has a math score of 0 or is 0 years old. To create a more meaningful intercept log odds, we may wish to center the continuous variables in the model on the mean of the sample. In this case, those variables would be age and math.

Once again, we can define the same model after recentering the estimates for age and math. We will grand-mean center each:

$$\eta_i = \beta_0 + \beta_1 lowSES + \beta_2 male + \beta_3 gmmath + \beta_4 gmage,$$ (7.4)

where $\eta_i = \log(\lambda)$.

TABLE 7.5 Model 1.1: Parameter Estimates

Parameter	β	Std. error	95% Wald confidence interval		Hypothesis test			Exp(β)	95% Wald confidence interval for exp(β)	
			Lower	Upper	Wald χ²	df	Sig.		Lower	Upper
(Intercept)	−2.473	0.5220	−3.496	−1.450	22.442	1	0.000	0.084	0.030	0.235
lowses	0.344	0.0372	0.271	0.417	85.532	1	0.000	1.411	1.312	1.518
male	0.168	0.0358	0.097	0.238	21.902	1	0.000	1.182	1.102	1.268
math	−0.007	0.0003	−0.007	−0.006	611.736	1	0.000	0.993	0.993	0.994
age	0.242	0.0375	0.169	0.316	41.721	1	0.000	1.274	1.184	1.372
(Scale)	1[a]									

Note: Dependent variable: fail; model: (intercept), lowses, male, math, age.
[a] Fixed at the displayed value.

Defining Model 1.2 With IBM SPSS Menu Commands

Note: Settings default to those used for Model 1.1.

1. Go to the tool-
bar and select
ANALYZE,
GENERALIZED
LINEAR
MODELS,
GENERALIZED
LINEAR
MODELS. This
command opens the
*Generalized Linear
Models* (GENLIN)
main dialog box.

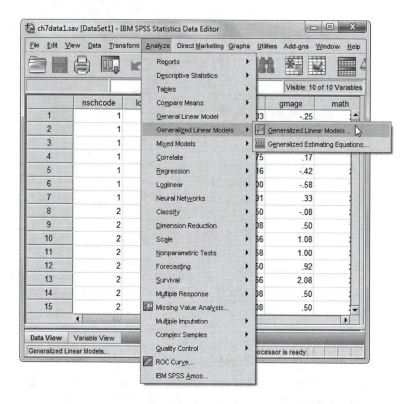

2a. The *Generalized
Linear Models* display
screen highlights the
Statistics tab, which
was the last com-
mand window used
for Model 1.1.

b. Click the *Predictors*
command tab.

c. We will change
the model by first
removing the
unwanted variables.
Click to select both
math and *age* from
the *Covariates* box
and then click the
left-arrow button to
remove them from
the model.

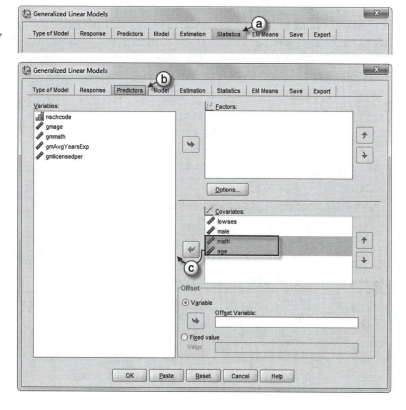

d. Now click to select the grand-mean variables (*gmage*, *gmmath*) from the *Variables* column and then click the right-arrow button to move them into the *Covariates* box.

e. To facilitate reading of the output tables, we will rearrange the variable order. Click to select *gmmath* and then click the up-arrow button on the right to place *gmmath* above *gmage*. The variable order is now *lowses, male, gmmath, gmage*.

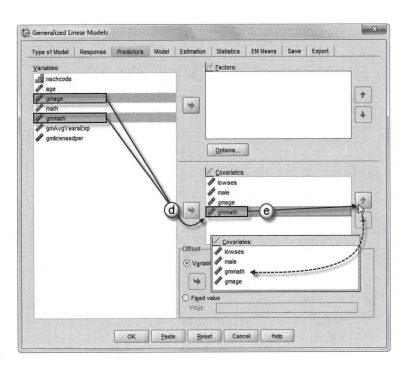

3a. Click the *Model* command tab so that the two grand-mean variables (*gmmath*, *gmage*) may be added as fixed effects into the model.

b. *Main effects* remains as the default setting.

c. Click to select *gmmath* and *gmage* from the *Factors and Covariates* column and then click the right-arrow button to move the two variables into the *Model* box.

Click the OK button to generate the model results.

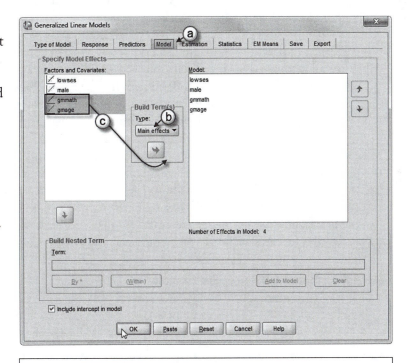

$$\eta_i = \beta_0 + \beta_1 lowSES + \beta_2 male + \boldsymbol{\beta_3 gmmath} + \boldsymbol{\beta_4 gmage} \quad \text{(Eq. 7.4)}$$

TABLE 7.6   Model 1.2: Continuous Variable Information

		N	Minimum	Maximum	Mean	Std. dev.
Dependent variable	fail	9956	0.00	4.00	0.60	1.09000
Covariate	lowses	9956	0.00	1.00	0.42	.49400
	male	9956	0.00	1.00	0.51	.50000
	gmmath	9956	−144.66	255.34	0.00	62.36500
	gmage	9956	−1.50	3.00	0.00	.43097

Interpreting the Output Results of Model 1.2

We can see in Table 7.3 that the mean for math is 244.66 and for age is 13.58 years. After we grand-mean center math and age (refer to Chapter 2), the new means will be 0 for math and age. We see the result of that in Table 7.6. Readers will notice that the standard deviations stay the same for the variables that have been grand-mean centered.

In Table 7.7, the new intercept log of the expected incident rate is now considerably different (−0.879, $p < .001$) from that in Table 7.5. The associated hypothesis test indicates that we can reject the null hypothesis that the intercept is zero, given the other predictors in the model, which are evaluated at zero. We can interpret the Poisson regression slopes as the amount that a one-unit change in the predictor increases or decreases the log odds of the expected event rate, holding the other predictors in the model constant.

First, we note in Table 7.7 that all the variables are significant in explaining students' likelihood to fail a course ($p < .001$). The estimated coefficients for dichotomous variables provide the change in the logs of expected counts. Because of the way the outcome is coded, positive coefficients indicate the individual is more likely to fail a course. For example, males are more likely to fail a course than females (0.168, $p < .001$), holding other variables constant. The difference in expected log counts is 0.168 units higher for males than for females. Similarly, holding other variables constant, low SES students are more likely to fail a core course than average or high SES students (0.344, $p < .001$). The difference in logs of expected counts is 0.344 higher for low SES students compared with more average or high SES students. For each one-unit increase in math, the difference in logs of expected counts would decrease slightly by 0.007. For age, for each added year, the difference in logs of expected counts would increase by 0.242.

Second, we can exponentiate the logs of the expected counts, which can then be interpreted as an incident rate ratio. For example, the logged incident rate for the intercept is −0.879 when all variables in the model are equal to 0. In this case, this can be interpreted as the rate for females (coded 0) who are of average or high SES (coded 0) and at the grand means (0) in age

TABLE 7.7   Model 1.2: Parameter Estimates

Parameter	β	Std. error	95% Wald confidence interval		Hypothesis test			Exp(β)	95% Wald confidence interval for exp(β)	
			Lower	Upper	Wald χ²	df	Sig.		Lower	Upper
(Intercept)	−0.879	0.0330	−0.944	−0.815	711.487	1	0.000	0.415	0.389	0.443
lowses	0.344	0.0372	0.271	0.417	85.532	1	0.000	1.411	1.312	1.518
male	0.168	0.0358	0.097	0.238	21.902	1	0.000	1.182	1.102	1.268
gmmath	−0.007	0.0003	−0.007	−0.006	611.736	1	0.000	0.993	0.993	0.994
gmage	0.242	0.0375	0.169	0.316	41.721	1	0.000	1.274	1.184	1.372
(Scale)	1[a]									

Note: Dependent variable: fail; model: (intercept), lowses, male, gmmath, gmage.
[a] Fixed at the displayed value.

and math. The corresponding predicted incident rate in Table 7.7 (e^β) is then 0.415. For males, the logged event rate will be the following:

$$\log(\hat{\lambda}) = -0.879 + 0.168(male),$$

which is a logged incident rate of −0.711.

Because the log of the expected value of Y is a linear function of explanatory variable(s), a unit change in X has a multiplicative effect on the expected event rate; that is, the expected incident rate is multiplied by the value $\exp(B)$ for each one unit increase in X. For males, coded 1, holding the other variables constant, the logged incident rate can also be expressed as a multiplicative model:

$$\hat{\lambda} = e^{-0.879} e^{0.168(1)} = (0.415)(1.183)$$

which corresponds to an expected incident rate of 0.491. This can be interpreted as the change in incident rate for a unit change in the predictor (i.e., from female coded 0 to male coded 1). The ratio of these two incident rates, which is the difference in event rates for males (coded 1) versus females (coded 0), is then the following:

$$\frac{males = 0.491}{females = 0.415} = 1.183$$

This is the same as the estimated difference in incident rate ratio of 1.182 reported for males in Table 7.7 (slightly different due to rounding). This suggests that males have an incident rate ratio of 1.18 times the incident rate ratio for females (or 18.3% more), holding the other variables constant.

For socioeconomic status, the intercept event rate for average/high SES students who are female is −0.879, holding the other variables constant (age and math at their grand means). The expected logged event rate for low SES students (holding the other variables constant) is then the following multiplicative model:

$$\hat{\lambda} = e^{-0.879} e^{0.344(1)} = (0.415)(1.411)$$

This corresponds to an expected event rate of 0.586. The ratio of the two expected event rates is then

$$\frac{lowSES = 0.586}{ave/hiSES = 0.415} = 1.412,$$

which is the difference in event rates for low SES students versus average or high SES students (1.411) in Table 7.7 (a small difference is due to rounding). We can interpret this as that, for low SES students, the expected course failure count is increased by a factor of 1.41 compared with average or high SES students (or about 41%). For students at the grand mean for age (0), the intercept logged event rate is −0.879 (expected event rate = 0.415), holding the other variables constant. An added year of age will increase the logged event rate by 0.242. The expected event rate will then be as follows:

$$\hat{\lambda} = e^{-0.879} e^{0.242(1)} = (0.415)(1.274)$$

This corresponds to the expected event rate of 0.529 when age is increased by 1 year. The ratio of these event rates will then be

$$\frac{1\,year = 0.529}{grandmean = 0.415} = 1.275,$$

which corresponds with the value in Table 7.7 (1.274). This suggests that the event rate count is increased by a factor of 1.274 (or 27.4%) for an added year of age. For each test score point increase, the expected count is decreased by a factor of 0.993.

Third, we can also use the EMMEANS command to obtain the predicted counts for the levels of our factors (*lowses* and *male*) and their interaction while holding the other continuous variables (gmmath, gmage) in the model at their grand means (which are 0). We provide the syntax to run this model in Appendix A. This can also help us understand the relationships present in Table 7.7. We can use this information (as well as the expected event rates in Table 7.7) to calculate the probability of failing one or more courses, given the variables in the model.

The results for the interaction between low SES and gender are shown in Table 7.8. We could test whether the cell means are significantly different by including the interaction term in the model, but for our purpose, we will simply examine the event rates for each of the subgroups implied in the 2 by 2 interaction table. From this table, we can see that the event rate for females who are average/high SES (cell 0, 0) is 0.415, holding the covariates fixed at their means. This is the same as the exponentiated intercept ($e^{-0.879}$) in Table 7.7. The event rate for males versus females in Table 7.7 is 1.182. If we multiply the event rate for females who are average/high SES, cell (0, 0), by the difference in event rates for males versus females [0.415(1.182)], we obtain the event rate (0.49053) for males who are low SES (cell 0, 1).

Readers will no doubt see that the logs of the expected counts in Table 7.7 basically represent another way of presenting the information contained in the cross-break table. Once we obtain the event rates for any particular groups of interest, we can calculate the resulting probability of failing one or more courses as in Equation 6.2, controlling for the other variables in the model. For example, the probability of an average/high SES female failing a course (cell 0, 0) with an estimated event rate mean of 0.415 (the intercept event rate in Table 7.7) will be

$$\frac{e^{-(0.415)}0.415^1}{1!} = \frac{(0.660)(0.415)^1}{1} = 0.274$$

The probability of average/high SES males failing a course (cell 0, 1) would be

$$\frac{e^{-(0.491)}0.491^1}{1!} = \frac{(0.612)(0.491)^1}{1} = 0.300$$

TABLE 7.8 Marginal Means for lowses*male With Covariates at Their Grand Means

| Lowses | Male | Mean | Std. error | 95% Wald confidence interval | |
				Lower	Upper
1	1	0.693	0.022	0.65	0.74
	0	0.586	0.020	0.55	0.63
0	1	0.491	0.015	0.46	0.52
	0	0.415	0.014	0.39	0.44

Note: Covariates appearing in the model are fixed at the following values: gmage = 0.0014; gmmath = 0.0034.

**TABLE 7.9 Marginal Mean Estimates for lowses*male
With gmmath 1 SD Above the Grand Mean**

Lowses	Male	Mean	Std. error	95% Wald confidence interval	
				Lower	Upper
1	1	0.449	0.022	0.42	0.48
	0	0.380	0.020	0.35	0.41
0	1	0.318	0.015	0.30	0.34
	0	0.269	0.014	0.25	0.29

Note: Covariates appearing in the model are fixed at the following values: gmage = 0.000; gmmath = 62.365.

We note in passing that the ratio of the male to female event rates (0.491/0.415) for average/high SES standing is the event rate we calculated for males versus females (1.18), which corresponds to the difference in male and female rates in Table 7.7.

We can also adjust the predictors and see the corresponding effect on the event rates for the groups under consideration. For example, we could set gmmath to be one standard deviation above its grand mean (62.365 in Table 7.6). Recall that grand-mean centering creates a mean of 0 but maintains the standard deviation in the units of the raw metric. If we used a z-score for math, then it would be standardized such that the mean is 0 and the standard deviation is 1.0. In that case, a one-unit increase in either variable would then be interpreted as producing a 1-SD increase in the estimated event rate. When we adjust the level of gmmath to be one standard deviation above the mean, we see the results in Table 7.9. We provide the syntax statement to alter the means in Appendix A.

We can see that the event rates for the intercept group (i.e., average/high SES females) declines considerably (from 0.415 in Tables 7.7 and 7.8 to 0.269 in Table 7.9). We can determine that the ratio of these event rates (0.269/0.415) would be the expected difference (0.648) in rate ratio for a female of average/high SES one standard deviation above the grand mean in math versus at the grand mean. This suggests that if an individual increased her test score by one standard deviation, her rate ratio for course failures would be expected to decrease by a factor of 0.648, while holding all other variables in the model constant.

We can check our calculations by rerunning the model summarized in Table 7.7 using a z-score for math (zmath). The output confirms that the expected event rate for *zmath* (mean = 0, SD = 1) is indeed a rate ratio of 0.648. In Table 7.10, we can confirm that a 1-SD increase in math reduces the expected event rate by a factor of 0.648 (or 35.2%).

Considering Possible Overdispersion

We note also in Table 7.7 and Table 7.10 a scale factor that is set to 1.0 as a default. This is not a parameter in the model. The fixed scale factor assumes that the observed residuals follow the Poisson distribution exactly. Depending on particular characteristics of the data set, over- or underdispersion can be present. This can result from misspecification of the model or, in a multilevel data set, from small within-group sample sizes (Hox, 2010).

There is information in the model output that can be helpful in thinking about possible dispersion. The ratio of the model deviance to the degrees of freedom is one indicator. The dispersion is the deviance statistic divided by its degrees of freedom. If there is no overdispersion, the ratio will be close to 1. As we have noted, deviance is usually –2*log likelihood (–2LL) of the final model. In this case, however, SPSS calculates the deviance in a different manner. Some prefer the ratio of the Pearson chi-square to the degrees of freedom for this purpose (Hilbe, 2007). In Table 7.11, we can see that the estimate is a bit larger for the Pearson χ^2 than for the deviance.

TABLE 7.10   Parameter Estimates With *zmath*

Parameter	β	Std. error	Hypothesis test Wald χ²	df	Sig.	Exp(β)	95% Wald confidence interval for exp(β) Lower	Upper
(Intercept)	−0.879	0.0330	711.526	1	0.000	0.415	0.389	0.443
[lowses = 1]	0.344	0.0372	85.532	1	0.000	1.411	1.312	1.518
[lowses = 0]	0[a]					1		
[male = 1]	0.168	0.0358	21.902	1	0.000	1.182	1.102	1.268
[male = 0]	0[a]					1		
gmage	0.242	0.0375	41.721	1	0.000	1.274	1.184	1.372
zmath	−0.433	0.0175	611.736	1	0.000	0.648	0.627	0.671
(Scale)	1[b]							

Note: Dependent variable: fail; model: (intercept), lowses, male, gmage, zmath. Both lowses and male are treated as factors for the EM means procedure.

[a] Set to 0 because this parameter is redundant.

[b] Fixed at the displayed value.

When the observed variance is much larger than expected under the assumptions of the Poisson model, one option is to add a dispersion parameter. IBM SPSS does not support this option for the Poisson model. However, it allows estimation of counts using a negative binomial model, which extends the Poisson model by allowing extra variance in the counts (Hox, 2010). The negative binomial distribution can incorporate the overdispersion that is often present directly into the model by estimating an extra parameter in the variance function (McCullagh & Nelder, 1989, p. 373). This amounts to adding an error term to the model. The value of the ancillary parameter can be any number greater than or equal to zero. The user can set the ancillary parameter to a fixed value (e.g., 1.0) or allow it to be estimated by the program. We note in passing that the two distributions are the same under certain conditions (i.e., when the scale parameter in the negative binomial distribution is fixed at 0.0). In Model 1.3, we will begin by fixing the dispersion parameter to 0.0.

TABLE 7.11   Goodness of Fit[a]

	Value	df	Value/df
Deviance	14322.071	9951	1.439
Scaled deviance	14322.071	9951	
Pearson chi-square	17971.844	9951	1.806
Scaled Pearson chi-square	17971.844	9951	
Log likelihood[b]	−10926.549		
Akaike's information criterion (AIC)	21863.098		
Finite sample corrected AIC (AICC)	21863.104		
Bayesian information criterion (BIC)	21899.127		
Consistent AIC (CAIC)	21904.127		

Note: Dependent variable: fail; model: (intercept), lowses, male, gmage, gmmath.

[a] Information criteria are in small-is-better form.

[b] The full log likelihood function is displayed and used in computing information criteria.

Defining Model 1.3 With IBM SPSS Menu Commands

Note: Settings default to those used for Model 1.2.

1. Go to the toolbar and select ANALYZE, GENERALIZED LINEAR MODELS, GENERALIZED LINEAR MODELS. This command opens the *Generalized Linear Models* (GENLIN) main dialog box.

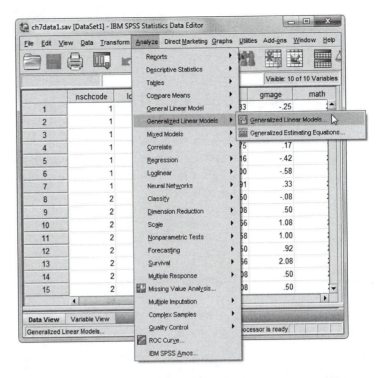

2a. The *Generalized Linear Models* display screen highlights the *Model* tab, which was the last command window used for Model 1.2.

b. Click the *Type of Model* command tab to access the distribution and link options.

c. We will use a custom setting to designate a negative binomial with log link distribution with a fixed dispersion value of 0. First, click to select *Custom*.

d. Click the *Distribution* pull-down menu and select *Negative binomial*.

e. Click the *Link function* pull-down menu and select *Log*.

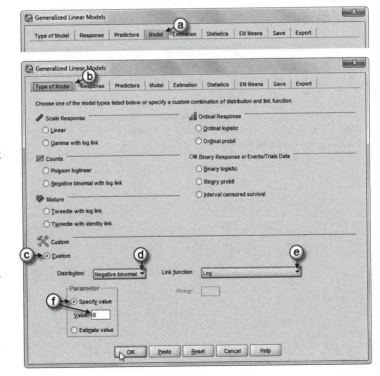

f. Within the *Parameter* option, use the *Specify value* setting but change the default value (1) to 0.

Click the OK button to generate the model results.

TABLE 7.12   Model 1.3: Model Information

Dependent variable	Fail
Probability distribution	Negative binomial (0)
Link function	Log

Interpreting the Output Results of Model 1.3

Table 7.12 provides the model information, summarizing that we have a negative binomial distribution with scale parameter fixed to zero and a log link function. We can see in Table 7.13 that the fixed-effects estimates confirm that the negative binomial distribution ancillary parameter set to 0 provides the same results as the previous Poisson model estimates (see Table 7.7). Next, we will fix the negative binomial dispersion parameter at 1.0. We provide the model commands for this particular step.

TABLE 7.13   Model 1.3: Parameter Estimates

Parameter	β	Std. error	95% Wald confidence interval		Hypothesis test			Exp(β)	95% Wald confidence interval for exp(β)	
			Lower	Upper	Wald χ^2	df	Sig.		Lower	Upper
(Intercept)	−0.879	0.0330	−0.944	−0.815	711.487	1	0.000	0.415	0.389	0.443
lowses	0.344	0.0372	0.271	0.417	85.532	1	0.000	1.411	1.312	1.518
male	0.168	0.0358	0.097	0.238	21.902	1	0.000	1.182	1.102	1.268
gmmath	−0.007	0.0003	−0.007	−0.006	611.736	1	0.000	0.993	0.993	0.994
gmage	0.242	0.0375	0.169	0.316	41.721	1	0.000	1.274	1.184	1.372
(Scale)	1ᵃ									
(Negative binomial)	0*									

Note: Dependent variable: fail; model: (intercept), lowses, male, gmmath, gmage.
ᵃ Fixed at the displayed value.

Defining Model 1.4 With IBM SPSS Menu Commands

Note: Settings default to those used for Model 1.3.

1. Go to the tool-bar and select ANALYZE, GENERALIZED LINEAR MODELS, GENERALIZED LINEAR MODELS. This command opens the *Generalized Linear Models* (GENLIN) main dialog box.

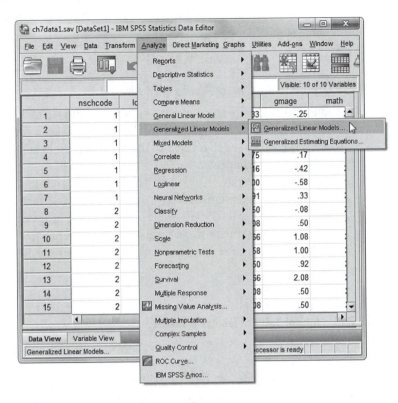

2a. The *Generalized Linear Models* display screen highlights the *Type of Model* tab which was the last command window used for Model 1.3.

b. We will continue using the. *Custom* setting with a negative binomial distribution and log link but now designate a fixed dispersion value of 1. Change the *Value* from 0 (used in the prior model) to 1.

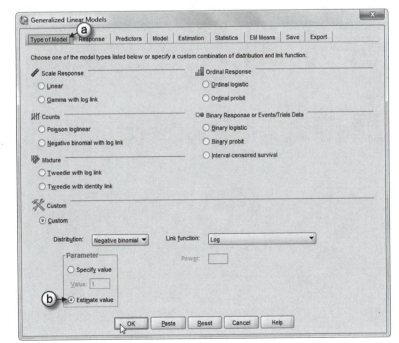

Click the OK button to generate the model results.

TABLE 7.14   Model 1.4: Model Information

Dependent variable	Fail
Probability distribution	Negative binomial (1)
Link function	Log

Interpreting the Output Results of Model 1.4

Table 7.14 confirms that we have set the parameter to 1.0. In Table 7.15, we can see that the fixed effects for the negative binomial distribution are a little different from the previous Poisson model. We will leave to readers to interpret any slight substantive differences in the coefficients. For example, the expected logged rate of course failures is slightly larger (−0.920) than the intercept logged event rate (−0.879) for the Poisson regression model in Table 7.7. The estimates also have slightly larger standard errors on most fixed-effect estimates. For example, the intercept standard error was 0.0330 in Table 7.7, and it increases slightly to 0.0340 in Table 7.15.

Next, if we estimate the overdispersion present (1.747), instead of fixing the value of the ancillary parameter at 1.0, the resulting estimates are only slightly different from the previous model.

TABLE 7.15   Model 1.4: Parameter Estimates

Parameter	β	Std. error	95% Wald confidence interval		Hypothesis test			Exp(β)	95% Wald confidence interval for exp(β)	
			Lower	Upper	Wald χ^2	df	Sig.		Lower	Upper
(Intercept)	−0.920	0.0340	−0.987	−0.853	733.613	1	0.000	0.399	0.373	0.426
lowses	0.367	0.0373	0.294	0.440	97.083	1	0.000	1.444	1.342	1.553
male	0.196	0.0366	0.125	0.268	28.871	1	0.000	1.217	1.100	1.307
gmmath	−0.008	0.0003	−0.008	−0.007	660.557	1	0.000	0.992	0.992	0.993
gmage	0.249	0.0390	0.172	0.325	40.768	1	0.000	1.282	1.188	1.384
(Scale)	1[a]									
(Negative binomial)	1									

Note: Dependent variable: fail; model: (intercept), lowses, male, gmmath, gmage.
[a] Fixed at the displayed value.

Defining Model 1.5 With IBM SPSS Menu Commands
Note: IBM SPSS settings default to those used for Model 1.4.

1. Go to the tool-bar and select ANALYZE, GENERALIZED LINEAR MODELS, GENERALIZED LINEAR MODELS. This command allows access to the *Generalized Linear Models* (GENLIN) main dialog box.

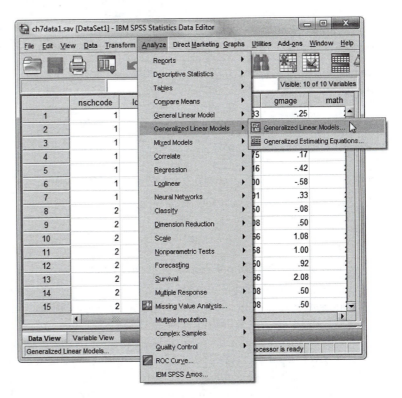

2a. The *Generalized Linear Models* display screen highlights the *Type of Model* tab as it was the last command window used for Model 1.4.

b. We will continue using the *Custom* setting with a negative binomial distribution and log link but allow the IBM SPSS software to estimate the over-dispersion. Click to select *Estimate value.*

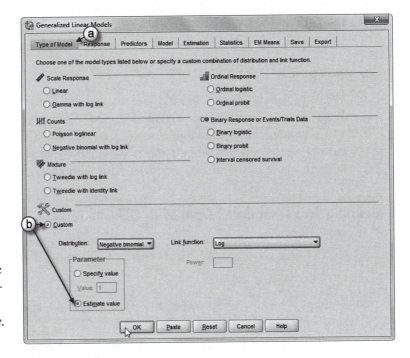

Click the OK button to generate the model results.

TABLE 7.16   Model 1.5: Parameter Estimates

Parameter	β	Std. error	95% Wald confidence interval		Hypothesis test			Exp(β)	95% Wald confidence interval for exp(β)	
			Lower	Upper	Wald χ²	df	Sig.		Lower	Upper
(Intercept)	−0.936	0.0346	−1.004	−0.868	732.880	1	0.000	0.392	0.367	0.420
lowses	0.376	0.0376	0.302	0.450	100.133	1	0.000	1.456	1.353	1.568
male	0.208	0.0371	0.135	0.281	31.458	1	0.000	1.231	1.145	1.324
gmmath	−0.008	0.0003	−0.009	−0.008	652.899	1	0.000	0.992	0.991	0.992
gmage	0.250	0.0395	0.173	0.328	40.279	1	0.000	1.285	1.189	1.388
(Scale)	1[a]									
(Negative binomial)	1.747	0.0710	1.614	1.892						

Note: Dependent variable: fail; model: (intercept), lowses, male, gmmath, gmage.
[a] Fixed at the displayed value.

Interpreting the Output Results of Model 1.5

Once again, in Table 7.16 the standard errors are slightly larger after estimating the overdispersion more directly. Because hypothesis tests are based on the ratio of the parameter to its standard error, including the ancillary parameter in the model improves the model fit and corrects the standard errors. This will result in more accurate hypothesis tests in the presence of overdispersion. As Hox (2010) cautions, if the scale factor is very different from 1.0, it may be desirable to locate the cause of the misfit and modify the model. Using the negative binomial distribution with terms for overdispersion results in a lower expected event rate of 0.392 (which represents the expected failure rate for females of average/high SES), holding gmmath and age at their grand means. This is a bit lower than the estimated event rate in the Poisson model (0.415).

Comparing the Fit

We can examine whether the added estimated dispersion parameter fits the data better than the Poisson model without a dispersion parameter. We can compare the fit of the negative binomial model against the Poisson model using the AIC and BIC (because the models are not nested). In this case, when we compare the two models, we find that the negative binomial model fits better because AIC and BIC are lower. This is summarized in Table 7.17.

Estimating Two-Level Count Data With GENLIN MIXED

We turn our attention now to defining the same model using GENLIN MIXED. We rerun the first model using a Poisson distribution. The single-level model is the same as Equation 7.4.

TABLE 7.17   Fit Comparison Results

	Poisson	Negative binomial
AIC	21,863.098	19,854.094
BIC	21,899.127	19,897,321

Defining Model 2.1 With IBM SPSS Menu Commands

Note: Refer to Chapter 4 for discussion concerning settings and modeling options available in the IBM SPSS GENLIN MIXED routine.

Continue using the *ch7data1.sav* data file.

1. Go to the toolbar and select ANALYZE, MIXED MODELS, GENERALIZED LINEAR MODELS. This command opens the *Generalized Linear Mixed Models* (GENLINMIXED) main dialog box.

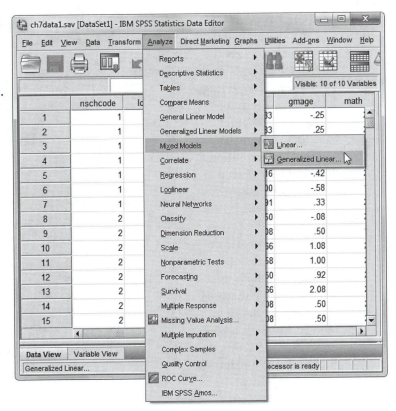

2. The *Generalized Linear Mixed Models* display screen shows four command tabs: *Data Structure, Fields & Effects, Build Options*, and *Model Options*.

a. The *Data Structure* command tab is the default screen display when creating a model for the first time. The options enable specifying structural relationships between data set records when observations are correlated.

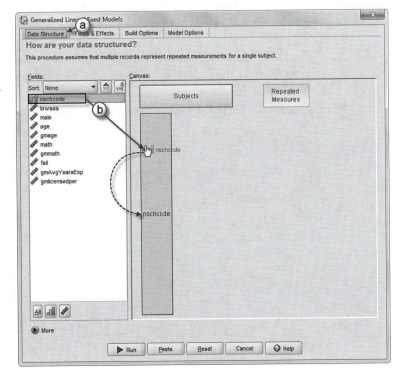

b. To build the model, begin by defining the subject variable. A "subject" is an observational unit that can be considered independently of other subjects. Click to select the *nschcode* variable from the left column and then "drag and drop" *nschcode* onto the canvas area beneath the *Subjects* box.

3a. Click the *Fields & Effects* command tab, which displays the default *Target* display screen. The settings allow specifying a model's target (dependent variable), distribution, and relationship to model predictors through the link functions.

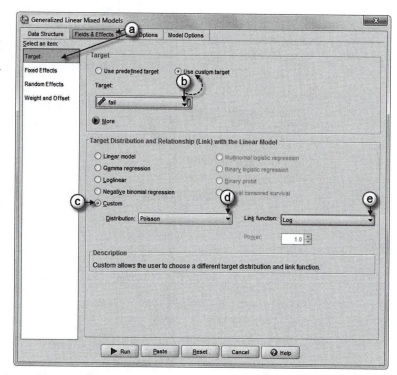

b. For this model we will use the number of courses that students fail during their ninth grade year of high school. Click the *Target* pull-down menu and select *fail*.

Note: Selecting the target variable will automatically change the setting to *Use custom input.* Also, the ordinal measurement setting of the dependent variable (*plans*) automatically designates the target distribution as *Multinomial logistic regression.*

c. For this first model we will designate a Poisson distribution and log link function. First, click to select the *Custom* option to designate a custom target distribution and link function.

d. Now click the *Distribution* pull-down menu and select *Poisson*.

e. Finally, click the *Link function* pull-down menu and select *Log*.

4a. Now click the *Fixed Effects* item to access the main dialog box.

b. Note that the *Include intercept in model* option is preselected. This is because a Level-1 model is intercept only (default), so other model effects must be specified.

c. We will use four predictor variables in this model. Click to select *lowses, male, gmage,* and *gmmath* and then drag and drop them into the *Effect builder* box within the *Main* section.

d. To facilitate reading of the output tables, we will rearrange the variable order. Click

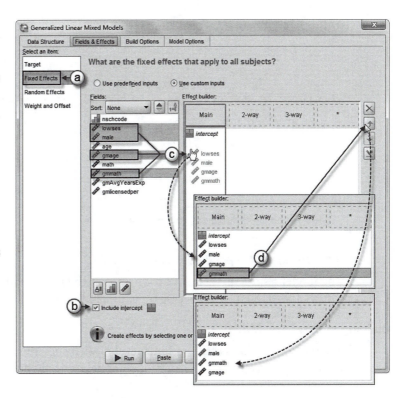

to select *gmmath* and then click the up-arrow button on the right to place *gmmath* above *gmage*. The variable order is now *lowses, male, gmmath, gmage*.

$$\eta_i = \beta_0 + \beta_1 lowSES + \beta_2 male + \beta_3 gmmath + \beta_4 gmage \quad \text{(Eq. 7.4)}$$

5a. Click the *Build Options* command tab located at the top of the screen to access advanced criteria for building the model. Skip over the *Sorting Order* option as the settings have no effect on count data. We will retain the default settings for all except *Tests of fixed effects and Coefficients*.

b. For *Tests of fixed effects and coefficients*, click to select *Use robust estimation to handle violations of model assumptions*.

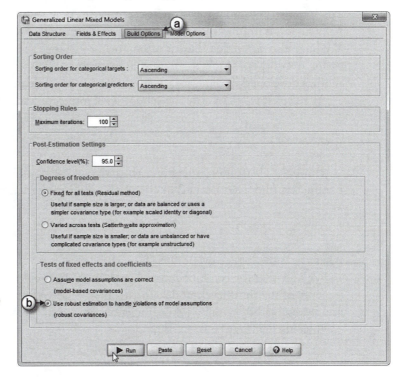

Click the RUN button to generate the model results.

TABLE 7.18   Model 2.1: Fixed Effects

Model term	Coefficient	Std. error	t	Sig.	Exp(coefficient)	95% Confidence interval for exp(coefficient)	
						Lower	Upper
(Intercept)	−0.879	0.033	−26.674	0.000	0.415	0.389	0.443
lowses	0.344	0.037	9.248	0.000	1.411	1.312	1.518
male	0.168	0.036	4.680	0.000	1.182	1.102	1.269
gmmath	−0.007	0.000	−24.733	0.000	0.993	0.993	0.994
gmage	0.242	0.038	6.459	0.000	1.274	1.184	1.372

Note: Probability distribution: Poisson; link function: log.

Interpreting the Output Results of Model 2.1

The model results in Table 7.18 confirm that the same model can be specified and estimated in GENLIN MIXED, in this case by not including a random intercept at Level 2.

Building a Two-Level Model

We will now build a series of multilevel models, beginning with an unconditional model with a random intercept. The unconditional model for individual i in high school j will simply be the following at Level 1:

$$\eta_{ij} = \log(\lambda_{ij}) = \beta_{0j}, \tag{7.5}$$

where η_{ij} is equal to the natural logarithm of the event rate λ_{ij}.

We reiterate that in a Poisson distribution there is no separate residual variance modeled at Level 1. The lowest level residuals are scaled to the variance of the particular distribution (for the Poisson distribution, it is the predicted mean). As we have noted, because the scale parameter is fixed at 1.0, it implies that the observed residuals follow the expected error distribution exactly. We have also noted that there is often greater variance than the expected scale value of 1.0 (indicating overdispersion). We can use the preliminary Poisson models and negative binomial (NB) models developed in the last section to identify possible dispersion.

At Level 2, we have only the school-level intercept and random variance:

$$\beta_{0j} = \gamma_{00} + u_{0j}. \tag{7.6}$$

Through substitution, therefore, we arrive at the combined single-equation model with two parameters (the fixed Level-2 intercept and the Level-2 variance):

$$\eta_{ij} = \gamma_{00} + u_{0j}, \tag{7.7}$$

where η_{ij} is the logged event rate (λ).

We note that the combined model contains both the between-groups and within-group components of the multilevel model. In this model, as well as subsequent models, the fixed effects are represented by gamma (γ) coefficients because in the combined model they are group-level coefficients that may be fixed or random.

Defining Model 2.2 With IBM SPSS Menu Commands
Note: IBM SPSS settings default to those used for Model 2.1.

1. Go to the toolbar and select ANALYZE, MIXED MODELS, GENERALIZED LINEAR MODELS. This command opens the *Generalized Linear Mixed Models* (GENLINMIXED) main dialog box.

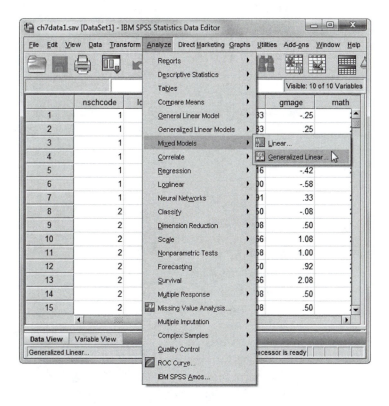

2a. The *Generalized Linear Mixed Models* dialog box displays the *Build Options* command tab, which was the last setting used in the prior model.

   b. Click to select the *Fields & Effects* command tab. Note that the *Fixed Effects* item is preselected as it was the last setting used from the *Fields & Effects* command tab in the prior model.

   c. Creating the null model requires removing the predictor variables from the model. To remove the variables, click to select *lowses, male, gmage,* and *gmmath* and then click the "X" button on the right. The *Effect builder* now displays a predictor-free (unconditional/null) model.

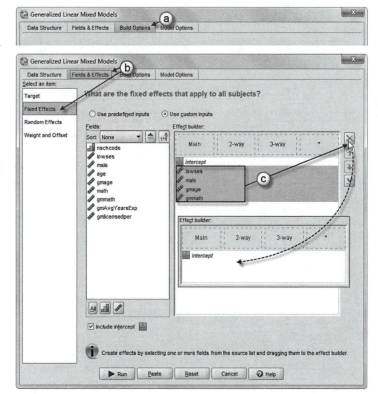

$$\eta_{ij} = \log(\lambda_{ij}) = \beta_{0j} \quad \text{(Eq. 7.5)}$$

3a. We will add a random effect (intercept) to the model. First, click the *Random Effects* item.

b. Click the ADD BLOCK button to access the *Random Effect Block* main dialog box.

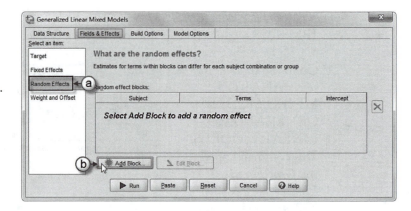

c. Click to select the *include intercept* option, which changes the *no intercept* default to *intercept*. Because there is only one random effect (the intercept), we can use the default variance component setting.

d. We need to identify the *Subject combination* by clicking the pull-down menu and selecting *nschcode*. Click the OK button to close the *Random Effect Block* dialog box and return to the *Random Effects* main screen.

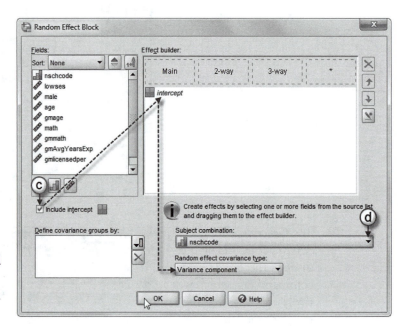

$$\eta_{ij} = \gamma_{00} + u_{0j} \quad \text{(Eq. 7.7)}$$

e. In the main screen notice that the *Random effect blocks* displays a summary of the block just created: *nschcode* (Subject) and *intercept* (Yes).

Click the RUN button to generate the model results.

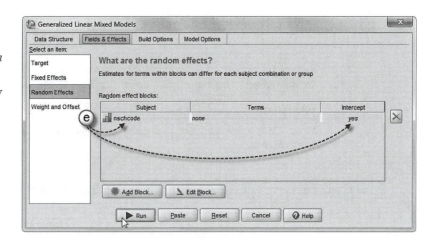

Interpreting the Output Results of Model 2.2

The model output in Table 7.19 confirms that we are estimating the model we intended. There are one fixed effect to be estimated and one random effect at Level 2. We can also see 44 units (high schools) in the Level-2 sample. Note that we subsequently will eliminate the information provided below the table.

As noted in Table 7.20, the estimated logged event rate for failing a course at the high school level is −0.701 (where the random effect is held constant at 0 or at the average variability among Level-2 units). The corresponding exponentiated Poisson regression coefficient is a rate ratio corresponding to a one-unit difference in the predictor. In this case, at Level 2, with no predictors in the model, the expected event rate is about 0.496.

TABLE 7.19 Model 2.2: Covariance Parameters

Covariance parameters	
Residual effect	0
Random effects	1
Design matrix columns	
Fixed effects	1
Random effects	1[a]
Common subjects	44

Note: Common subjects are based on the subject specifications for the residual and random effects and are used to chunk the data for better performance.

[a] This is the number of columns per common subject.

TABLE 7.20 Model 2.2: Fixed Effects

Model term	Coefficient	Std. error	*t*	Sig.	Exp(coefficient)	95% Confidence interval for exp(coefficient)	
						Lower	Upper
(Intercept)	−0.701	0.074	−9.455	0.000	0.496	0.429	0.574

Note: Probability distribution: Poisson; link function: log.

TABLE 7.21 Model 2.2: Variance Components

Random effect	Estimate	Std. error	z-Test	Sig.	95% Confidence interval	
					Lower	Upper
Var(Intercept)	0.220	0.055	3.977	0.000	0.134	0.360

Note: Covariance structure: variance components; subject specification: nschcode.

From this event rate, we can estimate the probability of failing one course as follows:

$$P(Y = 1) = \frac{e^{-\lambda}\lambda^c}{c!} = \frac{e^{-(0.496)}0.496^1}{1!} = \frac{(0.609)(0.496)^1}{1} = 0.302.$$

Readers can calculate the other probabilities as desired.

We will also make note of the Level-2 variance components ($\sigma^2_{Between} = 0.220$; SE = 0.055). The accompanying z-test is 3.997. We can conclude that we can build a Level-2 model to explain variance in course failures at the group level (see Table 7.21).

Within-Schools Model

We will next add the four Level-1 predictors to the proposed model. At Level 1, we have the following:

$$\eta_{ij} = \beta_{0j} + \beta_{1j}lowSES_{ij} + \beta_{2j}male_{ij} + \beta_{3j}gmmath_{ij} + \beta_{4j}gmage_{ij}, \tag{7.8}$$

where $h_{ij} = \log(1_{ij})$. At Level 2, we have the same intercept model (Equation 7.6):

$$\beta_{0j} = \gamma_{00} + u_{0j}.$$

In addition, we now have the following added slope parameters, which are defined as fixed because there are no corresponding random school variance parameters ($u_{1j} - u_{4j}$):

$$
\begin{aligned}
\beta_{1j} &= \gamma_{10}, \\
\beta_{2j} &= \gamma_{20}, \\
\beta_{3j} &= \gamma_{30}, \\
\beta_{4j} &= \gamma_{40}.
\end{aligned}
\tag{7.9}
$$

The combined single-equation model suggests that there is one random effect in the model (the intercept variance u_{0j}) and five fixed effects (i.e., the preceding intercept and four fixed slopes). We can see that the fixed effects for the within-group predictors are now fixed Level-2 coefficients:

$$\eta_{ij} = \gamma_{00} + \gamma_{10}lowSES_{ij} + \gamma_{20}male_{ij} + \gamma_{30}gmmath_{ij} + \gamma_{40}gmage_{ij} + u_{0j}. \tag{7.10}$$

Defining Model 2.3 With IBM SPSS Menu Commands

Note: IBM SPSS settings default to those used for Model 2.2.

1. Go to the tool-bar and select ANALYZE, MIXED MODELS, GENERALIZED LINEAR MODELS. This command opens the *Generalized Linear Mixed Models* (GENLINMIXED) main dialog box.

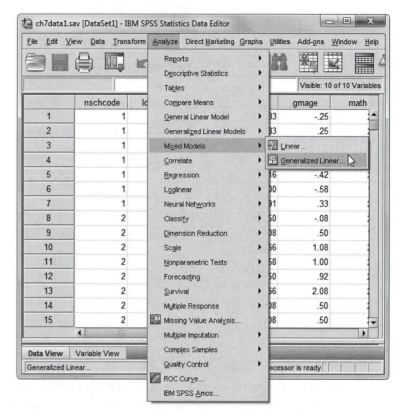

2a. The GENLINMIXED dialog box displays the *Build Options* command tab and *Random Effects* item, which was the last setting used in the prior model.

   Note: The *Random Effects* item is preselected (represented in the illustration by the dashed-line box but appears on-screen colored blue) as it was the last setting used from the prior model.

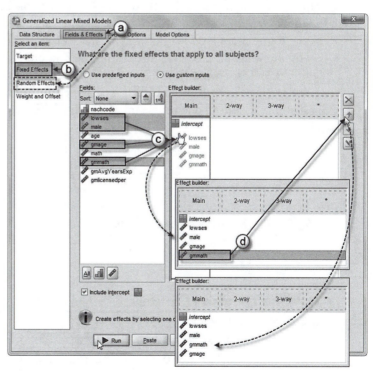

$$\eta_{ij} = \beta_{0j} + \beta_{1j}lowSES_{ij} + \beta_{2j}male_{ij} + \beta_{3j}gmmath_{ij} + \beta_{4j}gmage_{ij} \quad \text{(Eq. 7.8)}$$

TABLE 7.22 Model 2.3: Covariance Parameters

Covariance parameters	
Residual effect	0
Random effects	1
Design matrix columns	
Fixed effects	5
Random effects	1
Common subjects	44

b. We will begin to build the model by adding predictor variables. First, click to select the *Fixed Effects* item.

c. We will now add four variables to the model: *lowses*, *male*, *gmage*, and *gmmath*. Click to select the four variables and then drag and drop them into the *Main* (effect) column of the *Effect builder*.

d. To facilitate reading of the output tables, we will rearrange the variable order. Click to select *gmmath* and then click the up-arrow button on the right to place *gmmath* above *gmage*. The variable order is now *lowses*, *male*, *gmmath*, *gmage*.

Click the RUN button to generate the model results.

Interpreting the Output Results of Model 2.3

The model dimensions in Table 7.22 confirm that we are estimating five fixed effects (i.e., the intercept and slopes for the four Level-1 predictors) and one random effect at Level 2 (the intercept). We also see again that there are 44 high schools in the sample.

In Table 7.23, we can immediately see the estimated logged event rate for failing a course at the high school level is now –1.060. Recall that the intercept is the log odds of the expected event rate when all predictors are 0 (and the random effect held constant at 0). Once again, this logged event rate would correspond to a subject who is average/high SES, female, and at the grand means for math and age. The expected event rate for this individual is 0.346. The expected probability of failing one course for this individual is as follows:

$$P(Y = 1) = \frac{e^{-\lambda}\lambda^c}{c!} = \frac{e^{-(0.346)}0.346^1}{1!} = \frac{(0.708)(0.346)^1}{1} = 0.245$$

TABLE 7.23 Model 2.3: Fixed Effects

Model term	Coefficient	Std. error	t	Sig.	Exp(coefficient)	95% Confidence interval for exp(coefficient)	
						Lower	Upper
(Intercept)	–1.060	0.095	–11.215	0.000	0.346	0.288	0.417
lowses	0.297	0.057	5.245	0.000	1.346	1.204	1.504
male	0.165	0.032	5.128	0.000	1.179	1.107	1.256
gmmath	–0.007	0.000	–16.792	0.000	0.993	0.992	0.993
gmage	0.197	0.045	4.400	0.000	1.218	1.115	1.330

Note: Probability distribution: Poisson; link function: log.

When there are a number of variables in a model, our primary focus is on the predictors. All four Level-1 predictors are significantly related to students' probability of failing a course during ninth grade. Because the intercept is the expected event rate for high/average SES students who are female (holding age and math at their grand means), for low SES females the logged event rate is as follows:

$$\text{Log}(\hat{\lambda}) = -1.06 + 0.297 \text{ (low SES)},$$

which is a logged event rate of −0.763. The expected event rate ($e^{-0.763}$) can also be expressed as the following multiplicative model:

$$\lambda = e^{-1.06} e^{0.297(1)} = (0.346)(1.346) = 0.466.$$

The ratio of the difference in expected event rates between low SES students and average/high SES students can then be expressed as follows:

$$\frac{lowses = 0.466}{ave. / highSES = 0.346} = 1.3468,$$

which is the same as the event rate ratio of 1.346 for low SES (with slight difference for rounding) in Table 7.23.

This suggests that for low SES students, the estimated count increases by a factor of 1.35 (or about 35%) compared with the estimated count of their more average or high SES peers. Holding the other variable constant in the model, males are expected to have an event rate 1.179 times greater (or about 18%) for course failures during ninth grade compared to females. For each year of added age, the predicted count of failing increases by a factor of 1.218 (or 21.8%). For a 1-point increase in math, the predicted odds of failing a course would be expected to decrease by a factor of 0.993 (or 0.7%).

The Level-2 variance components for the intercept are summarized in Table 7.24. After adding the within-school predictors, there is still considerable variance to explain at Level 2. Notice that the variance component is actually slightly larger than in the null model. This is often the case because the Level-1 variance is rescaled each time variables are added to the model. This can also affect variance components at Level 2 (Hox, 2010).

Examining Whether the Negative Binomial Distribution Is a Better Choice

We have noted that the NB distribution with log link provided a better fit to the data than the Poisson distribution with the single-level model. Therefore, we might take time to investigate whether that holds for the two-level model we are building. We can change the distribution to negative binomial.

TABLE 7.24 Model 2.3: Variance Components

Random effect	Estimate	Std. error	z-Test	Sig.	95% Confidence interval	
					Lower	Upper
Var(Intercept)	0.237	0.060	3.927	0.000	0.144	0.391

Note: Covariance structure: variance components; subject specification: nschcode.

Defining Model 2.4 With IBM SPSS Menu Commands

Note: IBM SPSS settings default to those used for Model 2.3.

1. Go to the toolbar and select ANALYZE, MIXED MODELS, GENERALIZED LINEAR MODELS. This command opens the *Generalized Linear Mixed Models* (GENLINMIXED) main dialog box.

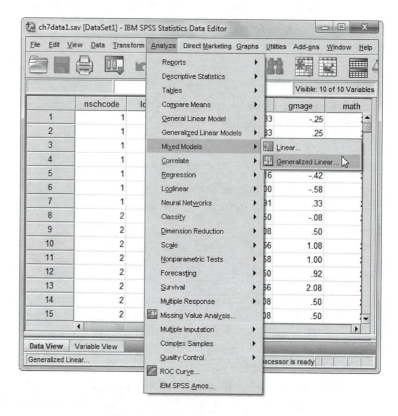

2a. The *Generalized Linear Mixed Models* dialog box displays the *Build Options* command tab and *Fixed Effects* item, which was the last setting used in the prior model.

   Note: The *Fixed Effects* item is preselected (represented in the illustration by the dashed-line box but appears on-screen colored blue) as it was the last setting used from the prior model.

b. Click to select the *Target* item.

c. Change the distribution setting by clicking the pull-down menu to select *Negative binomial.*

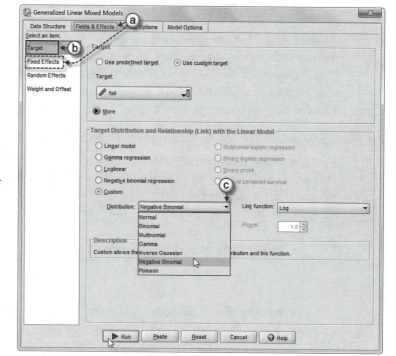

   Click the RUN button to generate the model results.

TABLE 7.25   Model 2.4: Fixed Effects

Model term	Coefficient	Std. error	t	Sig.	Exp(coefficient)	95% Confidence interval for exp(coefficient) Lower	Upper
(Intercept)	−1.124	0.094	−11.979	0.000	0.325	0.270	0.391
lowses	0.340	0.058	5.910	0.000	1.405	1.255	1.573
male	0.202	0.036	5.625	0.000	1.224	1.141	1.313
gmmath	−0.009	0.000	−19.599	0.000	0.991	0.990	0.992
gmage	0.231	0.052	4.479	0.000	1.260	1.139	1.394
Negative binomial	1.747						

Note: Probability distribution: negative binomial; link function: log; dependent variable: fail.

Interpreting the Output Results of Model 2.4

The GENLIN MIXED program will automatically estimate the ancillary parameter (at 1.747). When we do this, we note the following fixed effects. The standard errors are a little larger (e.g., *gmage*, *gmmath*) after we account for overdispersion in the model (see Table 7.25). The estimate of the Level-2 variance after rescaling is almost the same as the Poisson model, with slightly larger standard error (see Table 7.26).

We can again compare the fit of each distribution against the data to see whether we should have a preference for the Poisson distribution or the NB distribution. Keep in mind that model tests should be interpreted cautiously because they represent approximations of the likelihood function. The results support the view that the NB distribution seems to fit the data somewhat better, likely because of accounting for the presence of some overdispersion at Level 1. In this case, we will proceed with the negative binomial distribution with log link (see Table 7.27).

Does the SES-Failure Slope Vary Across Schools?

The two-level model typically provides unit-specific results—that is, the expected outcome for a Level-2 unit conditional on the set of random effects that have been specified (Raudenbush, Bryk, Cheong, & Congdon, 2004). Often, as we have noted, the concern is with a random intercept. At other times, we may have a particular theoretical interest in a within-school slope relationship. We might decide to see if there is evidence that a within-school slope varies randomly

TABLE 7.26   Model 2.4: Variance Components

Random effect	Estimate	Std. error	z-Test	Sig.	95% Confidence interval Lower	Upper
Var(Intercept)	0.238	0.063	3.773	0.000	0.142	0.401

Note: Covariance structure: variance components; subject specification: nschcode.

TABLE 7.27   Fit Comparison Results

	Poisson	Negative binomial
AIC	42,588.644	41,708.086
BIC	42,595.849	41,715.291

across schools. We may propose that the effect of student SES (β_{1j}) on the probability of failing a course varies across schools. At Level 1, the model remains the same as Equation 7.8.

At Level 2, we have the model for random intercepts, and we will add a random SES slope (β_{1j}) as follows:

$$\beta_{0j} = \gamma_{00} + u_{0j}$$
$$\beta_{1j} = \gamma_{10} + u_{1j}. \tag{7.11}$$

The other slopes remain fixed at Level 2. In the combined, single-equation model, we will now be estimating the same number (five) of fixed effects and two random effects at Level 2:

$$\eta_{ij} = \gamma_{00} + \gamma_{10} lowSES_{ij} + \gamma_{20} male_{ij} + \gamma_{30} gmmath_{ij} + \gamma_{40} gmage_{ij} + u_{1j} lowses_{ij} + u_{0j} \tag{7.12}$$

Defining Model 2.5 With IBM SPSS Menu Commands

Note: IBM SPSS settings default to those used for Model 2.4.

1. Go to the tool-bar and select ANALYZE, MIXED MODELS, GENERALIZED LINEAR MODELS. This command opens the *Generalized Linear Mixed Models* (GENLINMIXED) main dialog box.

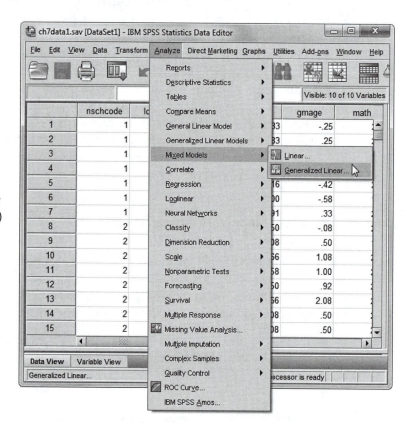

2a. The *Generalized Linear Mixed Models* dialog box displays the *Fields & Effects* command tab and *Target* item, which was the last setting used in the prior model.

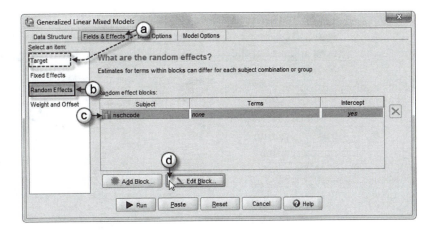

Note: The *Target* item is preselected (represented in the illustration by the dashed-line box but appears on-screen colored blue) as it was the last setting used from the prior model.

b. Click to select the *Random Effects* item.
c. Click on *nschcode* to activate the *Random effects block*.
d. Click the EDIT BLOCK button.

e. Designate *lowses* as a random effect. Click to select the variable (*lowses*) and then drag and drop it into the *Main* (effect) column of the *Effect builder*.

f. Confirm that the *Include intercept* option and *nschcode* are selected. Click the OK button to close the window and return to the *Random Effects* main screen.

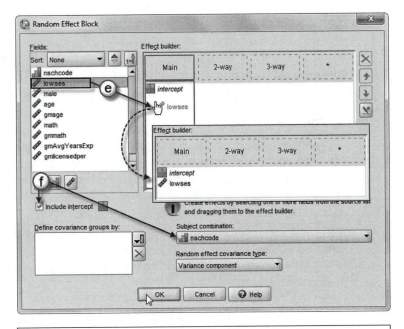

$$\beta_{1j} = \gamma_{10} + u_{1j} \quad (\text{Eq. 7.11})$$

g. In the main screen notice that *lowses* now appears in the *Random effects block* summary.

Click the RUN button to generate the model results.

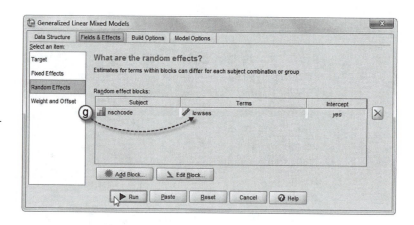

Interpreting the Output Results of Model 2.5

The model dimensions table (Table 7.28) confirms that the pattern of fixed and random effects estimated are as we intended. Table 7.29 provides the corresponding variance components at Level 2. The residual variance at Level 1 is reported as 1.0, which is the scale parameter. The estimates are slightly different from the same model estimated with a Poisson distribution (likely due to including the overdispersion directly into the negative binomial model). Once again, given the small nature of our sample size, we should treat the results cautiously at the group level. As Hox (2010) has noted, because variance cannot be negative, a one-way test can actually be used (which suggests dividing its p-value by 2). In this case, the ratio of the slope variance (0.04) to its standard error (0.026) with a one-tailed test appears to indicate that the slope might vary across groups in this small data set ($z = 1.567$, $p = .0585$). We are likely a little short on power to detect the slope variance due to the small sample size.

Modeling Variability at Level 2

Raudenbush and Bryk (2002) suggest building slope models with the same set of variables as the intercept model because the slope and intercept models can affect each other. Only after a variable is shown to be insignificant should it then be dropped for a more parsimonious model. The Level-1 model remains the same (as in Equation 7.8). We will add two Level-2 predictors: the average years of teaching experience of the staff and the proportion of teachers who were fully certified (using the default variance components matrix). We have grand-mean centered both variables.

At Level 2, we will first build the intercept model by adding two fixed effects to explain the random intercepts:

$$\beta_{0j} = \gamma_{00} + \gamma_{01}gmlicensedper_j + \gamma_{02}gmavgyearsexp_j + u_{0j}. \tag{7.13}$$

TABLE 7.28   Model 2.5: Covariance Parameters

Covariance parameters	
Residual effect	0
Random effects	2
Design matrix columns	
Fixed effects	5
Random effects	2
Common subjects	44

TABLE 7.29   Model 2.5: Variance Components

Random effect	Estimate	Std. error	z-Test	Sig.	95% Confidence interval Lower	Upper
Var(Intercept)	0.247	0.067	3.683	0.000	0.145	0.420
Var(lowses)	0.040	0.026	1.567	0.117	0.011	0.140

Note: Covariance structure: variance components; subject specification: nschcode; probability distribution: negative binomial; link function: log.

For the combined model, building on Equation 7.12, we will be estimating nine parameters (seven fixed effects and two random effects):

$$\eta_{ij} = \gamma_{00} + \gamma_{01}gmlicensedper_j + \gamma_{02}gmavgyearsexp_j + \gamma_{10}lowSES_{ij} + \gamma_{20}male_{ij}$$

$$+ \gamma_{30}gmmath_{ij} + \gamma_{40}gmage_{ij} + u_{1j}lowses_{ij} + u_{0j}.$$

(7.14)

Defining Model 2.6 With IBM SPSS Menu Commands

Note: IBM SPSS settings default to those used for Model 2.5.

1. Go to the toolbar and select ANALYZE, MIXED MODELS, GENERALIZED LINEAR MODELS. This command opens the *Generalized Linear Mixed Models* (GENLINMIXED) main dialog box.

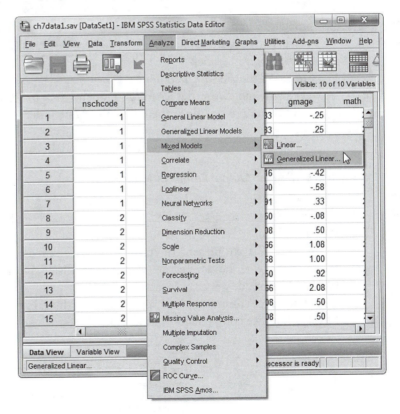

2a. The *Generalized Linear Mixed Models* dialog box displays the *Fields & Effects* command tab and *Random Effects* item, which was the last setting used in the prior model.

Note: The *Random Effects* item is preselected (represented in the illustration by the dashed-line box but appears onscreen colored blue) as it was the last setting used from the prior model.

$$\beta_{0j} = \gamma_{00} + \gamma_{01}gmlicensedper_j + \gamma_{02}gmavgyearsexp_j + u_{0j} \quad \text{(Eq. 7.13)}$$

b. We will modify the model by adding additional predictor variables. First, click to select the *Fixed Effects* item.

c. We will add two variables to the model. Click to select *gmAvgYearsExp* and *gmlicensedper* and then drag and drop them into the *Main* (effect) column of the *Effect builder*.

Click the RUN button to generate the model results.

Interpreting the Output Results of Model 2.6

First, we present the model dimensions table (Table 7.30). We can see seven fixed effects (intercept and two between-school and four within-school predictors) and two random effects (intercept and slope) at Level 2.

In Table 7.31 we present the results for the intercept part of the model first (adding three fixed effects) with random slope. We hold off discussing them until we provide the results for the complete model.

The variance components are very similar to those in the previous model. The corresponding Level-2 variance components are found in Table 7.32.

TABLE 7.30 Model 2.6: Covariance Parameters

Covariance parameters	
Residual effect	0
Random effects	2
Design matrix columns	
Fixed effects	7
Random effects	2
Common subjects	44

TABLE 7.31   Model 2.6: Fixed Effects

Model term	Coefficient	Std. error	*t*	Sig.	Exp(coefficient)	95% Confidence interval for exp(coefficient)	
						Lower	Upper
(Intercept)	−1.091	0.087	−12.553	0.000	0.336	0.283	0.398
lowses	0.344	0.056	6.189	0.000	1.410	1.265	1.572
male	0.203	0.036	5.614	0.000	1.225	1.141	1.314
gmmath	−0.009	0.000	−19.539	0.000	0.991	0.990	0.992
gmage	0.230	0.051	4.465	0.000	1.258	1.138	1.392
gmAvgYearsExp	0.090	0.032	2.770	0.006	1.094	1.027	1.166
gmlicensedper	−1.019	1.219	−0.836	0.403	0.361	0.033	3.938
Negative binomial	1.731						

Note: Probability distribution: negative binomial; link function: log; dependent variable: fail.

Adding the Cross-Level Interactions

In the next model (2.7) we add the two cross-level interactions to explain the random slope. The slope model will be the following:

$$\beta_{1j} = \gamma_{10} + \gamma_{11} gmAvgYearsExp_j + \gamma_{12} gmlicensedper_j + u_{1j}. \qquad (7.15)$$

The complete single-equation model that we have built in several steps will have a total of 11 parameters, which include nine fixed effects and two random covariance parameters:

$$\eta_{ij} = \gamma_{00} + \gamma_{01} gmlicensedper_j + \gamma_{02} gmavgyearsexp_j + \gamma_{10} lowSES_{ij} + \gamma_{11} gmAvgYearsExp_j * lowSES_{ij}$$
$$+ \gamma_{12} gmlicensedper_j * lowSES_{ij} + \gamma_{20} male_{ij} + \gamma_{30} gmmath_{ij} + \gamma_{40} gmage_{ij} + u_{1j} lowses_{ij} + u_{0j}. \qquad (7.16)$$

Finally, we will also add an unstructured covariance matrix to investigate the relationship between the intercept and slope, which will add a third covariance parameter at Level 2 and one more estimated parameter:

$$\begin{bmatrix} \sigma_I^2 & \sigma_{IS} \\ \sigma_{IS} & \sigma_S^2 \end{bmatrix}. \qquad (7.17)$$

A TABLE 7.32   Model 2.6: Variance Components

Random effect	Estimate	Std. error	*z*-Test	Sig.	95% Confidence interval	
					Lower	Upper
Var(Intercept)	0.233	0.064	3.642	0.000	0.136	0.398
Var(lowses)	0.042	0.026	1.649	0.099	0.013	0.139

Note: Covariance structure: variance components; subject specification: nschcode; probability distribution: negative binomial; link function: log.

Defining Model 2.7 With IBM SPSS Menu Commands

Note: IBM SPSS settings default to those used for Model 2.6.

1. Go to the toolbar and select ANALYZE, MIXED MODELS, GENERALIZED LINEAR MODELS. This command opens the *Generalized Linear Mixed Models* (GENLINMIXED) main dialog box.

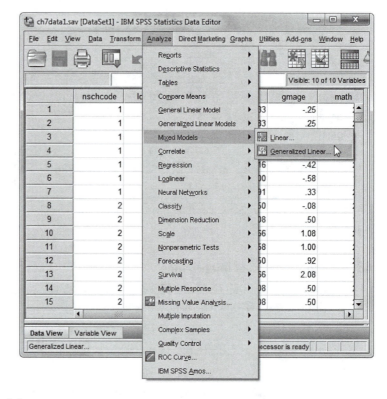

Adding Two Interactions to Model 2.7

2a. We will add the first interaction to the model. First, click to select *lowses* and *gmAvgYearsExp* and then drag both variables to the *2-way* column to create *lowses*gmAvgYearsExp*.

b. To create the second interaction, click to select *lowses* and *gmlicensedper* and then drag both variables to the *2-way* column to create *lowses*gmlicensedper*. The sequence of variables is now *lowses, male, gmmath, gmage, gmAvgYearsExp, gmlicensedper, lowses*gmAvgYearsExp, lowses*gmlicensedper*.

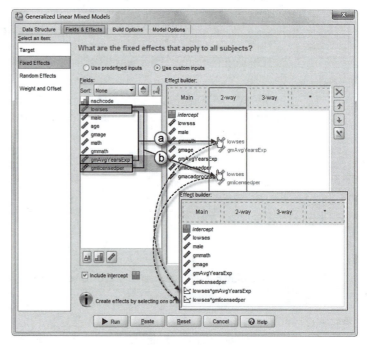

$$\eta_{tij} = \gamma_{00} + \gamma_{01}gmlicensedper_j + \gamma_{02}gmavgyearsexp_j + \gamma_{10}lowSES_{ij}$$
$$+ \gamma_{11}\mathbf{gmAvgYearsExp_j * lowSES_{ij}} + \gamma_{12}\mathbf{gmlicensedper_j * lowSES_{ij}}$$
$$+ \gamma_{20}male_{ij} + \gamma_{30}gmmath_{ij} + \gamma_{40}gmage_{ij} + u_{1j}lowses_{ij} + u_{0j}. \qquad \text{(Eq. 7.16)}$$

3a. Next we will change the covariance type from a variance components matrix used in the prior model to an unstructured matrix. First, click the *Random Effects* item to access the main screen.

b. Click on *nschcode* to activate the random effect block to enable editing the block.

c. Click the EDIT BLOCK button.

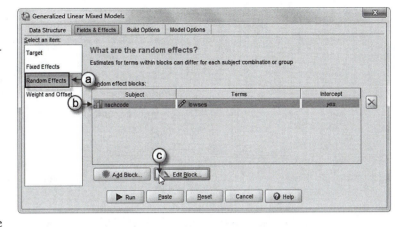

d. Change the random effect covariance structure by clicking the pull-down menu and selecting *Unstructured.*

Note: The unstructured matrix is a completely general covariance matrix.

e. Confirm that the *Include intercept* option and *nschode* are selected. Click the OK button to close the window and return to the *Random Effects* main screen.

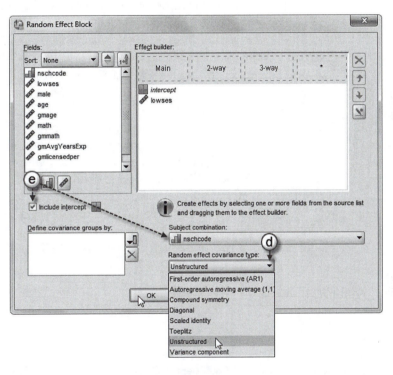

Click the RUN button to generate the model results

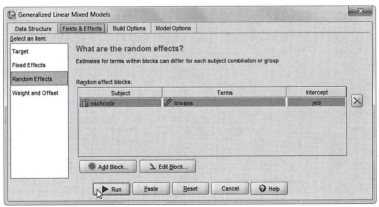

TABLE 7.33   Model 2.7: Covariance Parameters

Covariance parameters	
Residual effect	0
Random effects	3
Design matrix columns	
Fixed effects	9
Random effects	2
Common subjects	44

Interpreting the Output Results of Model 2.7

We can see in Table 7.33 that we have added the two fixed effects (increasing from seven to nine fixed effects). We also added a covariance between the intercept and slope.

The complete model is presented in Table 7.34. We note that the intercept rate (−1.106) is very similar to that of the previous model (−1.091). Moreover, the within-school effects change very little, as we would expect because the variables were added at Level 2. We note that we can interpret a within-school variable such as gmage (0.231) as the expected logged count difference in course failures for two students who attend the same school but differ by 1 year in age (holding the school random effects constant at 0). The expected incident rate (1.26) suggests that an added year of age increases the expected failure rate for two students in the same school by a factor of 1.26 (or 26%), holding the other variables and random effects constant. For this model, we will concentrate our attention on the addition of the school-level variables.

For the intercept model, we can see that the school-level average proportion of fully licensed teachers does not affect logged expected counts (−1.057, $p > .10$). In contrast, average years of teaching experience of the staff was positively related to logged expected counts; that is, an added year of experience can be expected to increase the logged event count by 0.102 ($p < .05$).

We might want to explore in a bit more detail what the meaning of this result is within the school portion of the model. For the reference group (i.e., females who are average/high SES)

TABLE 7.34   Model 2.7: Fixed Effects

Model term	Coefficient	Std. error	t	Sig.	Exp(coefficient)	95% Confidence interval for exp(coefficient) Lower	Upper
(Intercept)	−1.106	0.088	−12.534	0.000	0.331	0.278	0.393
lowses	0.367	0.051	7.186	0.000	1.443	1.306	1.595
male	0.202	0.036	5.571	0.000	1.224	1.140	1.314
gmmath	−0.009	0.000	−19.497	0.000	0.991	0.990	0.992
gmage	0.231	0.051	4.502	0.000	1.260	1.139	1.393
gmAvgYearsExp	0.102	0.042	2.438	0.015	1.108	1.020	1.203
gmlicensedper	−1.057	1.474	−0.717	0.473	0.348	0.019	6.245
lowses*gmAvgYearsExp	−0.037	0.035	−1.046	0.295	0.964	0.900	1.033
lowses*gmlicensedper	0.022	0.924	0.023	0.981	1.022	0.167	6.251
Negative binomial	1.731						

Note: Probability distribution: negative binomial; link function: log; dependent variable: fail.

who are in a school of average teacher experience (and holding the other covariates at their grand means), the logged event rate is –1.106. For similarly situated students (females who are average/high SES) who attend a neighboring school that differs by 1 year of added average teacher experience, the expected logged event rate would then be –1.004 (–1.106 + 0.102 = –1.004). This would be an expected event rate of

$$\hat{\lambda} = e^{-1.004} = 0.366.$$

We can then estimate the increase in incident rate ratio due to an added year of staff teaching experience. The incident ratio of the intercept school is 0.331 (in Table 7.31), so the ratio of the two incident rates is then 0.366/0.331 = 1.106. This is approximately the expected incident rate ratio of 1.108 for average staff experience in Table 7.34, with the slight difference due to rounding.

Looking at the incident rate ratio in Table 7.34, we can therefore interpret the estimate as the expected difference in incident rates for two students who have the same values on the other variables in the model, but who attend schools differing by 1 year of teaching experience (and holding the school variability constant). An added year of teaching experience at the school level increases the expected failure rate ratio by a factor of 1.108 (or about 10.8%). One interpretation of this finding may be that in schools with more experienced teaching staffs, the academic expectations are somewhat higher than in schools that are more average in teaching experience. Turning to the slope model (which concerns the two interaction terms), Table 7.34 suggests that proportion of fully certified staff and average teacher experience did not moderate the expected SES–course failure logged counts ($p > .10$). There might be other school context variables that would moderate the Level-1 slope effect under consideration.

Table 7.35 presents the corresponding Level-2 variance components. Given the small number of high schools in the sample, we would likely conclude that there is not enough slope variance left to continue adding Level-2 predictors. For example, even for a one-tailed test, the significance level would be 0.078. We could continue to add variables explaining variability in intercepts (keeping in mind that the slopes and intercepts tend to affect each other).

Developing a Two-Level Count Model With an Offset Variable

We will next turn our attention to developing a two-level model with varied exposure. Our previous model looked at course failures within a particular time interval (1 year). It is often the case, however, that because events occur over time (and space perhaps), the length of time (or amount of space) can vary from observation to observation. Our estimates of incident rates should be able to take this into account. In the second case, we provide a similar analysis of course failures. This new data set (*ch7data2.sav*) has 10,991 postsecondary students in 169 institutions (2-year and 4-year institutions).

TABLE 7.35 Model 2.7: Variance Components

Random effect	Estimate	Std. error	z-Test	Sig.	95% Confidence interval	
					Lower	Upper
Var(Intercept)	0.277	0.080	3.454	0.001	0.157	0.489
Var(slope-intercept)	–0.052	0.037	–1.409	0.159	–0.124	0.020
Var(lowses)	0.039	0.027	1.455	0.146	0.010	0.150

Note: Covariance structure: unstructured; subject specification: nschcode.

We can adapt models with count data and fixed exposure rates like the last one to consider varying rates of exposure. An *offset* variable represents an additional variable that allows the analyst to consider a varying exposure rate or number of trials. Here we will just refer to the varied exposure as a number of semesters, or time (t). The expected value $E(Y/t)$ is therefore the following:

$$E(Y/t) = \frac{1}{t}E(Y) = \lambda/t. \tag{7.18}$$

This suggests that the expected individual count (λ) is divided by the exposure. This implies that the expected count is an event rate; the difference is that, in this example, the exposure time is varied rather than constant. The Level-1 model for individual i in school j with a single X predictor is as follows:

$$\eta_{ij} = \log(\lambda_{ij}) = \beta_{0j} + \beta_{1j}X_{ij} + \log(t_{ij}). \tag{7.19}$$

This reflects that the offset parameter corresponding to the varied exposure rate is often logged so that it is in the same metric as the outcome.

Readers can refer to Figure 7.2 for the offset feature in IBM SPSS. The offset is a structural parameter whose coefficient is usually constrained to 1.0 (McCullagh & Nelder, 1989). It is therefore not directly estimated but, rather, is an individual adjustment term. In this case, our response variable is the count of course failures that postsecondary students experience during their undergraduate educational careers. The offset variable is the number of semesters that the student has been enrolled in a 2-year or 4-year institution. The number of trials should be greater than or equal to the number of events for each case (IBM SPSS, 2010). Events should be non-negative integers (because a count cannot be less than 0), and trials should be positive integers. It is important to note that trials must be defined as scale (continuous) variables in IBM SPSS.

The Data

The postsecondary data set summarized in Table 7.36 focuses on students who are pursuing undergraduate degrees. The students entered either a 2-year ($N = 4,735$, 43.1%) or a 4-year ($N = 6,256$, 56.9%) institution. They ranged from being in their second through 11th semesters as undergraduates. Regarding the outcome, 61.4% of the students pursuing a bachelor's degree did not fail any course during their postsecondary studies. In contrast, 38.6% failed one or more courses, with a few students failing as many as eight courses over the 10-semester period of data collection. Because over half the individuals did not fail an undergraduate course, the data are likely not normally distributed because there are too many zeros.

Research Questions

We may be interested in determining what background and institutional variables might explain students' probability of failing an undergraduate course. We can first ask whether background variables increase or decrease students' probability of failing a course. In particular, we may be interested whether the individual's choice to enter a 2-year or 4-year institution increases the probability of failing one or more courses, after controlling for other background factors known to affect student academic outcomes such as socioeconomic status, grade point

TABLE 7.36: Data Definition of *ch7data2.sav* (*N* = 10,991)

Variable	Level[a]	Description	Values	Measurement
schcode	School	School identifier (169 schools)	Integer	Ordinal
male	Individual	Demographic predictor variable representing student's gender	0 = female 1 = male	Scale
gmmath	Individual	State math test score (grand-mean centered) to describe student's achievement during Grade 8	−140.60 to 259.40	Scale
gmgpa	Individual	Students' cumulative grade point average (grand-mean centered)	−4.35 to 1.35	Scale
twoyear	Individual	The type of institution in which the student was enrolled	0 = 4-year institution 1 = 2-year college	Scale
fail	Individual	The number of courses that students fail during college	0 to 8	Scale
ses	Individual	Predictor variable (z-score) measuring student socioeconomic status	−2.89 to 2.30	Scale
gmacadprocess	School	Predictor variable (grand-mean centered factor score) assessing perceptions about the overall quality of the school's educational experiences and academic expectations for students	−1.67 to 3.17	Scale
gminsqual	School	Predictor variable (factor score) measuring students' perceptions of quality of classroom instruction in the school	−6.54 to 6.76	Scale
semester	Individual	The number of semesters in which students were enrolled	2 to 11	Scale
lnsemester	Offset	The log of the number of semesters a student is enrolled in college while pursuing a degree	0.69 to 2.40	Scale

[a] Individual = Level 1; school = Level 2.

average (GPA), and high school math achievement. Second, we may explore whether academic features of students' present postsecondary institutions have any impact on their likelihood to fail a course. In this case we examine whether perceived features of the quality of schools' academic experiences and classroom instruction have any impact on students' probability to fail one or more courses.

Offset Variable

The offset variable (Figure 7.2) is the number of semesters that the students have been enrolled. The model specification assumes that course failure has a Poisson distribution. This seems justified because we noted that the course failure counts are not normally distributed. For example, the outcome is positively skewed (1.535). Variables that measure the amount of exposure to the risk (or varied trials) are handled as offset variables in generalized linear models. We can see from Equation 7.19 that the expected counts will therefore depend on the exposure rate (t) and X, both of which are observations in the data set.

As we noted, the assumption is typically that the scale parameter of the offset is fixed at 1.0 (McCullagh & Nelder, 1989). We can also relax that restriction and estimate an overdispersed Poisson regression model. In some cases, modeling raw cell counts can be misleading if the

TABLE 7.37 **Frequency Distribution of Fail Variable**

		Frequency	Percent	Valid percent	Cumulative percent
Valid	0	6,746	61.4	61.4	61.4
	1	1,207	11.0	11.0	79.3
	2	898	8.2	8.2	88.4
	3	844	7.7	7.7	93.6
	4	786	7.2	7.2	95.7
	5	328	3.0	3.0	98.9
	6	102	0.9	0.9	99.4
	7	49	0.4	0.4	99.8
	8	31	0.3	0.3	100.0
	Total	10,991	100.0	100.0	

aggregate offset variable (in this case, semesters enrolled) varies with variables that measure the amount of exposure to the risk. As we have noted in this chapter, the Poisson regression model assumes that the log of the dependent variable (course failures) is linear with respect to the predictors. In some cases, it may be necessary to use the log of the offset (e.g., if the offset varies by some other particular factor such as the type of institutions that students entered). This has the effect of placing the offset and the dependent variable in the same metric scale (Hox, 2010).

Table 7.37 presents a distribution of the frequency of course occurrences over the ten semesters of data collection. We can see that 61.4% did not fail any course, 11% failed one course, and increasingly smaller percentages failed from two to eight courses.

The offset variable makes it possible to consider how length of time attending the institution might influence the incident rates. For example, the probability of failing a course may decrease as a student successfully navigates general education requirements and the requirements of his or her major over a number of semesters enrolled during college. We found that we needed to take the log of the number of semesters students were enrolled in order to obtain a more satisfactory model fit. Modeling the raw counts can be misleading because there were differing course failures associated with the types of institutions students first entered and the number of semesters students were enrolled. More specifically, failure rates were considerably higher in 2-year institutions than in 4-year institutions. Failure rates were also higher in the beginning semesters of the students' postsecondary careers.

Specifying a Single-Level Model

We will first specify a single-level model using GENLIN with five student background predictors only, without the offset parameter. We have found that a few more features are currently available in this program for preliminary model building (e.g., fit indices, alternative types of model specifications). These can be useful in "getting the horse in the corral." The specification is similar to Equation 7.3. The model will be as follows:

$$\eta_i = \beta_0 + \beta_1 ses + \beta_2 male + \beta_3 gmmath + \beta_4 gmgpa + \beta_5 twoyear. \tag{7.20}$$

Note that we have specified the dichotomous predictors as *scale* in this formulation so that the reference category will be coded 0. Alternatively, we could also enter the dichotomous predictors as nominal and set the categories for the predictors as *descending*, which means that the first category will be the reference category. We note that, with a count outcome, it does not make any difference whether we specify the outcome categories as *descending* or *ascending* because counts must be specified as continuous (scale).

Defining Model 3.1 With IBM SPSS Menu Commands

Note: Refer to Chapter 3 for discussion concerning settings and modeling options available in the IBM SPSS GENLIN routine.

Launch the IBM SPSS application program and select the *ch7data2.sav* data file.

1. Go to the tool-
   bar and select
   ANALYZE,
   GENERALIZED
   LINEAR
   MODELS,
   GENERALIZED
   LINEAR
   MODELS. This
   command opens the
   *Generalized Linear
   Models* (GENLIN)
   main dialog box.

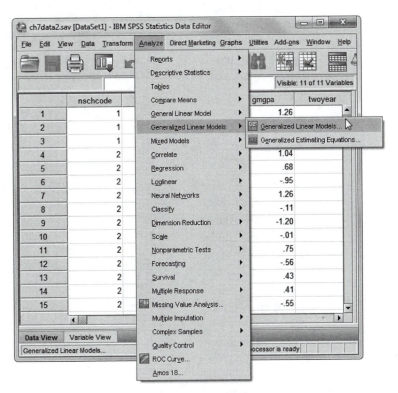

2a. The *Generalized Linear Models* screen displays the *Type of Model* command tab, which is the default setting when creating a model for the first time.

b. We will designate a Poisson distribution and log link function for this model. Beneath the *Counts* section, click the radio button to select *Poisson loglinear.*

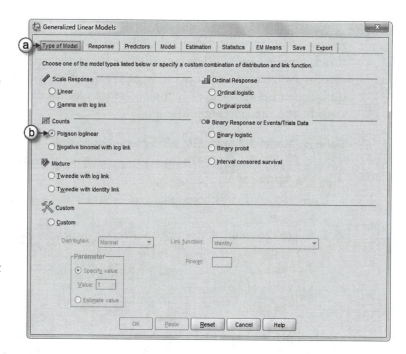

3a. Click the *Response* command tab, where we will select the dependent variable (*fail*) for the model. (*Fail* represents the number of courses that students fail during their undergraduate years.)

b. Click to select *fail* from the *Variables* list and then click the right-arrow button to move the variable into the *Dependent Variable* box.

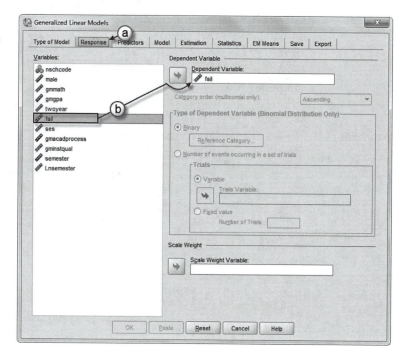

4a. Click the *Predictors* command tab, which enables selecting factors and covariates to use in predicting the dependent variable. To facilitate ease in reading the output tables, we will enter the variables in a different sequence from the way they are displayed.

b. First, click to select *ses* and then click the right-arrow button to move the variable into the *Covariates* box.

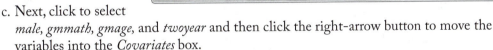

c. Next, click to select *male, gmmath, gmage,* and *twoyear* and then click the right-arrow button to move the variables into the *Covariates* box.

Note: An alternative method for entering the variables is to select all five variables at one time and then click the right-arrow button to move them into the *Covariates* box. Then, select *ses* and use the adjacent up-arrow button to move *ses* above *male.*

5a. Click the *Model* command tab to specify model effects based on the model's covariate variables (*ses, male, gmmath, gmage, twoyear*).

b. The default model is intercept only, so other model effects must be specified.

c. We will use the default *Main effects.*

d. Click to select all of the variables (*ses, male, gmmath, gmage, twoyear*) and then click the right-arrow button to move them into the *Model* box. Note that five effects are specified for the model (*Number of Effects in Model: 5*).

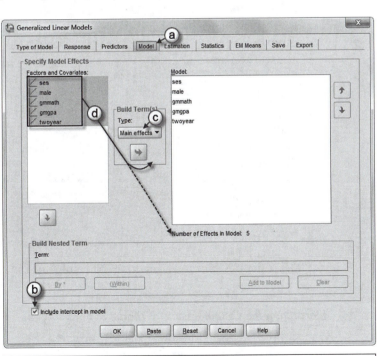

$$\eta_i = \beta_0 + \beta_1 ses + \beta_2 male + \beta_3 gmmath + \beta_4 gmgpa + \beta_5 twoyear \quad \text{(Eq. 7.20)}$$

6a. Click the *Estimation* command tab, which enables specifying estimation methods and providing initial values for the parameter estimates.

b. For this model we will use the default *Hybrid* estimation method setting.

c. We will change the *Covariance Matrix* by clicking to select the *Robust estimator*.

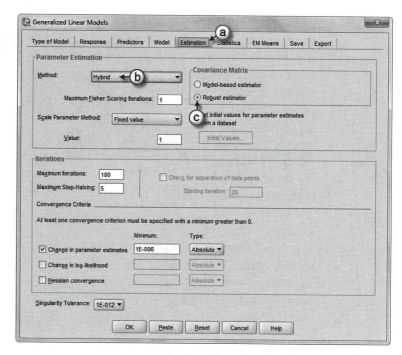

7a. Finally, click the *Statistics* command tab, which enables specifying estimation methods and providing initial values for the parameter estimates.

b. We will add an additional print option to display exponentiated parameter estimates (incident rate ratios) in addition to the raw parameter estimates. Click to select the option *Include exponential parameter estimates.*

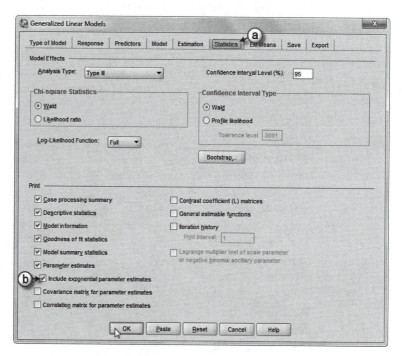

Click the OK button to generate the results for the model.

Interpreting the Output Results of Model 3.1

One can see that background variables similarly affect students' likelihood to fail a course in this postsecondary data set. The intercept in this case represents the logged count (−0.263) if students are female (coded 0), entered a 4-year institution (coded 0), and are at the means (0) for the other predictors. Controlling for the other covariates, females entering 2-year institutions have a logged incident rate of −0.263, which corresponds to an incident rate

TABLE 7.38 Model 3.1: Parameter Estimates

Parameter	β	Std. error	95% Wald confidence interval		Hypothesis test			Exp(β)	95% Wald confidence interval for exp(β)	
			Lower	Upper	Wald χ²	df	Sig.		Lower	Upper
(Intercept)	−0.263	0.0263	−0.314	−0.211	100.287	1	0.000	0.769	0.730	0.809
ses	0.004	0.0181	−0.032	0.039	0.043	1	0.837	1.004	0.969	1.040
male	0.168	0.0284	0.112	0.223	34.728	1	0.000	1.182	1.118	1.250
gmmath	−0.006	0.0002	−0.007	−0.006	819.927	1	0.000	0.994	0.993	0.994
gmgpa	−0.327	0.0176	−0.362	−0.293	344.916	1	0.000	0.721	0.696	0.746
twoyear	0.195	0.0290	0.138	0.251	44.961	1	0.000	1.215	1.148	1.286
(Scale)	1[a]									

Note: Dependent variable: fail; model: (intercept), ses, male, gmmath, gmgpa, twoyear.
[a] Fixed at the displayed value.

ratio of 0.769 in Table 7.38. For females who entered 2-year institutions, the logged count rate is as follows:

$$\eta_{ij} = \log(\lambda) = -0.263 + 0.195,$$

which is −0.068.

If we exponentiate this ($e^{-0.068}$), we obtain an expected incident rate of 0.934. The ratio of these incident rates for 2-year entrance to 4-year entrance is then $0.934/0.769 = 1.215$ ($p < .001$), which is the expected incident rate ratio for 2-year entrance in Table 7.38. We can interpret this as that entering a 2-year institution increases the expected failure rate by a factor of 1.215 or 21.5%. Except for SES, the other predictors are also statistically significant in explaining expected failure rates ($p < .001$). Controlling for the other variables, we can see that increasing student GPA by 1 point decreases the expected failure rate by a factor of 0.721 (or 27.9%). Each point added for the students' twelfth grade math level reduces the incident rate ratio by a factor of 0.994 (or about 0.6%). Males have higher expected rate ratios than females by a factor of 1.182 (or about 18.2%).

Adding the Offset

We may want to adjust the previous estimates for varying exposure rates to the probability of failing a course. For purposes of demonstration, we will use the number of semesters a student has been enrolled, which ranged from 2 to 11 (mean = 5.5; SD = 2.6). We will enter the logged version of semesters as the offset variable (*Lnsemester*), as shown in Figure 7.2. Keep in mind that the offset must be continuous and defined as *scale*, or one will receive the output message shown in Figure 7.3. Note that the offset also cannot be a predictor variable in the model simultaneously.

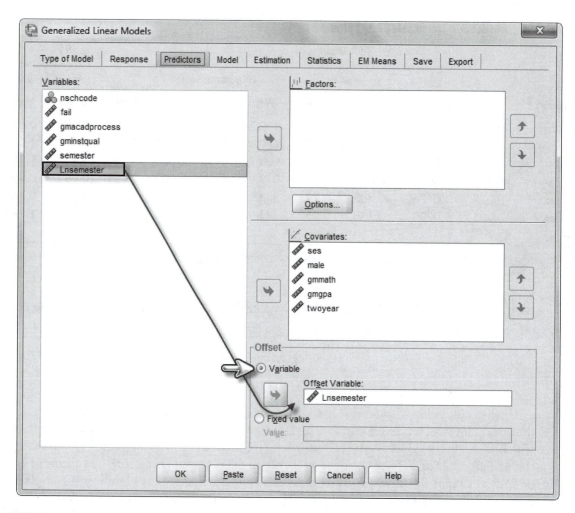

FIGURE 7.2 Selecting an offset variable (*Lnsemester*) within the GENLIN routine's *Predictors* display screen.

Warnings

The offset must have a continuous measurement level. Execution of this command stops.

FIGURE 7.3 Offset function warning message.

Defining Model 3.2 With IBM SPSS Menu Commands

Note: IBM SPSS settings default to those used for Model 3.1.

1. Go to the tool-bar and select ANALYZE, GENERALIZED LINEAR MODELS, GENERALIZED LINEAR MODELS. This command opens the *Generalized Linear Models* (GENLIN) main dialog box.

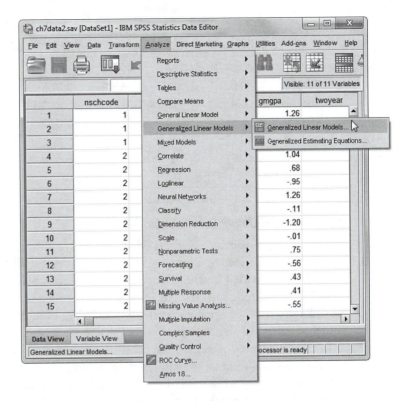

2a. The *Generalized Linear Mixed Models* main screen highlights the *Statistics* tab as it was the last command window used for Model 3.1.

b. We will add an offset variable (*Lnsemester*) into the model. First, click the *Predictors* command tab.

c. Click to select *Lnsemester* and then click the right-arrow button to move the variable into the *Offset Variable* box.

Click the OK button to generate the model results.

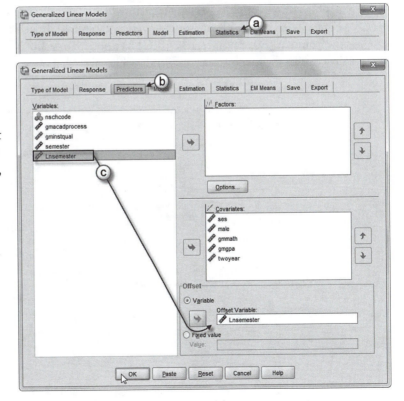

TABLE 7.39 Model 3.2: Parameter Estimates

Parameter	β	Std. error	95% Wald confidence interval		Hypothesis test			Exp(β)	95% Wald confidence interval for exp(β)	
			Lower	Upper	Wald χ²	df	Sig.		Lower	Upper
(Intercept)	−1.972	0.0261	−2.024	−1.921	5726.878	1	0.000	0.139	0.132	0.146
ses	0.027	0.0183	−0.009	0.063	2.203	1	0.138	1.028	0.991	1.065
male	0.158	0.0286	0.102	0.214	30.460	1	0.000	1.171	1.107	1.238
gmmath	−0.007	0.0002	−0.007	−0.007	893.082	1	0.000	0.993	0.993	0.993
gmgpa	−0.264	0.0162	−0.295	−0.232	265.136	1	0.000	0.768	0.744	0.793
twoyear	0.220	0.0292	0.162	0.277	56.614	1	0.000	1.245	1.176	1.319
(Scale)	1[a]									

Note: Dependent variable: fail; model: (intercept), ses, male, gmmath, gmgpa, twoyear, offset = lnsemester.

Interpreting the Output Results of Model 3.2

In Table 7.39 we provide the fixed effects. With the offset covariate added to the model, we can see that intercept logged count (−1.972) is considerably different. This is because the expected count is now divided by the offset value (λ_{ij}/t_{ij}); however, most of the other estimates in the model are consistent with the previous model. The logged count for student GPA, at −0.264, is a bit different from the previous model (−0.327). We note that the estimate of the intercept in the model will depend on how the offset is specified (i.e., it is generally fixed at 1.0).

In the single-level model, we can relax the assumption that the scale parameter in Table 7.39 is 1.0. This allows us to fit an overdispersed Poisson regression model. We note also that the Pearson χ^2 method can be used to estimate the scale parameter, with the other model-fitting criteria left to their default values (i.e., using the model-based estimator rather than the robust estimator). This can be used to obtain more conservative variance estimates and significance levels (McCullagh & Nelder, 1989) (see Figure 7.4). These results are summarized in Table 7.40. This does not change the other overall fit criteria in the model (e.g., deviance, AIC, BIC). We note that, in GENLIN MIXED, this can be accomplished in a similar manner by using an NB distribution so that the scale parameter is estimated.

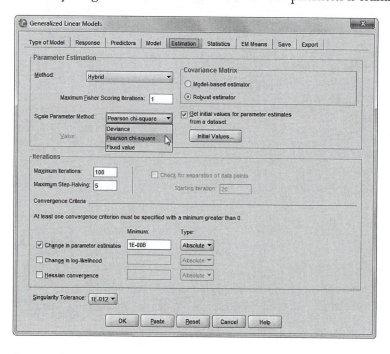

FIGURE 7.4 Selecting Pearson chi-square in the GENLIN routine's *Estimation* main screen.

TABLE 7.40 Model 3.3: Parameter Estimates

Parameter	β	Std. error	95% Wald confidence interval		Hypothesis test			Exp(β)	95% Wald confidence interval for exp(β)	
			Lower	Upper	Wald χ²	df	Sig.		Lower	Upper
(Intercept)	−1.972	0.0261	−2.024	−1.921	5726.878	1	0.000	0.139	0.132	0.146
ses	0.027	0.0183	−0.009	0.063	2.203	1	0.138	1.028	0.991	1.065
male	0.158	0.0286	0.102	0.214	30.460	1	0.000	1.171	1.107	1.238
gmmath	−0.007	0.0002	−0.007	−0.007	893.082	1	0.000	0.993	0.993	0.993
gmgpa	−0.264	0.0162	−0.295	−0.232	265.136	1	0.000	0.768	0.744	0.793
twoyear	0.220	0.0292	0.162	0.277	56.614	1	0.000	1.245	1.176	1.319
(Scale)	2.558[a]									

Note: Dependent variable: fail; model: (intercept), ses, male, gmmath, gmgpa, twoyear, offset = lnsemester.
[a] Computed based on the Pearson chi-square.

Defining Model 3.3 With IBM SPSS Menu Commands

Note: IBM SPSS settings default to those used for Model 3.2.

1. Go to the tool-bar and select ANALYZE, GENERALIZED LINEAR MODELS, GENERALIZED LINEAR MODELS. This command opens the *Generalized Linear Models* (GENLIN) main dialog box.

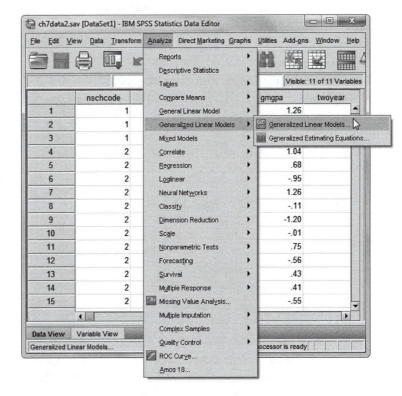

2a. The *Generalized Linear Models* main screen highlights the *Predictors* tab as it was the last command window used for Model 3.2.

b. We will change the *Scale Parameter Method*. First, click the *Estimation* command tab.

c. Now change the *Scale Parameter Method* by clicking the pull-down menu to select *Pearson chi square*.

Click the OK button to generate the model results.

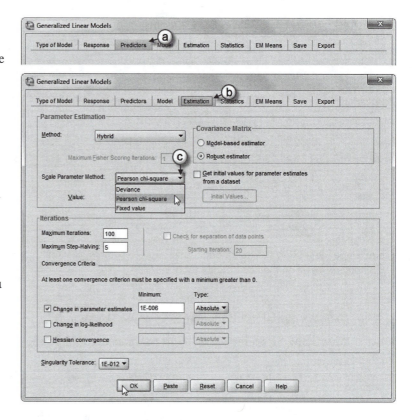

Interpreting the Output Results of Model 3.3

In Table 7.40, we note that when the scale is estimated (computed based on the Pearson χ^2), the result suggests the presence of overdispersion. The scale parameter in Table 7.41 (computed based on the Pearson χ^2) is the ratio of the deviance to the degrees of freedom summarized in the goodness of fit indices in Table 7.41.

TABLE 7.41   Model 3.3: Goodness of Fit[a]

	Value	df	Value/df
Deviance	22726.791	10985	2.069
Scaled deviance	8884.876	10985	
Pearson chi-square	28098.737	10985	2.558
Scaled Pearson chi-square	10985.000	10985	
Log likelihood[b,c]	−17201.622		
Adjusted log likelihoo[d]	−6724.851		
Akaike's information criterion (AIC)	34415.243		
Finite sample corrected AIC (AICC)	34415.251		
Bayesian information criterion (BIC)	34459.072		
Consistent AIC (CAIC)	34465.072		

Note: Dependent variable: fail; model: (intercept), ses, male, gmmath, gmgpa, twoyear, offset = lnsemester.
[a] Information criteria are in small-is-better form.
[b] The full log likelihood function is displayed and used in computing information criteria.
[c] The log likelihood is based on a scale parameter fixed at 1.
[d] The adjusted log likelihood is based on an estimated scale parameter and is used in the model fitting omnibus test.

We can also consider what the logged count rates might look like if we entered the varied exposure variable (*Lnsemester*) into the model as a covariate. We specify this as Model 3.4. Hox (2010) suggests that when the software program does not provide an offset feature, one can instead enter it into the model as a covariate to control for the possible effects of varied exposure rates on the outcome.

Defining Model 3.4 With IBM SPSS Menu Commands

Note: IBM SPSS settings default to those used for Model 3.3.

1. Go to the tool-bar and select ANALYZE, GENERALIZED LINEAR MODELS, GENERALIZED LINEAR MODELS. This command opens the *Generalized Linear Models* (GENLIN) main dialog box.

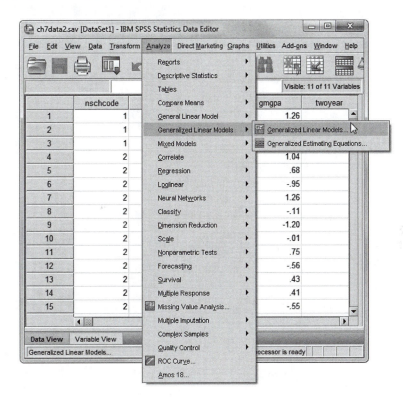

2a. The *Generalized Linear Models* main screen highlights the *Estimation* tab as it was the last command window used for Model 3.3.

b. Click the *Predictors* command tab.

c. We will add an offset variable into the model. First, remove *Lnsemester* from the *Offset* variable box and then click the left-arrow button. This returns *Lnsemester* to the *Variables* column.

d. Now click to select *Lnsemester* from the *Variables* column and then click the right-arrow button to place it into the *Covariates* box.

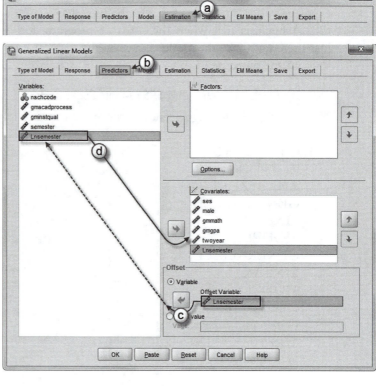

3a. Next we will add *Lnsemester* as a fixed effect into the model. First, click the *Model* command tab.

b. *Main effects* is the default setting.

c. We will add the offset variable (*Lnsemester*) into the model by clicking to select it and then clicking the right-arrow button to move it into the *Model* box. Notice that the number of fixed effects in the model has increased to six effects (*Number of Effects in Model: 6*).

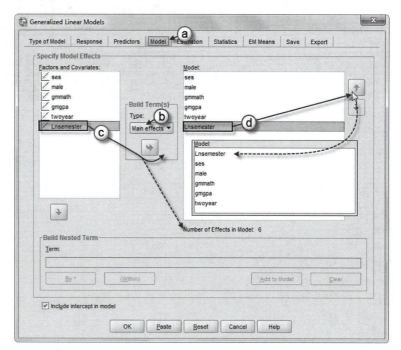

d. To facilitate reading of the output tables, we will change the sequence order of *Lnsemester* by clicking the up-arrow button until the variable is above *ses*.

4a. Click the *Estimation* command tab so that we may change the *Scale Parameter Method*.

b. Click the pull-down menu for *Scale Parameter Method* to select *Fixed value*.

Click the OK button to generate the model results.

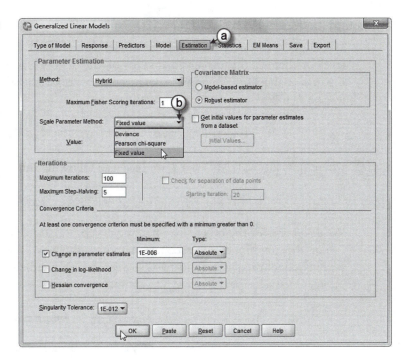

Interpreting the Output Results of Model 3.4

When we enter the logged semester variable into the model as a predictor (i.e., instead of as part of the event rate), we can see that as the number of semesters a student is enrolled increases by 1.0, the logged course failure rate is also increased (0.412, $p < .01$). If we calculate a measure of association (gamma) between the number of semesters students are enrolled and number of courses failed (not tabled), we find that the association between semesters and courses failed is weak but statistically significant (0.07, $p < .05$). Adjusting the model outcome for the number of semesters a student is enrolled, therefore, is likely to improve the accuracy of our results (see Table 7.42).

TABLE 7.42   Model 3.4 Parameter Estimates

Parameter	β	Std. error	95% Wald confidence interval Lower	95% Wald confidence interval Upper	Wald χ^2	df	Sig.	Exp(β)	95% Wald confidence interval for exp(β) Lower	95% Wald confidence interval for exp(β) Upper
(Intercept)	−0.939	0.0524	−1.042	−0.836	320.711	1	0.000	0.391	0.353	0.433
lnsemester	0.412	0.0278	0.357	0.466	218.777	1	0.000	1.510	1.429	1.594
ses	0.013	0.0179	−0.022	0.048	0.556	1	0.456	1.013	0.979	1.050
male	0.164	0.0280	0.109	0.219	34.349	1	0.000	1.179	1.116	1.245
gmmath	−0.007	0.0002	−0.007	−0.006	889.968	1	0.000	0.993	0.993	0.994
gmgpa	−0.302	0.0165	−0.334	−0.270	334.384	1	0.000	0.739	0.716	0.764
twoyear	0.206	0.0286	0.150	0.262	51.792	1	0.000	1.228	1.162	1.299
(Scale)	1[a]									

Note: Dependent variable: fail; model: (intercept), lnsemester, ses, male, gmmath, gmgpa, twoyear.
[a] Fixed at the displayed value.

Estimating the Model With GENLIN MIXED

We can now estimate the complete model using GENLIN MIXED. We will add two school-level predictors: the perceived quality of the school's academic processes and its classroom instructional quality. We grand-mean center both variables. The model specification for Level 1 will be as follows:

$$\eta_{ij} = \beta_{0j} + \beta_{1j}ses_{ij} + \beta_{2j}male_{ij} + \beta_{3j}gmmath_{ij} + \beta_{4j}gmgpa_{ij} + \beta_{5j}twoyear_{ij}, \qquad (7.21)$$

where $\eta_{ij} = \log(\lambda_{ij})$. At Level 2, we have the following intercept model:

$$\beta_{0j} = \gamma_{00} + \gamma_{01}gmacadprocess_j + \gamma_{02}gminstqual_j + u_{0j}. \qquad (7.22)$$

We also consider the other slope parameters ($\beta_1 - \beta_5$) as fixed at Level 2 as in Equation 7.9.

For the combined single-equation model, there will be eight fixed effects (intercept, five within-school predictors, and two between-school predictors) and one random effect (random intercept variance parameter u_{0j}) to estimate:

$$\eta_{ij} = \gamma_{00} + \gamma_{01}gmacadprocess_j + \gamma_{02}gminstqual_j + \gamma_{10}ses_{ij} + \gamma_{20}male_{ij}$$
$$+ \gamma_{30}gmmath_{ij} + \gamma_{40}gmgpa_{ij} + \gamma_{50}twoyear_{ij} + u_{0j} \qquad (7.23)$$

As in the single-level situation, we will estimate a Poisson regression model with log link function.

Defining Model 4.1 With IBM SPSS Menu Commands

Note: Refer to Chapter 3 for discussion concerning settings and modeling options available in the IBM SPSS GENLIN MIXED routine.

Continue using the *ch7data2.sav* data file.

1. Go to the toolbar and select ANALYZE, MIXED MODELS, GENERALIZED LINEAR MODELS. This command opens the *Generalized Linear Mixed Models* (GENLINMIXED) main dialog box.

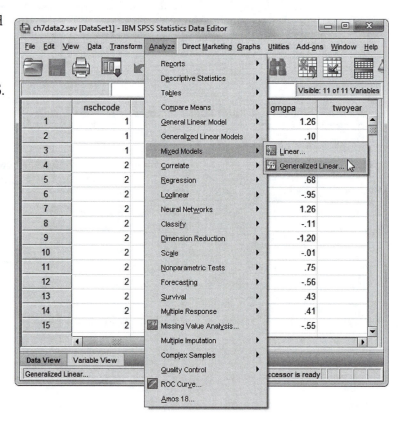

2a. The *Generalized Linear Mixed Models* dialog box displays the default *Data Structure* command tab. This tab allows specifying structural relationships between data set records when observations are correlated.

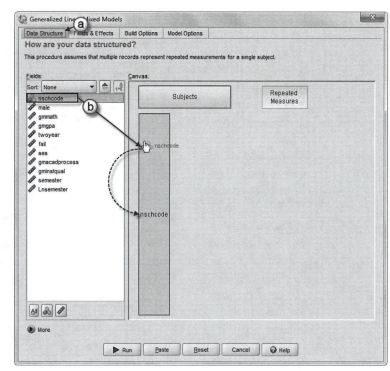

b. To build our model, begin by defining the subject variable. A "subject" is an observational unit that can be considered independently of other subjects. Click to select the *nschcode* variable from the left column and then drag and drop *nschcode* onto the canvas area beneath the *Subjects* box.

3a. Click the *Fields & Effects* command tab, which displays the default *Target* display screen that enables specifying a model's target (dependent variable), its distribution, and its relationship to its predictors through the link functions.

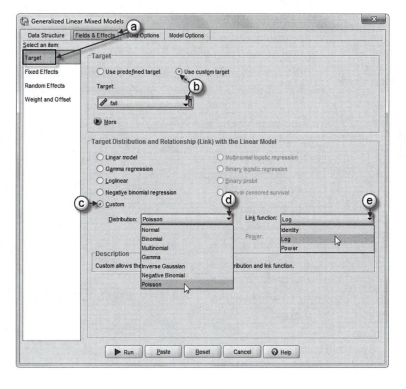

b. Click the *Target* pull-down menu and select the dependent variable *fail*. (*Fail* represents the number of courses that students fail during their ninth grade year of school.) Note that the selection of the variable automatically changes the target option to *Use custom target*.

c. For this first model we will designate a Poisson distribution and log link function. Click to select the *Custom* option to designate a custom target distribution and link function.

d. Now click the pull-down menu and designate the custom *Distribution* option as *Poisson*.

e. Finally, click the pull-down menu and designate the custom *Link function* option as *Log*.

4a. Next click the *Fixed Effects* item to access the main dialog box.

b. The *Include intercept* option is preselected as Level-1 models are intercept only by default.

c. We will add seven predictor variables in this model. Click to select *male, gmmath, gmgpa, twoyear, ses, gmacadprocess,* and *gminstqual*. Then, drag and drop them into the *Effect builder* box beneath the *Main* section.

d. To facilitate reading of the output tables, we will rearrange the variable sequence. Click to select *ses* and then click the up-arrow button on the right to move

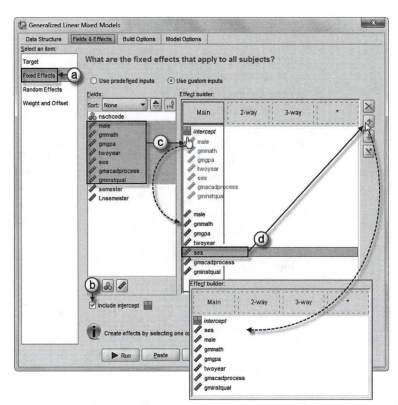

the variable so that it appears above *male*. The variable sequence now appears as *ses, male, gmmath, gmgpa, twoyear, gmacadprocess, gminstqual*.

$$\eta_{ij} = \beta_{0j} + \beta_{1j}ses_{ij} + \beta_{2j}male_{ij} + \beta_{2}gmmath_{ij} + \beta_{4}gmgpa_{ij} + \beta_{5j}twoyear_{ij}, \quad \text{(Eq. 7.21)}$$

$$\beta_{0j} = \gamma_{00} + \gamma_{01}gmacadprocess_{j} + \gamma_{02}gminstqual_{j} + u_{0j}. \quad \text{(Eq. 7.22)}$$

5. We will now add a random effect where the subject is *nschcode* (no effects, intercept only).

a. Click the *Random Effects* item.

b. Click the ADD BLOCK button to access the *Random Effect Block* main dialog box.

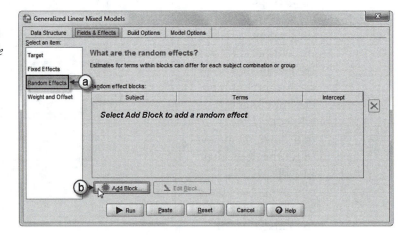

c. Click to select the *include intercept* option, which changes the *no intercept* default to *intercept*. Because there is only one random effect (the intercept), we can use the default variance component setting.

d. We need to identify the *Subject combination* by clicking the pull-down menu and selecting *nschcode*. Click the OK button to close the *Random Effect Block* dialog box and return to the *Random Effects* main screen.

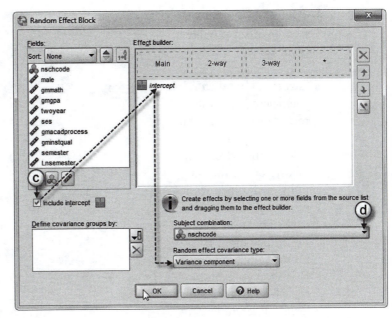

e. In the main screen notice that the *Random Effect Block* displays a summary of the block just created *nschcode* (Subject) and *intercept* (Yes).

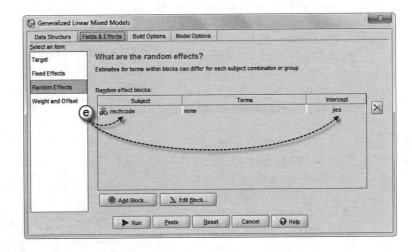

6. We will now designate an offset variable (*Lnsemester*). The offset term is a "structural" predictor. Its coefficient is not estimated by the model but is assumed to have the value 1; the offset is an adjustment term; thus individual values of the offset are simply added to the linear predictor of the target. This is especially useful in Poisson regression models, where each case may have a different level of exposure to the event of interest. (IBM SPSS, 2010).

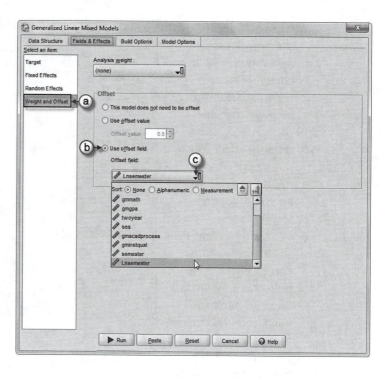

a. Click the *Weight and Offset* item to access the main dialog box.
b. Click to select the *Use offset field* option.
c. Click the pull-down menu then locate and select: *Lnsemester*.

7a. Finally, click the *Build Options* tab located at the top of the screen to access advanced criteria for building the model. Skip over the *Sorting Order* option as the settings have no effect on count data. We will, however, retain most of the default settings except for the *Tests of Fixed Effects and Coefficients*.

b. For *Tests of fixed effects and coefficients*, click to select *Use robust estimation to handle violations of model assumptions*.

Click the RUN button to generate the model results.

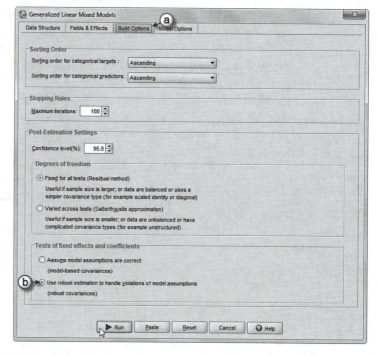

TABLE 7.43 Model 4.1: Fixed Effects

Model term	Coefficients	Std. error	t	Sig.	Exp(coefficient)	95% Confidence interval for exp(coefficient)	
						Lower	Upper
(Intercept)	−2.087	0.056	−37.517	0.000	0.124	0.111	0.138
ses	0.008	0.027	0.294	0.769	1.008	0.956	1.063
male	0.154	0.030	5.202	0.000	1.166	1.101	1.236
gmmath	−0.007	0.000	−19.786	0.000	0.993	0.992	0.993
gmgpa	−0.223	0.017	−13.519	0.000	0.800	0.774	0.826
twoyear	0.162	0.033	4.961	0.000	1.175	1.103	1.253
gmacadprocess	−0.056	0.030	−1.856	0.064	0.945	0.891	1.003
gminstqual	−0.054	0.017	−3.117	0.002	0.947	0.916	0.980

Note: Probability distribution: Poisson; link function: log; dependent variable: fail; offset variable: lnsemester.

Interpreting the Output Results of Model 4.1

At the school level, the intercept logged incident rate in this model (−2.087) is similar to the previous intercept logged count rate (−1.972) in Table 7.39 with just the individual background variables, institutional type, and offset parameter. In this case, the reference group is female students who entered a 2-year institution, with other variables (and school random effect) held constant at their means (0). The expected incident rate ratio for this group is 0.124. We note that the individual-level variables are similar to models that we have already discussed (with SES not a significant predictor of logged course failure counts) (see Table 7.43).

Turning to the school predictors, perceptions of stronger classroom instructional practices are related to reduced log counts (−0.054, $p < .01$). Holding the other predictors constant, for two female students who both entered 2-year institutions (coded 0), but who attend institutions that differ by one standard deviation in perceived instructional quality, the incident rate ratio is expected to decrease by a factor of 0.947 (or a decline of 5.3%) in the higher quality school (holding the school random effect u_{0j} constant). Similarly, there was some evidence that in institutions where students rated the academic focus more strongly, student logged counts would be expected to decline (−0.056, $p < .07$), holding the other variables constant (and the school random effect); that is, a 1-SD increase in perceived quality of academic processes would be expected to decrease the incident rate ratio by a factor of 0.945 (or 5.5%). These results suggest that institutional academic contexts do have some effects on students' likelihoods to experience academic success.

Table 7.44 presents the variance components for the random intercept at Level 2. We can see considerable variability in intercepts at Level 2. We can also note the overall fit of this model in Table 7.45 and we could compare it with our final model estimated with the NB distribution to address overdispersion (keeping in mind that estimates are approximate only in the multilevel setting).

TABLE 7.44 Model 4.1: Variance Components

Random effect	Estimate	Std. error	z-Test	Sig.	95% Confidence interval	
					Lower	Upper
Var(Intercept)	0.269	0.037	7.248	0.000	0.206	0.353

Note: Covariance structure: variance components; subject specification: nschcode.

TABLE 7.45   Model 4.1: Fit Results

	Poisson
AIC	48,429.654
BIC	48,436.958

Note: Target: fail; probability distribution: Poisson; link function: log. Information criteria are based on the –2 log pseudolikelihood (48,427.654) and are used to compare models. Models with smaller information criterion values fit better. When comparing models using pseudolikelihood values, caution should be used because different data transformations may be used across the models.

Finally, we could calculate the probability of failing one or more courses in this model. The adjusted intercept event rate is 0.124. For example, holding the other variables constant, this estimated incident rate suggests that females entering 2-year institutions would have an expected probability of 0.11 of failing one course:

$$\frac{e^{-(0.124)}0.124^1}{1!} = \frac{(0.883)(0.124)^1}{1} = 0.110$$

We leave it to readers to calculate other probabilities of course failures, as desired. We can also reestimate the model using an NB distribution to account for likely overdispersion at Level 1.

Defining Model 4.2 With IBM SPSS Menu Commands

Note: IBM SPSS settings default to those used for Model 4.1.

1. Go to the toolbar and select ANALYZE, MIXED MODELS, GENERALIZED LINEAR MODELS. This command opens the *Generalized Linear Mixed Models* (GENLINMIXED) main dialog box.

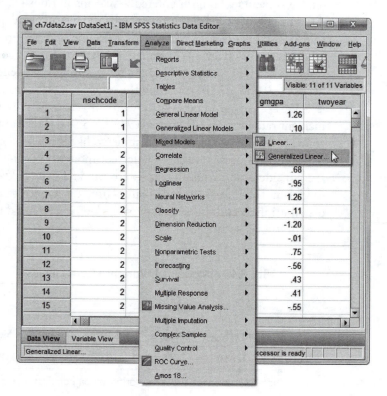

2a. The *Generalized Linear Mixed Models* dialog box displays the *Build Options* command tab, which was the last setting used in the prior model.

b. First, click to select the *Fixed Effects* command tab, which displays the *Weight and Offset* display screen that was the last setting used in the prior model.

Note: The *Weight and Offset* item is preselected (represented in the illustration by the dashed-line box but appears on-screen colored blue) as it was the last setting used from the prior model.

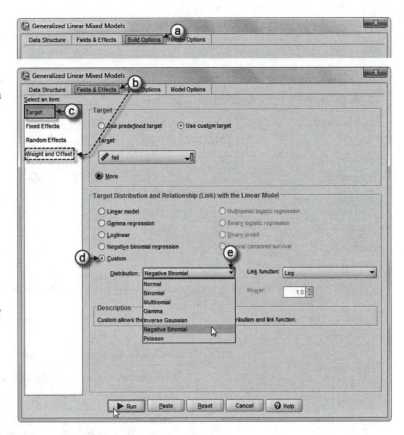

c. Click to select the *Target* item.

d. We will change the model's Poisson distribution to a negative binomial. We will retain the *Custom* setting used in the prior model but change the distribution.

e. Change the distribution setting by clicking the pull-down menu to select *Negative binomial*.

Click the RUN button to generate the model results.

Interpreting the Output Results of Model 4.2

The fixed-effect results using an NB distribution are consistent with the results we obtained previously in Table 7.43. Table 7.46 suggests that most variables have similar effects. The one change is that the academic process variable is definitely not significant ($p > .10$). As we have noted, this is likely because of the addition of the error term to the model, which tends to increase the size of the standard errors, making hypothesis tests of parameters more conservative.

In Table 7.47, the negative binomial appears to fit the data better than the Poisson distribution, as summarized by the fit indices (negative binomial AIC = 44,367.314; Poisson AIC = 48,429.654). This is likely because of the overdispersion at Level 1. The Level-2 variance component for the random intercept was similar to the last model, so we do not reproduce it here. We could, of course, continue to make subtle changes to the model, perhaps by examining whether there might be a random slope of theoretical interest that might be investigated.

TABLE 7.46   Model 4.2: Fixed Effects[a]

Model term	Coefficients	Std. error	t	Sig.	Exp(coefficient)	95% Confidence interval for exp(coefficient)	
						Lower	Upper
(Intercept)	−2.116	0.056	−38.086	0.000	0.120	0.108	0.134
ses	0.010	0.029	0.338	0.735	1.010	0.954	1.069
male	0.244	0.034	7.114	0.000	1.277	1.194	1.366
gmmath	−0.009	0.000	−28.288	0.000	0.991	0.990	0.991
gmgpa	−0.229	0.025	−9.217	0.000	0.795	0.757	0.835
twoyear	0.201	0.037	5.462	0.000	1.223	1.138	1.315
gmacadprocess	−0.057	0.035	−1.610	0.107	0.945	0.882	1.012
gminstqual	−0.049	0.018	−2.811	0.005	0.952	0.919	0.985
Negative binomial	1.914						

Note: Probability distribution: negative binomial; link function: log; dependent variable: fail; offset variable: lnsemester.

TABLE 7.47   Fit Comparison Results

	Poisson	Negative binomial
AIC	48,429.654	44,367.314
BIC	48,436.958	44,374.618

Note: Target: fail; probability distribution: Poisson; link function: log; information criteria are based on the −2 log pseudolikelihood (48,427.654) and are used to compare models. Models with smaller information criterion values fit better. When comparing models using pseudolikelihood values, caution should be used because different data transformations may be used across the models.

Summary

In this chapter we presented two extended examples examining count outcomes with fixed and varied exposure rates. We illustrated that the NB model is often a useful extension of the Poisson regression model because it can address overdispersion (by adding an error term) that can be present in Poisson regression models. This is similar to an added dispersion parameter in the Poisson regression model (Hox, 2010). We can often make use of the AIC and BIC indices to help make decisions about whether the negative binomial model is effective in addressing model dispersion.

Readers should keep in mind, however, that in the multilevel setting, these comparisons can be more tenuous because model estimation depends on quasilikelihood estimation. When the data show an excess of zeros compared to the Poisson or NB distribution, there may actually be two populations (one that is always zeros and one that produces counts similar to a Poisson distribution) (Hox, 2010). These types of mixtures are often referred to as zero inflated Poisson (ZIP) models (Lee et al., 2006).

CHAPTER 8

Concluding Thoughts

In our 2009 book, *An Introduction to Multilevel Modeling Techniques* (second edition), we sought to expand the application of multilevel modeling techniques to a new range of research questions. The modeling we presented in that book for continuous and categorical outcomes used two popular multilevel statistical programs: HLM and Mplus. Although we were, and remain, very satisfied with those excellent programs, the cost and the learning curve associated with their use present barriers to many of our colleagues who want to incorporate multilevel modeling techniques in their graduate research courses. Moreover, relatively few social science computing labs make these programs available for student and classroom use. We believe that these types of software-related challenges represent a significant barrier for instructors hoping to introduce multilevel modeling techniques to their students.

Several companies providing mainstream statistical programs have developed a multilevel functionality in more recent versions of their software. Among these are IBM SPSS, Stata, and SAS. We have observed that among these mainstream packages, SPSS enjoys widespread use in many social and behavioral science departments and in graduate schools of education. Beyond its widespread availability, the IBM SPSS software has one of the most user-friendly interfaces available. For these reasons, we have chosen to use SPSS as the vehicle for the lessons contained in our workbooks.

In our 2010 workbook we provided an applied introduction to multilevel modeling along with instruction for managing multilevel data, specifying a range of multilevel models, and interpreting output generated through the IBM SPSS MIXED procedure. After using the MIXED modeling procedure for the past few years (and having several students complete dissertations using it), we feel that it provides a powerful tool for investigating a variety of multilevel and longitudinal models with both discrete and continuous outcomes. In compiling this new IBM SPSS workbook for single-level and multilevel categorical outcomes, we also triangulated many of our results with other multilevel specific software programs. As might be expected, we found small differences in individual parameters but, overall, a high degree of substantive convergence.

Our goal in producing these workbooks was to widen exposure to and understanding of some general multilevel modeling principles and applications. Each workbook has been designed as a complement to our more in-depth treatment of the statistical and conceptual issues surrounding multilevel modeling provided in our 2009 introductory multilevel book. With an eye toward the "how to" aspect of setting up a study to investigate a research problem, in the first chapter of this workbook we introduced a number of key conceptual and methodological issues for categorical outcomes. We noted that in IBM SPSS MIXED, one can formulate models using either syntax statements or menu commands. We suspect most users are more familiar with the menu-based approach because they have likely used that to examine single-level models (e.g., analysis of variance, multiple regression, factor analysis). Accordingly, as in our first workbook, we chose the menu-based approach for presenting our model-building procedures.

Although our examples are built through the menu system in IBM SPSS, we also provide corresponding modeling syntax in Appendix A (and in electronic format via our website) for each chapter. We offer our instruction using the menu-based approach; however, we strongly encourage users to move to the syntax approach as they become more familiar with the multilevel modeling routines available in MIXED. Although the menu-based approach provides an intuitive way into the models initially, the syntax approach provides a detailed record of a set of models that have been developed and, in some instances, also facilitates conducting more nuanced types of analyses that cannot be directly conducted with the menu commands. The IBM SPSS program gives the user a good head start by providing the option for the user to generate the syntax corresponding to any models developed through the menus. Once this syntax is generated, the user can often begin to add or remove variables manually or to modify model options directly from the command line.

In our opening chapter, we detailed a number of important conceptual and mathematical differences between models for continuous and categorical outcomes. First, we noted that researchers in the social sciences often encounter variables that are not continuous because outcomes are often perceptual (e.g., defined on an ordinal scale), dichotomous (e.g., dropping out or persisting), or refer to membership in different unordered groups (e.g., religious affiliation, race/ethnicity). We also noted that categorical outcomes can be harder to report because they are in metrics (e.g., log odds and probit coefficients) different from those of the unstandardized and standardized multiple regression coefficients with which most readers are familiar.

Second, we found that many different types of models are available in IBM SPSS for working with categorical variables, beginning with basic contingency tables and related measures of association, loglinear models, discriminant analysis, logistic and ordinal regression, probit regression, and survival models, as well as multilevel formulations of many of these basic single-level models. This flexibility opens up many different opportunities for investigating different types of outcomes. The application of multilevel methods to categorical outcomes is relatively recent, however, and still being developed. At present, this provides some challenges and necessary compromises in empirically representing complex theoretical processes. Third, we also devoted attention to tying up a few loose ends from our first workbook—one being the use of the add-on multiple imputation routine available in IBM SPSS to deal with missing data in a more satisfactory manner.

In Chapter 2, we presented some adjustments to our chapter on managing cross-sectional and longitudinal data within IBM SPSS. We felt the inclusion of this chapter was necessary for people interested in learning about single-level and multilevel models with categorical outcomes who were perhaps not familiar with our other workbook. Data management (e.g., preparation of variables for analysis, manipulation of variables between levels) is an important skill in making multilevel and longitudinal modeling techniques more accessible to beginning researchers.

Chapter 3 is where we connected a set of concepts unique to models for categorical outcomes. We introduced and described a range of probability distributions, link functions, and structural models for categorical outcomes. This chapter provided an important segue to the range of models we developed in the remaining chapters. This third chapter was a capstone to the first part of the book where we offered an introductory-level consideration of the conceptual and statistical bases for categorical models.

Each of the remaining chapters in this workbook dealt with different types of probability distributions and link functions often associated with a particular type of categorical outcome. Our primary focus was on the meaning and interpretation of the output that accompanies the models that are specified. Although we did not attempt to cover every possible model, we do cover a broad number of research possibilities. It is important to note that although important omissions exist in the types of categorical models we present in this workbook (e.g., cross-classified models), we believe that enough information has been provided through our two workbooks for the reader to make the connections necessary to extend these models to cover most research possibilities that are likely to be encountered. Ultimately, appropriate data analysis returns to issues related to research questions, design, and data collection procedures. These all help produce a credible explanation of a phenomenon under study.

Throughout we have tried to direct a good portion of our analyses to the explanation of the output within the specific mathematical models that were proposed. It is our hope that making these modeling capabilities available through SPSS will make them more accessible to a wider audience, even though we only scratch the surface regarding the use of these techniques with regard to single-level, multilevel, and longitudinal models with categorical outcomes. As the reader can tell, multilevel models can get quite complex. It is one thing to dump a set of variables blindly into a single-level model, but quite another to adopt a recklessly exploratory approach in the multilevel framework. Although exploratory analysis is an important and necessary feature of empirical research, we emphasize that the models we cover in this workbook typically demand a more disciplined approach to model conceptualization and specification. With multiple levels of analysis and the possibility for numerous cross-level interactions (as well as possible mediating effects), choices about fixed or random slopes and intercepts, centering, weighting, and estimation algorithms, the interpretation of the results from these models can quickly become bogged down and rendered useless or, worse, misleading.

In our multilevel book and in these workbooks, we began with the principle that quantitative analysis represents the translation of abstract theories into concrete models and that theoretical frameworks are essential guides to empirical investigation. The statistical model is therefore designed to represent a set of proposed theoretical relationships that are thought to exist in a population from which we have a sample of data. With this as a guiding principle, the researcher makes conscious decisions about the analysis that define research questions, design, data structures, and methods of analysis. The potential complexity of the multilevel model thus demands the careful attention of the analyst.

One specific issue to which we call attention in this workbook is that multilevel models are quite demanding of data, and the analyst should be very aware of the limitations that exist as model specification becomes more complicated. Models with random intercepts will often work with 30 or 40 units; however, the moment we decide to investigate random slopes, we often find that much larger data sets are generally needed to observe significant variation in slopes across groups. They often require 20–30 individuals within groups and 200 or more groups to be able to detect variability in the slopes.

Of course, relatively large effects can be detected in smaller samples. This did give us pause, however, to consider that many published multilevel studies may report nonsignificant results due primarily to making Type II errors (i.e., failure to reject the null hypothesis) related to not having adequate data required to detect the effects. This implies that the available data often may not be up to the task of sufficiently supporting the complexity of the desired analyses. The failure to find significant results may not always be a weakness in the proposed theoretical model but, rather, can result from the shortcomings of the data (e.g., sample size, measurement properties). Hence, we emphasize that just because a model can be complex does not mean it should be complex. Parsimony is one indicator of a good empirical application of theory.

Another issue we touch upon concerns the sampling scheme for the data used in multilevel studies. Researchers have for some time now been concerned about so-called design effects resulting from complex sample schemes (i.e., where a combination of multistage cluster sampling and disproportionate stratified sampling is used). The multilevel model actually capitalizes on such sample designs, incorporating the nesting of data that result from multistage clustering. Although the multilevel model incorporates this important sample design feature, it does not address the disproportionate stratified sampling that generally occurs with such sampling. This aspect of the design is usually dealt with through the application of sampling weights during the model specification stage. Weighting is pretty routine and straightforward in single-level analyses, but as we noted in Chapter 1, it can become more challenging in the multilevel framework. Although some multilevel software can now accommodate weights at multiple levels of analysis (e.g., student-level weights and school-level weights), IBM SPSS uses a global weight and forces the researcher to choose between weights that may exist across different levels.

We note that comparing differences in multilevel weighting schemes available in various software programs is an issue that needs further research. As we illustrated briefly in Chapter 1, applying or not applying sample weights can change model estimates of standard errors considerably. In turn, because standard errors are used in determining hypothesis tests about individual parameters, potentially they can be underestimated when weights are not applied; this can lead to more findings of significance than should be the case.

Asparouhov (2006) recommends that if sample weights are only available at Level 1, it may be best to perform the analysis as a single-level analysis with adjustments for design effects (e.g., using the complex sample survey routine in IBM SPSS). If that is the case, Level-2 estimates would then have to be approached with some caution because, depending on the sampling scheme, some types of units may be over-represented and the standard errors for Level-2 variables will be calculated as if they were measured at the individual level. When weights cannot be used or do not exist, one might choose to include variables that are known to be related to the sample design in an effort to control for some of the effect (Stapleton & Thomas, 2008). For example, if "institutional type" at Level 2 is a stratified control (e.g., public and private schools), the analyst could include a public/private indicator at Level 2 to minimize the stratification effect in the absence of a school-level weight.

A third issue we ask readers to consider is that of missing data, which can present serious challenges for the researcher and should be dealt with as forthrightly as possible. Rarely are data missing completely at random (MCAR). If one were to go with the IBM SPSS defaults regarding missing data, any case with even one piece of missing data would be eliminated from the analysis (i.e., when data are not vertically organized). We emphasize that researchers need to be aware of this and, when possible, seek viable solutions for data sets that contain considerable amounts of missing data. We favor situations where the software will estimate model parameters in the presence of the missing data (e.g., where variables are arranged vertically in IBM SPSS) because, in truth, there is no way of completely compensating for the fact that some data have been irreversibly lost (Raykov & Marcoulides, 2008). We can try to obtain further information about the patterns of missing data and use this information to help devise a manner for dealing with them. We do not recommend regression-based missing data substitution (an option within IBM SPSS).

For IBM SPSS MIXED, we instead encourage the analyst to employ multiple imputation strategies, which produce multiple imputed data sets by repeatedly drawing values for each missing datum from a distribution (Raykov & Marcoulides, 2008). We demonstrated this briefly also in Chapter 1. Unfortunately, this procedure (IBM SPSS Missing Values) is not available in the IBM SPSS base or advanced program and must be purchased separately. At a minimum, we encourage the analyst to become very familiar with the data being used, to identify where data are missing, and to develop a strategy for testing the effects of missing data on the results generated by the model. Our experience is that journals are becoming more aware of missing data issues and often ask authors to be explicit in defining their strategy for dealing with missing data.

Finally, as should be very clear from the models we have developed throughout the workbook, there is a logical process underlying the development of the models and a series of steps involved for moving from the partitioning of variance between levels to the specification of models that may have random intercepts and slopes. We illustrated that our focus of investigation may be slightly different when investigating multilevel models with continuous versus categorical outcomes, given differences in underlying probability distributions and link functions.

However, we believe that in either case the focus should be on necessary steps in proposing and testing theoretical propositions within these varied types of hierarchical data structures. One's focus should always be on interpreting the output in relation to what propositions are being advanced. The data structures and flexibility of the multilevel modeling methods permit examining many new types of theoretical relationships. In this process, we strongly encourage the reader to devise a logical naming system to keep track of the various models used in any given analysis. Without a clear history of model specification, it is sometimes very difficult to understand how one arrived at a final model; that is, oftentimes the journey is more telling than the final destination.

We noted several things to consider in working with types of probability distributions and link functions other than *normal* and *identity*, respectively. One was the need to consider the various components of the categorical outcome model—that is, the random (or probability distribution) component and the link function, as well as the particular structural model to estimate the underlying predictor of Y. Another issue raised was the challenge of accounting for variance in multilevel models with categorical outcomes because of the nature of the outcomes themselves. Although we note that it is certainly possible to rescale a set of models so that they can be compared in terms of variance explained, this is often not an important concern within multilevel categorical models.

A third related issue is lack of a separate variance component at Level 1 for models with categorical outcomes. Mostly, these are just issues that require some extra thought in specifying and implementing a strategy to examine single-level and multilevel models with categorical outcomes. We hope we have provided some insight about some of these issues in the various examples that we included in the workbook.

Multilevel modeling provides us with another powerful means for investigating the types of processes referred to by our theories in more refined ways. We note that these models are hardly new and, in fact, can be traced back to the 1970s. A great deal is known about the statistical aspects of these models. As in many areas, however, the jump between knowing and doing can be quite large, and the programming of software that will run these models has slowly evolved over the past 20 years. It is our judgment that all of the programs that we have used to estimate multilevel models could be improved. Understanding the strengths and limitations of each for particular types of data structures has helped us find a suitable approach and software program for particular research problems that we have been trying to investigate. Although these software improvements can be nontrivial from a programming perspective, they continue to be realized and the range of software available continues to evolve.

Knowledge and application of multilevel models has increased greatly over the past few decades; however, there are still considerable issues of concern. For example, model estimation under various conditions, the application of sample weighting, and the proper accommodation of missing data are all issues on which a great deal of work continues to be done. Our knowledge base in these areas and others continues to grow and solutions to some of the important problems we have raised in this workbook will no doubt be forthcoming. As those solutions become better understood, it is our hope that the reader will be able to incorporate that new knowledge readily into the set of examples and lessons we have provided here.

Although the models presented in this workbook were simplified for purposes of demonstrating the techniques, we believe that these beginning models serve as a foundation for the more thorough specifications that can be formulated and tested. We note that the specification of multilevel models opens up many new possibilities for examining more complex relationships, including cross-level moderators as well as mediation effects within and between levels of data hierarchy. We encourage the reader searching for more detail to consult the many other excellent resources available and referenced throughout this workbook.

References

Agresti, A. (1996). *An introduction to categorical data analysis.* New York, NY: John Wiley & Sons.

Agresti, A. (2002). *Categorical data analysis* (2nd ed.). New York, NY: Wiley-Interscience.

Agresti, A. (2007). *An introduction to categorical data analysis.* Hoboken, NJ: John Wiley & Sons, Inc.

Aitkin, M., Anderson, D. A., Francis, B. J., & Hinde, J. P. (1989). *Statistical modelling in GLIM.* Oxford, England: University Press.

Albright, J.J. & Marinova, D. M. (2010). *Estimating multilevel models using SPSS, Stata, SAS, and R.* Retrieved from http://www.indiana.edu/~statmath/stat/all/hlm/hlm.pdf

Ashford, J. R., & Sowden, R. R. (1970). Multivariate probit analysis. *Biometrics, 26*(3), 535–546.

Asparouhov, T. (2005). Sampling weights in latent variable modeling. *Structural Equation Modeling, 12*(3), 411–434.

Asparouhov, T. (2006). General multi-level modeling with sampling weights. *Communications in Statistics: Theory & Methods, 35*(3), 439–460.

Azen, R., & Walker, C. M. (2011). *Categorical data analysis for the behavioral and social sciences.* New York, NY: Routledge.

Berkhof, J., & Snijders, T. A. B. (2001). Variance component testing in multilevel models. *Journal of Educational and Behavioral Statistics, 26,* 133–152.

Bryk, A. S., & Raudenbush, S. W. (1992). Hierarchical linear models: Applications and data analysis methods. Newbury Park, CA: Sage.

Chantala, K., Blanchette, D., & Suchinindran, C. M. (2011, April 28). Software to compute sampling weights for multilevel analysis. Retrieved from http://www.cpc.unc.edu/research/tools/data_analysis/ml_sampling_weights

Cheng, T. T. (1949). The normal approximation to the Poisson distribution and a proof of a conjecture of ramanujan. *Bulletin of the American Mathematical Society, 55,* 396–401.

Chrisman, N. R. (1998). Rethinking levels of measurement for cartography. *Cartography and Geographic Information Science, 25*(4), 231–242.

Curtin, T. R., Ingels, S. J., Wu, S., & Heuer, R. (2002). *National educational longitudinal study of 1988: Base-year to fourth follow-up data file user's manual (NCES 2002-323).* Washington, DC: US Department of Education, National Center for Education Statistics. Retrieved from http://nces.ed.gov/pubsearch/pubsinfo.asp?pubid=2002323

Dobson, A. (1990). *An introduction to generalized linear models.* London, England: Chapman & Hall.

du Toit, M., & du Toit, S. (2001). *Interactive LISREL user's guide.* Lincolnwood, IL: Scientific Software.

Evans, M., Hastings, N., & Peacock, B. (2000). *Statistical distributions.* New York, NY: Wiley.

Folmer, H. (1981). Measurement of the effects of regional policy instruments by means of linear structural equation models and panel data. *Environment and Planning A, 13,* 1435–1448.

Gill, J., & King, G. (2004). What to do when your Hessia is not invertable: Alternatives to model specification in nonlinear estimation. *Sociological Methods & Research, 8*(1), 54–87.

Goldstein, H. (1991). Non-linear multilevel models with an application to discrete response data. *Biometrika, 78*(1), 45–51.

Goldstein, H. (1995). *Multilevel statistical models.* New York, NY: Halsted.

Goodman, L. A. (1970). The multivariate analysis of qualitative data: Interactions among multiple classifications. *Journal of the American Statistical Association, 65*(329), 226–256.

Goodman, L. A. (1972). A general model for the analysis of surveys. *American Journal of Sociology, 77*(6), 1035–1086.

Grilli, L., & Pratesi, M. (2004). Weighted estimation in multilevel ordinal and binary models in the presence of informative sampling designs. *Survey Methodology, 30,* 93–103.

Hamilton, L. C. (1992). *Regression with graphics.* Belmont, CA: Wadsworth, Inc.

Heck, R. H., & Thomas, S. L. (2009). *An introduction to multilevel modeling techniques* (2nd ed.). New York, NY: Psychology Press.

Heck, R. H., Thomas, S. L., & Tabata, L. N. (2010). *Multilevel and longitudinal modeling with IBM SPSS.* New York, NY: Routledge.

Hedeker, D. (2005). Generalized linear mixed models. In B. Everitt & D. Howell (Eds.), *Encyclopedia of statistics in behavioral science* (pp. 727–738). New York, NY: Wiley.

Hedeker, D. (2007). Multilevel models for ordinal and nominal variables. In J. de Leeuw & E. Meijers (Eds.), *Handbook of multilevel analysis* (pp. 341–376). New York, NY: Springer.

Hedeker, D., & Mermelstein, R.J. (1998). A multilevel thresholds of change model for analysis of stages of change data. *Multivariate Behavioral Research, 33,* 427–455.

Hilbe, J. (1994). Negative binomial regression. *Stata Technical Bulletin 3(*18), StataCorp LP.

Horton, N. J., & Lipsitz, S. R. (1999). Review of software to fit generalized estimating equation (GEE) regression models. *American Statistician, 53,* 160–169.

Hox, J. (2002). *Multilevel analysis: Techniques and applications.* Mahwah, NJ: Lawrence Erlbaum Associates.

Hox, J. J. (2010). *Multilevel analysis: Techniques and applications* (2nd ed.). New York, NY: Routledge.

IBM SPSS. (2010). *Generalized linear mixed models.* Retrieved from http://127.0.0.1:49215/help/advanced/print.jsp?topic=/com.ibm.spss.statistics.help/idh_glm

Jia, Y., Stokes, L., Harris, I., & Wang, Y. (2011, April). *The evaluation of bias of the weighted random effects model estimators* (ETS RR-11-13). Princeton, NJ: Educational Testing Service.

Knoke, D., & Burke, P. J. (1980). *Log-linear models.* Newbury Park, CA: Sage.

Kreft, I., & de Leeuw, J. (1998). *Introducing multilevel modeling.* Thousand Oaks, CA: Sage.

Lancaster, H. O. (1951). Complex contingency tables treated by the partition of chi-square. *Journal of the Royal Statistical Society. Series B (Methodological), B13*(2), 242–249.

Lau, M. S. K., Yue, S. P., Ling, K. V., & Maciejowski, J. M. (2009, August). A comparison of interior point and active set methods for FPGA implementation of model predictive control. *European Control Conference, ECC'09,* Retrieved from http://publications.eng.cam.ac.uk/16840/

Lee, A.H., Wang, K., Scott, J.A., Yau, K.K.W., & McLachlan, G.W. (2006). Multi-level zero-inflated Poisson regression modeling of correlated count data with excess zeros. *Statistical Methods in Medical Research, 15,* 47–61.

Leyland, A. H. (2004). A review of multilevel modelling in SPSS. Retrieved from http://stat.gamma.rug.nl/reviewspss.pdf

Liang, K.-L., & Zeger, S. L. (1986). Longitudinal analysis using generalized linear models. *Biometrika, 73*(1), 13–22.

Loh, W. (1987). Some modifications of Levene's test of variance homogeneity. *Journal of Statistical Computation and Simulation, 28*(3), 213–226.

Long, J. S. (1997). *Regression models for categorical and limited dependent variables.* Thousand Oaks, CA: Sage Publications.

Longford, N. T. (1993). *Random coefficients models.* Oxford, England: Clarendon Press.

Luce, R. D. (1997). Quantification and symmetry: Commentary on Michell "quantitative science and the definition of measurement in psychology." *British Journal of Psychology, 88,* 395–398.

MacKinnon, D. P. (2008). *Introduction to statistical mediation analysis.* New York, NY: Psychology Press.

Marcoulides, G. A., & Hershberger, S. L. (1997). *Multivariate statistical methods: A first course.* Mahwah, NJ: Lawrence Erlbaum Associates.

McCullagh, P., & Nelder, J. A. (1989). *Generalized linear models* (2nd ed.). New York, NY: Chapman & Hall.

McKelvey, R., & Zavoina, W. (1975). A statistical model for the analysis of ordinal level dependent variables. *Journal of Mathematical Sociology, 4,* 103–120.

Mehta, P. D., & Neale, M. C. (2005). People are variables too: Multilevel structural equation modeling. *Psychological Methods, 10*(3), 259–284.

Michell, J. (1986). Measurement scales and statistics: A clash of paradigms. *Psychological Bulletin, 3,* 398–407.

Mislevy, R.J., & Bock, R.D. (1989). A hierarchical item-response model for educational testing. In R. D. Bock (Ed.), *Multi-level analysis for educational data* (pp. 57–74). San Diego, CA: Academic Press.

Morris, C. N. (1995). Hierarchical models for educational data: An overview. *Journal of Educational and Behavioral Statistics, 20*(2), 190–200.

Mueller, J. H., Schuessler, K. F., & Costner, H. L. (1977). *Statistical reasoning in sociology* (3rd ed.). New York, NY: Houghton Mifflin.

Muthén, B. O., & Kaplan, D. (1985). A comparison of some methodologies for the factor analysis of non-normal likert variables. *British Journal of Mathematical and Statistical Psychology, 38,* 171–189.

Muthén, B. O., & Kaplan, D. (1992). A comparison of some methodologies for the factor analysis of non-normal likert variables: A note on the size of the model. *British Journal of Mathematical and Statistical Psychology, 45,* 19–30.

Muthén, B. O., & Satorra, A. (1995). Complex sample data in structural equation modeling. In P. V. Marsden (Ed.), *Sociological methodology 1995* (pp. 267–316). Boston, MA: Blackwell Publishers.

Muthén, L. K., & Muthén, B. O. (1998–2006). *Mplus user's guide* (4th ed.). Los Angeles, CA: Authors.

Nelder, J. A., & Wedderburn, R. W. M. (1972). Generalized linear models. *Journal of the Royal Statistical Society: Series A (General), 135*(3), 370–384.

Nocedal, J., & Wright, S. J. (2006). *Numerical optimization* (2nd ed.). New York, NY: Springer.

Nunnally, J. C. (1978). *Psychometric theory* (2nd ed.). New York, NY: McGraw–Hill.

O'Connell, A. A. (2006). *Logistic regression models for ordinal response variables.* Thousand Oaks, CA: Sage Publications.

Olsson, U. (1979). On the robustness of factor analysis against crude classification of the observations. *Multivariate Behavioral Research, 14,* 485–500.

Park, T. (1994). Multivariate regression models for discrete and continuous repeated measurements. *Communications in Statistics: Theory & Methods, 22,* 1547–1564.

Pearson, K., & Heron, D. (1913). On theories of association. *Biometrika, 9,* 159–315.

Pedhazur, E. J., & Schmelkin, L. P. (1991). *Measurement, design, and analysis: An integrated approach.* Hillsdale, NJ: Lawrence Erlbaum Associates.

Peugh, J. L., & Enders, C. K. (2004). Missing in educational research: A review of reporting practices and suggestions for improvement. *Review of Educational Research, 74,* 525–556.

Pfeffermann, D., Skinner, C. J., Holmes, D. J., Goldstein, H., & Rasbash, J. (1998). Weighting for unequal selection probabilities in multilevel models. *Journal of the Royal Statistical Society: Series B (Statistical Methodology), 60*(1), 23.

Pierce, D. A., & Sands, B. R. (1975). *Extra-Bernoulli variation in binary data* (technical report 46). Oregon State University Department of Statistics.

Prentice, R. L. (1988). Correlated binary regression with covariates specific to each binary observation. *Biometrics, 44,* 1033–1048.

Rabe-Hesketh, S. R., & Skrondal, A. (2006). Multilevel modeling of complex survey data. *Journal of the Royal Statistical Society, 169* (Part 4), 805–827.

Rasbash, J., Steele, F., Browne, W. J., & Goldstein, H. (2009). A user's guide to MLwiN, v. 210. Retrieved from http://www.bristol.ac.uk/cmm/software/mlwin/download/manuals.html

Raudenbush, S. W., & Bryk, A. S. (2002). *Hierarchical linear models: Applications and data analysis methods* (2nd ed.). Thousand Oaks, CA: Sage Publications.

Raudenbush, S. W., Bryk, A. S., Cheong, Y., & Congdon, R. T., Jr. (2004). *HLM 6: Hierarchical linear and nonlinear modeling.* Lincolnwood, IL: Scientific Software International.

Raykov, T., & Marcoulides, G. A. (2008). *An introduction to applied multivariate analysis.* New York, NY: Routledge Academic.

Sabatier, P. A. (1999). *Theories of the policy process.* Boulder, CO: Westview Press.

Sable, J., & Noel, A. (2008). Public elementary and secondary student enrollment and staff from the common core of data: School year 2006-07 (NCES 2009-305). Washington, DC: National Center for Education Statistics, Institute of Education Sciences, Department of Education. Retrieved from http://nces.ed.gov/pubsearch/pubsinfo.asp?pubid=2009305

Satterthwaite, F. E. (1946). An approximate distribution of estimates of variance components. *Biometrics Bulletin,* 110–114.

Shapiro, J. Z., & McPherson, R. B. (1987). State board desegregation policy: An application of the problem-finding model of policy analysis. *Educational Administration Quarterly, 23*(2), 60–77.

Singer, J. D., & Willett, J. B. (2003). *Applied longitudinal data analysis: Modeling change and event occurrence.* New York, NY: Oxford University Press.

Skinner, C. J. (2005). *The use of survey weights in multilevel modeling.* Paper presented at the Workshop on Latent Variable Models and Survey Data for Social Science Research, Montreal, Canada. (May 2005).

Snijders, T. A. B. (2005). Power and sample size in multilevel linear models. In B. S. Everitt & D. C. Howell (Eds.), *Encyclopedia of statistics in behavioral sciences* (Vol. 2, pp. 1570–1573). New York: Wiley.

Snijders, T. A. B., & Bosker, R. (1999). *An introduction to basic and advanced multilevel modeling.* Thousand Oaks, CA: Sage.

Sobel, M. E. (1982). Asymptotic confidence intervals for indirect effects in structural equation models. In S. Leinhardt (Ed.), *Sociological methodology 1982* (pp. 290–312). Washington, DC: American Sociological Association.

SPSS. (2002). Linear mixed effects modeling in SPSS: An introduction to the mixed procedure (technical report LMEMWP-1002). Chicago, IL: SPSS.

Stapleton, L. M. (2002). The incorporation of sample weights into multilevel structural equation models. *Structural Equation Modeling, 9*(4), 475–502.

Stapleton, L. & Thomas, S. L. (2008). The use of national data sets for teaching and research: Sources and issues. In A. O'Connell & B. McCoach (Eds.) *Multilevel modeling of educational data* (pp. 11–57). Charlotte, NC: Information Age Publishing.

Stevens, S. S. (1951). Mathematics, measurement, and psychophysics. In S. S. Stevens (Ed.), *Handbook of experimental psychology* (pp. 1–49). New York, NY: John Wiley & Sons.

Stevens, S. S. (1968). Measurement, statistics, and the schemapiric view. *Science, 161,* 849–856.

Tabachnick, B. G. (2008, March). Multivariate statistics: An introduction and some applications. Workshop presented to the American Psychology-Law Society, Jacksonville, FL.

Tabachnick, B. G., & Fidell, L. S. (2006). *Using multivariate statistics.* Boston, MA: Allyn & Bacon.

Velleman, P. F., & Wilkinson, L. (1993). Nominal, ordinal, interval, and ratio typologies are misleading. *American Statistician, 47*(1), 65–72.

Wong, G. Y., & Mason, W. M. (1989). *Ethnicity, comparative analysis and generalization of the hierarchical normal linear model for multilevel analysis* (Research Rep. No. 89–138). Ann Arbor, MI: Population Studies Center, University of Michigan.

Wright, D. B. (1997). Extra-binomial variation in multilevel logistic models with sparse structures. *British Journal of Mathematical and Statistical Psychology, 50,* 21–29.

Yule, G. U. (1912). *An introduction to the theory of statistics* (2nd ed.). London, England: Charles Griffin & Co.

Zeger, S. L., & Liang, K.-L. (1986). Longitudinal data analysis for discrete and continuous outcomes. *Biometrics, 42*(1), 121–130.

Ziegler, A., Kastner, C., & Blettner, M. (1998). The generalized estimating equations: An annotated bibliography. *Biometrical Journal, (2),* 115–139.

Appendix A: Syntax Statements

Please note that syntax statements may be obtained through the "paste" feature in the IBM SPSS menu interface. This function will insert an introductory comment line to describes the routine (e.g., *Generalized Linear Models). This comment line has been omitted for the models in this section but may be viewed in the syntax files (.sps) accompanying the data sets for each chapter and it is available for downloading from the publisher's website.

The syntax statements give information about

the grouping variable (SUBJECTS = schcode)
the error distribution (BINOMIAL)
the link (LOGIT)
the fixed effects in the model (e.g., female, the intercept)
the random effects (RANDOM USE INTERCEPT = TRUE)
the type of covariance matrix (TYPE=VARIANCE_COMPONENTS)
the outcome category order (TARGET_CATEGORY_ORDER=DESCENDING)
the inputs category order (DESCENDING)
the maximum number of iterations to find a solution (100)
the confidence intervals (95%)
the standard errors (ROBUST)

Using the syntax can be useful in making quick changes to the model as well as helping to keep a record of what models have been developed. They should be saved in a manner that allows easy referencing for future use (e.g., RprofMod1.sps, RprofMod2.sps).

Chapter 3: Specification of Generalized Linear Models (*ch3data.sav*)
Specifying the Model
Model 1.1: Tables 3.6, 3.8, 3.9, 3.14 (*ch3data.sav*)

```
GENLIN readprof (REFERENCE=FIRST)
/MODEL INTERCEPT=YES
DISTRIBUTION=BINOMIAL LINK=LOGIT
/CRITERIA METHOD=FISHER(1) SCALE=1 COVB=ROBUST MAXITERATIONS=100
    MAXSTEPHALVING=5
PCONVERGE=1E-006(ABSOLUTE) SINGULAR=1E-012 ANALYSISTYPE=3(WALD)
    CILEVEL=95 CITYPE=WALD
LIKELIHOOD=KERNEL
/MISSING CLASSMISSING=EXCLUDE
/PRINT CPS DESCRIPTIVES MODELINFO FIT SUMMARY SOLUTION (EXPONENTIATED)
    HISTORY(1).
```

Adding Gender to the Model
Model 1.2: Tables 3.11, 3.13 (ch3data.sav)

```
GENLIN readprof (REFERENCE=FIRST) BY female (ORDER=DESCENDING)
/MODEL female INTERCEPT=YES
DISTRIBUTION=BINOMIAL LINK=LOGIT
/CRITERIA METHOD=FISHER(1) SCALE=1 COVB=ROBUST MAXITERATIONS=100
   MAXSTEPHALVING=5
PCONVERGE=1E-006(ABSOLUTE) SINGULAR=1E-012 ANALYSISTYPE=3(WALD)
   CILEVEL=95 CITYPE=WALD
LIKELIHOOD=KERNEL
/MISSING CLASSMISSING=EXCLUDE
/PRINT CPS DESCRIPTIVES MODELINFO FIT SUMMARY SOLUTION (EXPONENTIATED)
   HISTORY(1).
```

Adding Additional Background Predictors
Model 1.3: Tables 3.16, 3.17 (ch3data.sav)

```
GENLIN readprof (REFERENCE=FIRST) BY female lowses minor
   (ORDER=DESCENDING)
/MODEL female lowses minor INTERCEPT=YES
DISTRIBUTION=BINOMIAL LINK=LOGIT
/CRITERIA METHOD=FISHER(1) SCALE=1 COVB=ROBUST MAXITERATIONS=100
   MAXSTEPHALVING=5
PCONVERGE=1E-006(ABSOLUTE) SINGULAR=1E-012 ANALYSISTYPE=3(WALD)
   CILEVEL=95 CITYPE=WALD
LIKELIHOOD=KERNEL
/MISSING CLASSMISSING=EXCLUDE
/PRINT CPS DESCRIPTIVES MODELINFO FIT SUMMARY SOLUTION (EXPONENTIATED)
   HISTORY(1).
```

Chapter 4: Multilevel Model With Dichotomous Outcomes (*ch4data1.sav, ch4data2.sav*)
The Unconditional (Null) Model
Model 1.1: Tables 4.2, 4.3, 4.4 (ch4data1.sav)

```
GENLINMIXED
/DATA_STRUCTURE SUBJECTS=schcode
/FIELDS TARGET=readprof TRIALS=NONE OFFSET=NONE
/TARGET_OPTIONS DISTRIBUTION=BINOMIAL LINK=LOGIT
/FIXED USE_INTERCEPT=TRUE
/RANDOM USE_INTERCEPT=TRUE SUBJECTS=schcode COVARIANCE_TYPE= VARIANCE_
   COMPONENTS
/BUILD_OPTIONS TARGET_CATEGORY_ORDER=DESCENDING INPUTS_ CATEGORY_
   ORDER=DESCENDING
MAX_ITERATIONS=100 CONFIDENCE_LEVEL=95 DF_METHOD=RESIDUAL COVB=ROBUST
/EMMEANS_OPTIONS SCALE=ORIGINAL PADJUST=LSD.
```

Defining the Within-School Variables
Model 1.2: Tables 4.5, 4.6, 4.7, 4.8 (ch4data1.sav)

```
GENLINMIXED
/DATA_STRUCTURE SUBJECTS=schcode
/FIELDS TARGET=readprof TRIALS=NONE OFFSET=NONE
/TARGET_OPTIONS DISTRIBUTION=BINOMIAL LINK=LOGIT
/FIXEDEFFECTS=female lowses minor USE_INTERCEPT=TRUE
/RANDOM USE_INTERCEPT=TRUE SUBJECTS=schcode COVARIANCE_TYPE=VARIANCE_
  COMPONENTS
/BUILD_OPTIONS TARGET_CATEGORY_ORDER=DESCENDING INPUTS_CATEGORY_
  ORDER=DESCENDING
MAX_ITERATIONS=100 CONFIDENCE_LEVEL=95 DF_METHOD=RESIDUAL COVB=ROBUST
/EMMEANS_OPTIONS SCALE=ORIGINAL PADJUST=LSD.
```

Examining Whether a Level-1 Slope Varies Between Schools
Model 1.3: Table 4.9 (ch4data1.sav)

```
GENLINMIXED
/DATA_STRUCTURE SUBJECTS=schcode
/FIELDS TARGET=readprof TRIALS=NONE OFFSET=NONE
/TARGET_OPTIONS DISTRIBUTION=BINOMIAL LINK=LOGIT
/FIXEDEFFECTS=female lowses minor USE_INTERCEPT=TRUE
/RANDOM EFFECTS=minor USE_INTERCEPT=TRUE SUBJECTS=schcode COVARIANCE_
  TYPE=VARIANCE_COMPONENTS
/BUILD_OPTIONS TARGET_CATEGORY_ORDER=DESCENDING INPUTS_CATEGORY_
  ORDER=DESCENDING
MAX_ITERATIONS=100 CONFIDENCE_LEVEL=95 DF_METHOD=RESIDUAL COVB=ROBUST
/EMMEANS_OPTIONS SCALE=ORIGINAL PADJUST=LSD.
```

Adding Level-2 Predictors to Explain Variability in Intercepts
Model 1.4: Tables 4.10, 4.11, 4.12 (ch4data1.sav)

```
GENLINMIXED
/DATA_STRUCTURE SUBJECTS=schcode
/FIELDS TARGET=readprof TRIALS=NONE OFFSET=NONE
/TARGET_OPTIONS DISTRIBUTION=BINOMIAL LINK=LOGIT
/FIXEDEFFECTS=schcomp smallsch female lowses minor USE_INTERCEPT=TRUE
/RANDOM EFFECTS=minor USE_INTERCEPT=TRUE SUBJECTS=schcode COVARIANCE_
  TYPE=VARIANCE_COMPONENTS
/BUILD_OPTIONS TARGET_CATEGORY_ORDER=DESCENDING INPUTS_CATEGORY_
  ORDER=DESCENDING
MAX_ITERATIONS=100 CONFIDENCE_LEVEL=95 DF_METHOD=RESIDUAL COVB=ROBUST
/EMMEANS_OPTIONS SCALE=ORIGINAL PADJUST=LSD.
```

Adding Level-2 Variables to Explain Variation in Level-1 Slopes (Cross-Level Interaction)
Model 1.5: Tables 4.13, 4.14 (ch4data1.sav)

```
GENLINMIXED
/DATA_STRUCTURE SUBJECTS=schcode
/FIELDS TARGET=readprof TRIALS=NONE OFFSET=NONE
/TARGET_OPTIONS DISTRIBUTION=BINOMIAL LINK=LOGIT
/FIXEDEFFECTS=schcomp smallsch female lowses minor minor*schcomp
  minor*smallsch
USE_INTERCEPT=TRUE
/RANDOM EFFECTS=minor USE_INTERCEPT=TRUE SUBJECTS=schcode COVARIANCE_
  TYPE=VARIANCE_COMPONENTS
/BUILD_OPTIONS TARGET_CATEGORY_ORDER=DESCENDING INPUTS_CATEGORY_
  ORDER=DESCENDING
MAX_ITERATIONS=100 CONFIDENCE_LEVEL=95 DF_METHOD=RESIDUAL COVB=ROBUST
/EMMEANS_OPTIONS SCALE=ORIGINAL PADJUST=LSD.
```

Estimating Means
Model Based on Equation 4.22: Tables 4.16, 4.17 (ch4data1.sav)

```
GENLINMIXED
/DATA_STRUCTURE SUBJECTS=schcode
/FIELDS TARGET=readprof TRIALS=NONE OFFSET=NONE
/TARGET_OPTIONS DISTRIBUTION=BINOMIAL LINK=LOGIT
/FIXEDEFFECTS=lowses minor lowses*minor USE_INTERCEPT=TRUE
/RANDOM EFFECTS=minor USE_INTERCEPT=TRUE SUBJECTS=schcode COVARIANCE_
  TYPE=VARIANCE_COMPONENTS
/BUILD_OPTIONS TARGET_CATEGORY_ORDER=DESCENDING INPUTS_CATEGORY_
  ORDER=DESCENDING
MAX_ITERATIONS=100 CONFIDENCE_LEVEL=95 DF_METHOD=RESIDUAL COVB=ROBUST
/EMMEANS TABLES=lowses COMPARE=lowses CONTRAST=SIMPLE
/EMMEANS TABLES=minor COMPARE=minor CONTRAST=SIMPLE
/EMMEANS TABLES=lowses*minor COMPARE=lowses CONTRAST=SIMPLE
/EMMEANS_OPTIONS SCALE=ORIGINAL PADJUST=SEQBONFERRONI.
```

Probit Link Function
Model 1.6: Table 4.20 (ch4data1.sav)

```
GENLINMIXED
/DATA_STRUCTURE SUBJECTS=schcode
/FIELDS TARGET=readprof TRIALS=NONE OFFSET=NONE
/TARGET_OPTIONS DISTRIBUTION=BINOMIAL LINK=PROBIT
/FIXEDEFFECTS=schcomp smallsch female lowses minor minor*schcomp
  minor*smallsch
USE_INTERCEPT=TRUE
/RANDOM EFFECTS=minor USE_INTERCEPT=TRUE SUBJECTS=schcode
  COVARIANCE_TYPE=VARIANCE_COMPONENTS
/BUILD_OPTIONS TARGET_CATEGORY_ORDER=DESCENDING INPUTS_CATEGORY_
  ORDER=DESCENDING
MAX_ITERATIONS=100 CONFIDENCE_LEVEL=95 DF_METHOD=RESIDUAL COVB=ROBUST
/EMMEANS_OPTIONS SCALE=ORIGINAL PADJUST=LSD.
```

The Unconditional Model
Model 2.1: Table 4.23 (ch4data2.sav)

```
GENLINMIXED
/DATA_STRUCTURE SUBJECTS=schcode*Rteachid
/FIELDS TARGET=readprof TRIALS=NONE OFFSET=NONE
/TARGET_OPTIONS DISTRIBUTION=BINOMIAL LINK=LOGIT
/FIXED USE_INTERCEPT=TRUE
/RANDOM USE_INTERCEPT=TRUE SUBJECTS=schcode COVARIANCE_TYPE= VARIANCE_
    COMPONENTS
/RANDOM USE_INTERCEPT=TRUE SUBJECTS=schcode*Rteachid COVARIANCE_TYPE=
    VARIANCE_COMPONENTS
/BUILD_OPTIONS TARGET_CATEGORY_ORDER=DESCENDING INPUTS_CATEGORY_
    ORDER=DESCENDING
MAX_ITERATIONS=100 CONFIDENCE_LEVEL=95 DF_METHOD=RESIDUAL COVB=ROBUST
/EMMEANS_OPTIONS SCALE=ORIGINAL PADJUST=LSD.
```

Defining the Three-Level Model
Model 2.2: Tables 4.24, 4.25, 4.26 (ch4data2.sav)

```
GENLINMIXED
/DATA_STRUCTURE SUBJECTS=schcode*Rteachid
/FIELDS TARGET=readprof TRIALS=NONE OFFSET=NONE
/TARGET_OPTIONS DISTRIBUTION=BINOMIAL LINK=LOGIT
/FIXEDEFFECTS=schcomp smallsch gmclasscomp gmclass_size zteacheff
    lowses female minor
USE_INTERCEPT=TRUE
/RANDOM USE_INTERCEPT=TRUE SUBJECTS=schcode COVARIANCE_TYPE= VARIANCE_
    COMPONENTS
/RANDOM USE_INTERCEPT=TRUE SUBJECTS=schcode*Rteachid COVARIANCE_TYPE=
    VARIANCE_COMPONENTS
/BUILD_OPTIONS TARGET_CATEGORY_ORDER=DESCENDING INPUTS_CATEGORY_
    ORDER=DESCENDING
MAX_ITERATIONS=100 CONFIDENCE_LEVEL=95 DF_METHOD=RESIDUAL COVB=ROBUST
/EMMEANS_OPTIONS SCALE=ORIGINAL PADJUST=LSD.
```

Chapter 5: Multilevel Models With a Categorical Repeated Measures Outcome (ch5data1.sav, ch5data1-ord.sav, ch5data2.sav)
Model 1.1: Tables 5.2, 5.3, 5.4, 5.5, 5.6 (ch5data1.sav)

```
GENLIN readprof (REFERENCE=FIRST) WITH time
/MODEL time INTERCEPT=YES
DISTRIBUTION=BINOMIAL LINK=LOGIT
/CRITERIA METHOD=FISHER(1) SCALE=1 MAXITERATIONS=100 MAXSTEPHALVING= 5
    PCONVERGE=1E-006(ABSOLUTE)
SINGULAR=1E-012 ANALYSISTYPE=3(WALD) CILEVEL=95 LIKELIHOOD=FULL
/REPEATED SUBJECT=id WITHINSUBJECT=time SORT=YES CORRTYPE= EXCHANGEABLE
    ADJUSTCORR=YES COVB=ROBUST
MAXITERATIONS=100 PCONVERGE=1e-006(ABSOLUTE) UPDATECORR=1
/MISSING CLASSMISSING=EXCLUDE
/PRINT CPS DESCRIPTIVES MODELINFO FIT SOLUTION (EXPONENTIATED) CORB.
```

Alternate Coding of the Time Variable
Model 1.2: Tables 5.9, 5.11, 5.12 (ch5data1-ord.sav)

```
GENLIN readprof (REFERENCE=FIRST) BY time (ORDER=DESCENDING)
/MODEL time INTERCEPT=YES
DISTRIBUTION=BINOMIAL LINK=LOGIT
/CRITERIA METHOD=FISHER(1) SCALE=1 MAXITERATIONS=100 MAXSTEPHALVING=5
  PCONVERGE=1E-006(ABSOLUTE)
SINGULAR=1E-012 ANALYSISTYPE=3(WALD) CILEVEL=95 LIKELIHOOD=FULL
/REPEATED SUBJECT=id WITHINSUBJECT=time SORT=YES CORRTYPE= EXCHANGEABLE
  ADJUSTCORR=YES COVB=ROBUST
MAXITERATIONS=100 PCONVERGE=1e-006(ABSOLUTE) UPDATECORR=1
/MISSING CLASSMISSING=EXCLUDE
/PRINT CPS DESCRIPTIVES MODELINFO FIT SUMMARY SOLUTION (EXPONENTIATED)
  WORKINGCORR.
```

Table 5.10: Correlations of Parameter Estimates (Recoded Time Variable; ch5data1-ord.sav)

```
GENLIN readprof (REFERENCE=FIRST) WITH timeend
/MODEL timeend INTERCEPT=YES
DISTRIBUTION=BINOMIAL LINK=LOGIT
/CRITERIA METHOD=FISHER(1) SCALE=1 MAXITERATIONS=100 MAXSTEPHALVING= 5
  PCONVERGE=1E-006(ABSOLUTE)
SINGULAR=1E-012 ANALYSISTYPE=3(WALD) CILEVEL=95 LIKELIHOOD=FULL
/REPEATED SUBJECT=id WITHINSUBJECT=timeend SORT=YES CORRTYPE=
  EXCHANGEABLE ADJUSTCORR=YES
COVB=ROBUST MAXITERATIONS=100 PCONVERGE=1e-006(ABSOLUTE) UPDATECORR=1
/MISSING CLASSMISSING=EXCLUDE
/PRINT CPS DESCRIPTIVES MODELINFO FIT SUMMARY SOLUTION (EXPONENTIATED)
  CORB.
```

Model 1.3: Table 5.13 (Unstructured Correlation Matrix; ch5data1.sav)

```
GENLIN readprof (REFERENCE=FIRST) WITH time
/MODEL time INTERCEPT=YES
DISTRIBUTION=BINOMIAL LINK=LOGIT
/CRITERIA METHOD=FISHER(1) SCALE=1 MAXITERATIONS=100 MAXSTEPHALVING=5
  PCONVERGE=1E-006(ABSOLUTE)
SINGULAR=1E-012 ANALYSISTYPE=3(WALD) CILEVEL=95 LIKELIHOOD=FULL
/REPEATED SUBJECT=id WITHINSUBJECT=time SORT=YES CORRTYPE= UNSTRUCTURED
  ADJUSTCORR=YES COVB=ROBUST
MAXITERATIONS=100 PCONVERGE=1e-006(ABSOLUTE) UPDATECORR=1
/MISSING CLASSMISSING=EXCLUDE
/PRINT CPS DESCRIPTIVES MODELINFO FIT SUMMARY SOLUTION (EXPONENTIATED)
  WORKINGCORR.
```

Adding a Predictor

Model 1.4: Table 5.14 (ch5data1.sav)

```
GENLIN readprof (REFERENCE=FIRST) BY female (ORDER=DESCENDING) WITH
  time
/MODEL female time INTERCEPT=YES
DISTRIBUTION=BINOMIAL LINK=LOGIT
/CRITERIA METHOD=FISHER(1) SCALE=1 MAXITERATIONS=100 MAXSTEPHALVING=5
  PCONVERGE=1E-006(ABSOLUTE)
SINGULAR=1E-012 ANALYSISTYPE=3(WALD) CILEVEL=95 LIKELIHOOD=FULL
/REPEATED SUBJECT=id WITHINSUBJECT=time SORT=YES CORRTYPE= EXCHANGEABLE
  ADJUSTCORR=YES COVB=ROBUST
MAXITERATIONS=100 PCONVERGE=1e-006(ABSOLUTE) UPDATECORR=1
/MISSING CLASSMISSING=EXCLUDE
/PRINT CPS DESCRIPTIVES MODELINFO FIT SUMMARY SOLUTION (EXPONENTIATED)
  WORKINGCORR.
```

Alternative Model 1.4: Table 5.15 (Time Treated as Ordinal; data5-1ord.sav)

```
GENLIN readprof (REFERENCE=FIRST) BY female time (ORDER=DESCENDING)
/MODEL female time INTERCEPT=YES
DISTRIBUTION=BINOMIAL LINK=LOGIT
/CRITERIA METHOD=FISHER(1) SCALE=1 MAXITERATIONS=100 MAXSTEPHALVING=5
  PCONVERGE=1E-006(ABSOLUTE)
SINGULAR=1E-012 ANALYSISTYPE=3(WALD) CILEVEL=95 LIKELIHOOD=FULL
/REPEATED SUBJECT=id WITHINSUBJECT=time SORT=YES CORRTYPE= EXCHANGEABLE
  ADJUSTCORR=YES COVB=ROBUST
MAXITERATIONS=100 PCONVERGE=1e-006(ABSOLUTE) UPDATECORR=1
/MISSING CLASSMISSING=EXCLUDE
/PRINT CPS DESCRIPTIVES MODELINFO FIT SUMMARY SOLUTION
  (EXPONENTIATED).
```

Adding an Interaction Between Female and the Time Parameter

Model 1.5: Table 5.16 (ch5data1.sav)

```
GENLIN readprof (REFERENCE=FIRST) BY female (ORDER=DESCENDING) WITH
  time
/MODEL female time female*time INTERCEPT=YES
DISTRIBUTION=BINOMIAL LINK=LOGIT
/CRITERIA METHOD=FISHER(1) SCALE=1 MAXITERATIONS=100 MAXSTEPHALVING=5
  PCONVERGE=1E-006(ABSOLUTE)
SINGULAR=1E-012 ANALYSISTYPE=3(WALD) CILEVEL=95 LIKELIHOOD=FULL
/REPEATED SUBJECT=id WITHINSUBJECT=time SORT=YES CORRTYPE= EXCHANGEABLE
  ADJUSTCORR=YES COVB=ROBUST
MAXITERATIONS=100 PCONVERGE=1e-006(ABSOLUTE) UPDATECORR=1
/MISSING CLASSMISSING=EXCLUDE
/PRINT CPS DESCRIPTIVES MODELINFO FIT SUMMARY SOLUTION (EXPONENTIATED)
  WORKINGCORR.
```

Alternate Model 1.5: Table 5.17 (Time Variable Is Ordinal; ch5data1-ord.sav)

```
GENLIN readprof (REFERENCE=FIRST) BY female time (ORDER=DESCENDING)
/MODEL female time female*time INTERCEPT=YES
DISTRIBUTION=BINOMIAL LINK=LOGIT
/CRITERIA METHOD=FISHER(1) SCALE=1 MAXITERATIONS=100 MAXSTEPHALVING=5
   PCONVERGE=1E-006(ABSOLUTE)
SINGULAR=1E-012 ANALYSISTYPE=3(WALD) CILEVEL=95 LIKELIHOOD=FULL
/REPEATED SUBJECT=id WITHINSUBJECT=time SORT=YES CORRTYPE= EXCHANGEABLE
   ADJUSTCORR=YES COVB=ROBUST
MAXITERATIONS=100 PCONVERGE=1e-006(ABSOLUTE) UPDATECORR=1
/MISSING CLASSMISSING=EXCLUDE
/PRINT CPS DESCRIPTIVES MODELINFO FIT SUMMARY SOLUTION (EXPONENTIATED)
   WORKINGCORR.
```

Specifying a GEE Model Within GENLIN MIXED
Model 2.1: Tables 5.18, 5.19, 5.20 (ch5data1.sav)

```
GENLINMIXED
/DATA_STRUCTURE SUBJECTS=id REPEATED_MEASURES=time COVARIANCE_
   TYPE=COMPOUND_SYMMETRY
/FIELDS TARGET=readprof TRIALS=NONE OFFSET=NONE
/TARGET_OPTIONS DISTRIBUTION=BINOMIAL LINK=LOGIT
/FIXEDEFFECTS=female time USE_INTERCEPT=TRUE
/BUILD_OPTIONS TARGET_CATEGORY_ORDER=DESCENDING INPUTS_
 CATEGORY_ ORDER=DESCENDING
MAX_ITERATIONS=100 CONFIDENCE_LEVEL=95 DF_METHOD=RESIDUAL COVB=ROBUST
/EMMEANS_OPTIONS SCALE=ORIGINAL PADJUST=LSD.
```

Examining a Random Intercept at the Between-Student Level
Model 2.2: Tables 5.21, 5.22, 5.23 (ch5data1.sav)

```
GENLINMIXED
/DATA_STRUCTURE SUBJECTS=id REPEATED_MEASURES=time COVARIANCE_ TYPE=AR1
/FIELDS TARGET=readprof TRIALS=NONE OFFSET=NONE
/TARGET_OPTIONS DISTRIBUTION=BINOMIAL LINK=LOGIT
/FIXEDEFFECTS=time USE_INTERCEPT=TRUE
/RANDOM USE_INTERCEPT=TRUE SUBJECTS=id COVARIANCE_TYPE= VARIANCE_
   COMPONENTS
/BUILD_OPTIONS TARGET_CATEGORY_ORDER=DESCENDING INPUTS_CATEGORY_
   ORDER=DESCENDING
MAX_ITERATIONS=100 CONFIDENCE_LEVEL=95 DF_METHOD=RESIDUAL COVB=ROBUST
/EMMEANS_OPTIONS SCALE=ORIGINAL PADJUST=LSD.
```

What Variables Affect Differences in Proficiency Across Individuals?

Model 2.3: Tables 5.24, 5.25, 5.26 (ch5data1.sav)

```
GENLINMIXED
/DATA_STRUCTURE SUBJECTS=id REPEATED_MEASURES=time COVARIANCE_ TYPE=AR1
/FIELDS TARGET=readprof TRIALS=NONE OFFSET=NONE
/TARGET_OPTIONS DISTRIBUTION=BINOMIAL LINK=LOGIT
/FIXEDEFFECTS=female lowses time female*time time*lowses USE_
   INTERCEPT=TRUE
/RANDOM USE_INTERCEPT=TRUE SUBJECTS=id COVARIANCE_TYPE=VARIANCE_
   COMPONENTS
/BUILD_OPTIONS TARGET_CATEGORY_ORDER=DESCENDING INPUTS_CATEGORY_
   ORDER=DESCENDING
MAX_ITERATIONS=100 CONFIDENCE_LEVEL=95 DF_METHOD=RESIDUAL COVB=ROBUST
/EMMEANS_OPTIONS SCALE=ORIGINAL PADJUST=LSD.
```

Alternative Model (GEE Procedure) 2.3: Table 5.27 (ch5data1.sav)

```
GENLIN readprof (REFERENCE=FIRST) BY female lowses (ORDER=DESCENDING)
   WITH time
/MODEL female lowses time female*time lowses*time INTERCEPT=YES
DISTRIBUTION=BINOMIAL LINK=LOGIT
/CRITERIA METHOD=FISHER(1) SCALE=1 MAXITERATIONS=100 MAXSTEPHALVING=5
   PCONVERGE=1E-006(ABSOLUTE)
SINGULAR=1E-012 ANALYSISTYPE=3(WALD) CILEVEL=95 LIKELIHOOD=FULL
/REPEATED SUBJECT=id WITHINSUBJECT=time1 SORT=YES CORRTYPE=
   EXCHANGEABLE ADJUSTCORR=YES
COVB=ROBUST MAXITERATIONS=100 PCONVERGE=1e-006(ABSOLUTE) UPDATECORR=1
/MISSING CLASSMISSING=EXCLUDE
/PRINT CPS DESCRIPTIVES MODELINFO FIT SUMMARY SOLUTION
   (EXPONENTIATED).
```

The Beginning Model

Model 3.1: Tables 5.28, 5.29, 5.30 (ch5data1.sav)

```
GENLINMIXED
/DATA_STRUCTURE SUBJECTS=nschcode*Rid REPEATED_MEASURES=time
   COVARIANCE_TYPE=AR1
/FIELDS TARGET=readprof TRIALS=NONE OFFSET=NONE
/TARGET_OPTIONS DISTRIBUTION=BINOMIAL LINK=LOGIT
/FIXEDEFFECTS=time USE_INTERCEPT=TRUE
/RANDOM EFFECTS=time USE_INTERCEPT=TRUE SUBJECTS=nschcode COVARIANCE_
   TYPE=UNSTRUCTURED
/RANDOM USE_INTERCEPT=TRUE SUBJECTS=nschcode*Rid COVARIANCE_TYPE=
   VARIANCE_COMPONENTS
/BUILD_OPTIONS TARGET_CATEGORY_ORDER=DESCENDING INPUTS_CATEGORY_
   ORDER=DESCENDING
MAX_ITERATIONS=100 CONFIDENCE_LEVEL=95 DF_METHOD=RESIDUAL COVB=ROBUST
/EMMEANS_OPTIONS SCALE=ORIGINAL PADJUST=LSD.
```

Adding Student and School Predictors

Model 3.2: Tables 5.31, 5.32, 5.33 (ch5data1.sav)

```
GENLINMIXED
/DATA_STRUCTURE SUBJECTS=nschcode*Rid REPEATED_MEASURES=time
  COVARIANCE_TYPE=AR1
/FIELDS TARGET=readprof TRIALS=NONE OFFSET=NONE
/TARGET_OPTIONS DISTRIBUTION=BINOMIAL LINK=LOGIT
/FIXEDEFFECTS=Zses_mean smallsch female lowses time Zses_mean*time
  time*smallsch female*time
time*lowses USE_INTERCEPT=TRUE
/RANDOM EFFECTS=time USE_INTERCEPT=TRUE SUBJECTS=nschcode COVARIANCE_
  TYPE=UNSTRUCTURED
/RANDOM USE_INTERCEPT=TRUE SUBJECTS=nschcode*Rid COVARIANCE_TYPE=
  VARIANCE_COMPONENTS
/BUILD_OPTIONS TARGET_CATEGORY_ORDER=DESCENDING INPUTS_ CATEGORY_
  ORDER=DESCENDING
MAX_ITERATIONS=100 CONFIDENCE_LEVEL=95 DF_METHOD=RESIDUAL COVB=ROBUST
/EMMEANS_OPTIONS SCALE=ORIGINAL PADJUST=LSD.
```

An Example Experimental Design

Model 4.1: Table 5.34 (ch5data2.sav)

```
GENLINMIXED
/DATA_STRUCTURE SUBJECTS=id REPEATED_MEASURES=time COVARIANCE_
  TYPE=IDENTITY
/FIELDS TARGET=readprof TRIALS=NONE OFFSET=NONE
/TARGET_OPTIONS DISTRIBUTION=BINOMIAL LINK=LOGIT
/FIXEDEFFECTS=time treatment treatment*time USE_INTERCEPT=TRUE
/RANDOM USE_INTERCEPT=TRUE SUBJECTS=id COVARIANCE_TYPE= VARIANCE_
  COMPONENTS
/BUILD_OPTIONS TARGET_CATEGORY_ORDER=DESCENDING INPUTS_CATEGORY_
  ORDER=DESCENDING
MAX_ITERATIONS=100 CONFIDENCE_LEVEL=95 DF_METHOD=RESIDUAL COVB=ROBUST
/EMMEANS_OPTIONS SCALE=ORIGINAL PADJUST=LSD.
```

Table 5.34: GEE Fixed Effects Model (ch5data2.sav)

```
GENLIN readprof (REFERENCE=FIRST) WITH treatment time
/MODEL time treatment treatment*time INTERCEPT=YES
DISTRIBUTION=BINOMIAL LINK=LOGIT
/CRITERIA METHOD=FISHER(1) SCALE=1 MAXITERATIONS=100 MAXSTEPHALVING=5
  PCONVERGE=1E-006(ABSOLUTE)
SINGULAR=1E-012 ANALYSISTYPE=3(WALD) CILEVEL=95 LIKELIHOOD=FULL
/REPEATED SUBJECT=id WITHINSUBJECT=time SORT=YES CORRTYPE= INDEPENDENT
  ADJUSTCORR=YES COVB=ROBUST
MAXITERATIONS=100 PCONVERGE=1e-006(ABSOLUTE) UPDATECORR=1
/MISSING CLASSMISSING=EXCLUDE
/PRINT CPS DESCRIPTIVES MODELINFO FIT SUMMARY SOLUTION
  (EXPONENTIATED).
```

Chapter 6: Two-Level Models With Multinomial and Ordinal Outcomes (*ch6data1.sav, ch6data2.sav*)
Defining a Preliminary Single-Level Model
Model 1.1: Table 6.3 (ch6data1.sav)

```
GENLINMIXED
/DATA_STRUCTURE SUBJECTS=schcode
/FIELDS TARGET=plans TRIALS=NONE OFFSET=NONE
/TARGET_OPTIONS REFERENCE=2 DISTRIBUTION=MULTINOMIAL LINK= LOGIT
/FIXEDEFFECTS=gmgpa lowses male USE_INTERCEPT=TRUE
/BUILD_OPTIONS TARGET_CATEGORY_ORDER=ASCENDING INPUTS_CATEGORY_
   ORDER=ASCENDING MAX_ITERATIONS=100
CONFIDENCE_LEVEL=95 DF_METHOD=RESIDUAL COVB=ROBUST
/EMMEANS_OPTIONS SCALE=ORIGINAL PADJUST=LSD.
```

Unconditional Two-Level Model
Model 2.1: Tables 6.4, 6.5, 6.6 (ch6data1.sav)

```
GENLINMIXED
/DATA_STRUCTURE SUBJECTS=schcode
/FIELDS TARGET=plans TRIALS=NONE OFFSET=NONE
/TARGET_OPTIONS REFERENCE=2 DISTRIBUTION=MULTINOMIAL LINK= LOGIT
/FIXED USE_INTERCEPT=TRUE
/RANDOM USE_INTERCEPT=TRUE SUBJECTS=schcode COVARIANCE_TYPE= VARIANCE_
   COMPONENTS
/BUILD_OPTIONS TARGET_CATEGORY_ORDER=ASCENDING INPUTS_CATEGORY_
   ORDER=ASCENDING MAX_ITERATIONS=100
CONFIDENCE_LEVEL=95 DF_METHOD=RESIDUAL COVB=ROBUST
/EMMEANS_OPTIONS SCALE=ORIGINAL PADJUST=LSD.
```

Level-1 Model
Model 2.2: Tables 6.7, 6.9 (ch6data1.sav)

```
GENLINMIXED
/DATA_STRUCTURE SUBJECTS=schcode
/FIELDS TARGET=plans TRIALS=NONE OFFSET=NONE
/TARGET_OPTIONS REFERENCE=2 DISTRIBUTION=MULTINOMIAL LINK=LOGIT
/FIXEDEFFECTS=gmgpa lowses male USE_INTERCEPT=TRUE
/RANDOM USE_INTERCEPT=TRUE SUBJECTS=schcode COVARIANCE_TYPE= VARIANCE_
   COMPONENTS
/BUILD_OPTIONS TARGET_CATEGORY_ORDER=ASCENDING INPUTS_CATEGORY_
   ORDER=ASCENDING MAX_ITERATIONS=100
CONFIDENCE_LEVEL=95 DF_METHOD=RESIDUAL COVB=ROBUST
/EMMEANS_OPTIONS SCALE=ORIGINAL PADJUST=LSD.
```

Adding School-Level Predictors
Model 2.3: Tables 6.10, 6.11 (ch6data1.sav)

```
GENLINMIXED
/DATA_STRUCTURE SUBJECTS=schcode
/FIELDS TARGET=plans TRIALS=NONE OFFSET=NONE
/TARGET_OPTIONS REFERENCE=2 DISTRIBUTION=MULTINOMIAL LINK=LOGIT
/FIXEDEFFECTS=gmacadpress gmlicense gmperlowses gmgpa lowses male USE_
    INTERCEPT=TRUE
/RANDOM USE_INTERCEPT=TRUE SUBJECTS=schcode COVARIANCE_TYPE= VARIANCE_
    COMPONENTS
/BUILD_OPTIONS TARGET_CATEGORY_ORDER=ASCENDING INPUTS_CATEGORY_
    ORDER=ASCENDING MAX_ITERATIONS=100
CONFIDENCE_LEVEL=95 DF_METHOD=RESIDUAL COVB=ROBUST
/EMMEANS_OPTIONS SCALE=ORIGINAL PADJUST=LSD.
```

Investigating a Random Slope
Model 2.4: Table 6.12 (ch6data1.sav)

```
GENLINMIXED
/DATA_STRUCTURE SUBJECTS=schcode
/FIELDS TARGET=plans TRIALS=NONE OFFSET=NONE
/TARGET_OPTIONS REFERENCE=2 DISTRIBUTION=MULTINOMIAL LINK= LOGIT
/FIXEDEFFECTS=gmacadpress gmlicense gmperlowses gmgpa lowses male USE_
    INTERCEPT=TRUE
/RANDOM EFFECTS=gmgpa USE_INTERCEPT=TRUE SUBJECTS=schcode COVARIANCE_
    TYPE=VARIANCE_COMPONENTS
/BUILD_OPTIONS TARGET_CATEGORY_ORDER=ASCENDING INPUTS_CATEGORY_
    ORDER=ASCENDING MAX_ITERATIONS=100
CONFIDENCE_LEVEL=95 DF_METHOD=RESIDUAL COVB=ROBUST
/EMMEANS_OPTIONS SCALE=ORIGINAL PADJUST=LSD.
```

Preliminary Analysis
Model 3.1 (persistcat1): Table 6.18 (ch6data2.sav)

```
GENLIN persistcat1 (REFERENCE=FIRST) WITH zses hiabsent zabsent zqual
    zenroll zcomp
/MODEL zses hiabsent zabsent zqual zenroll zcomp INTERCEPT=YES
DISTRIBUTION=BINOMIAL LINK=LOGIT
/CRITERIA METHOD=FISHER(1) SCALE=1 COVB=ROBUST MAXITERATIONS=100
    MAXSTEPHALVING=5
PCONVERGE=1E-006(ABSOLUTE) SINGULAR=1E-012 ANALYSISTYPE=3(WALD)
    CILEVEL=95 CITYPE=WALD
LIKELIHOOD=FULL
/MISSING CLASSMISSING=EXCLUDE
/PRINT CPS DESCRIPTIVES MODELINFO FIT SUMMARY SOLUTION
    (EXPONENTIATED).
```

Model 3.1 (persistcat1for2): Table 6.18 (ch6data2.sav)

```
GENLIN persistcat1for2 (REFERENCE=FIRST) WITH zses hiabsent zabsent
  zqual zenroll zcomp
/MODEL zses hiabsent zabsent zqual zenroll zcomp INTERCEPT=YES
DISTRIBUTION=BINOMIAL LINK=LOGIT
/CRITERIA METHOD=FISHER(1) SCALE=1 COVB=ROBUST MAXITERATIONS=100
  MAXSTEPHALVING=5
PCONVERGE=1E-006(ABSOLUTE) SINGULAR=1E-012 ANALYSISTYPE=3(WALD)
  CILEVEL=95 CITYPE=WALD
LIKELIHOOD=FULL
/MISSING CLASSMISSING=EXCLUDE
/PRINT CPS DESCRIPTIVES MODELINFO FIT SUMMARY SOLUTION
  (EXPONENTIATED).
```

Adding Student Background Predictors
Model 3.2: Table 6.21 (ch6data2.sav)

```
GENLIN persist3 (ORDER=ASCENDING) WITH zses hiabsent
/MODEL zses hiabsent
DISTRIBUTION=MULTINOMIAL LINK=CUMLOGIT
/CRITERIA METHOD=FISHER(1) SCALE=1 COVB=ROBUST MAXITERATIONS=100
  MAXSTEPHALVING=5
PCONVERGE=1E-006(ABSOLUTE) SINGULAR=1E-012 ANALYSISTYPE=3(WALD)
  CILEVEL=95 CITYPE=WALD
LIKELIHOOD=FULL
/MISSING CLASSMISSING=EXCLUDE
/PRINT CPS DESCRIPTIVES MODELINFO FIT SUMMARY SOLUTION
  (EXPONENTIATED).
```

Testing an Interaction
Model 3.3: Table 6.22, 6.23 (ch6data2.sav)

```
GENLIN persist3 (ORDER=ASCENDING) WITH zses hiabsent
/MODEL zses hiabsent zses*hiabsent
DISTRIBUTION=MULTINOMIAL LINK=CUMLOGIT
/CRITERIA METHOD=FISHER(1) SCALE=1 COVB=ROBUST MAXITERATIONS=100
  MAXSTEPHALVING=5
PCONVERGE=1E-006(ABSOLUTE) SINGULAR=1E-012 ANALYSISTYPE=3(WALD)
  CILEVEL=95 CITYPE=WALD
LIKELIHOOD=FULL
/MISSING CLASSMISSING=EXCLUDE
/PRINT CPS DESCRIPTIVES MODELINFO FIT SUMMARY SOLUTION
  (EXPONENTIATED).
```

Multilevel Ordinal Unconditional Model

Model 4.1: Tables 6.24, 6.25, 6.26, 6.27 (ch6data2.sav)

```
GENLINMIXED
/DATA_STRUCTURE SUBJECTS=schcode
/FIELDS TARGET=persist3 TRIALS=NONE OFFSET=NONE
/TARGET_OPTIONS REFERENCE=2 DISTRIBUTION=MULTINOMIAL LINK=LOGIT
/FIXED USE_INTERCEPT=TRUE
/RANDOM USE_INTERCEPT=TRUE SUBJECTS=schcode COVARIANCE_TYPE=VARIANCE_
   COMPONENTS
/BUILD_OPTIONS TARGET_CATEGORY_ORDER=ASCENDING INPUTS_CATEGORY_
   ORDER=ASCENDING MAX_ITERATIONS=100
CONFIDENCE_LEVEL=95 DF_METHOD=RESIDUAL COVB=ROBUST
/EMMEANS_OPTIONS SCALE=ORIGINAL PADJUST=LSD.
```

Within-School Predictor

Model 4.2: Tables 6.28, 6.29, 6.30 (ch6data2.sav)

```
GENLINMIXED
/DATA_STRUCTURE SUBJECTS=schcode
/FIELDS TARGET=persist3 TRIALS=NONE OFFSET=NONE
/TARGET_OPTIONS REFERENCE=2 DISTRIBUTION=MULTINOMIAL LINK=LOGIT
/FIXEDEFFECTS=zses hiabsent USE_INTERCEPT=TRUE
/RANDOM USE_INTERCEPT=TRUE SUBJECTS=schcode COVARIANCE_TYPE=VARIANCE_
   COMPONENTS
/BUILD_OPTIONS TARGET_CATEGORY_ORDER=ASCENDING INPUTS_CATEGORY_
   ORDER=ASCENDING MAX_ITERATIONS=100
CONFIDENCE_LEVEL=95 DF_METHOD=RESIDUAL COVB=ROBUST
/EMMEANS_OPTIONS SCALE=ORIGINAL PADJUST=LSD.
```

Adding the School-level Predictors

Model 4.3: Tables 6.31, 6.32, 6.33 (ch6data2.sav)

```
GENLINMIXED
/DATA_STRUCTURE SUBJECTS=schcode
/FIELDS TARGET=persist3 TRIALS=NONE OFFSET=NONE
/TARGET_OPTIONS REFERENCE=2 DISTRIBUTION=MULTINOMIAL LINK=LOGIT
/FIXEDEFFECTS=zqual zabsent zenroll zcomp zses hiabsent USE_
   INTERCEPT=TRUE
/RANDOM USE_INTERCEPT=TRUE SUBJECTS=schcode COVARIANCE_TYPE=VARIANCE_
   COMPONENTS
/BUILD_OPTIONS TARGET_CATEGORY_ORDER=ASCENDING INPUTS_CATEGORY_
   ORDER=ASCENDING MAX_ITERATIONS=100
CONFIDENCE_LEVEL=95 DF_METHOD=RESIDUAL COVB=ROBUST
/EMMEANS_OPTIONS SCALE=ORIGINAL PADJUST=LSD.
```

Alternative Model 4.3: Table 6.34: Fixed Effects With Complementary Log–Log (ch6data2.sav)

```
GENLINMIXED
/DATA_STRUCTURE SUBJECTS=schcode
/FIELDS TARGET=persist3 TRIALS=NONE OFFSET=NONE
/TARGET_OPTIONS REFERENCE=2 DISTRIBUTION=MULTINOMIAL LINK= CLOGLOG
/FIXEDEFFECTS=zqual zabsent zenroll zcomp zses hiabsent USE_
  INTERCEPT=TRUE
/RANDOM USE_INTERCEPT=TRUE SUBJECTS=schcode COVARIANCE_TYPE= VARIANCE_
  COMPONENTS
/BUILD_OPTIONS TARGET_CATEGORY_ORDER=ASCENDING INPUTS_CATEGORY_
  ORDER=ASCENDING MAX_ITERATIONS=100
CONFIDENCE_LEVEL=95 DF_METHOD=RESIDUAL COVB=ROBUST
/EMMEANS_OPTIONS SCALE=ORIGINAL PADJUST=LSD.
```

Interpreting a Categorical Predictor
Tables 6.35, 6.36 (ch6data2.sav)

```
GENLINMIXED
/DATA_STRUCTURE SUBJECTS=schcode
/FIELDS TARGET=persist3 TRIALS=NONE OFFSET=NONE
/TARGET_OPTIONS REFERENCE=2 DISTRIBUTION=MULTINOMIAL LINK= LOGIT
/FIXEDEFFECTS=zqual zabsent zcomp enroll3 zses hiabsent USE_
  INTERCEPT=TRUE
/RANDOM USE_INTERCEPT=TRUE SUBJECTS=schcode COVARIANCE_TYPE= VARIANCE_
  COMPONENTS
/BUILD_OPTIONS TARGET_CATEGORY_ORDER=ASCENDING INPUTS_CATEGORY_
  ORDER=DESCENDING
MAX_ITERATIONS=100 CONFIDENCE_LEVEL=95 DF_METHOD=RESIDUAL COVB=ROBUST
/EMMEANS_OPTIONS SCALE=ORIGINAL PADJUST=LSD.
```

Examining a Mediating Effect at Level 1
Model 4.4: Table 6.37 (ch6data2.sav)

```
GENLINMIXED
/DATA_STRUCTURE SUBJECTS=schcode
/FIELDS TARGET=persist3 TRIALS=NONE OFFSET=NONE
/TARGET_OPTIONS REFERENCE=0 DISTRIBUTION=MULTINOMIAL LINK= LOGIT
/FIXEDEFFECTS=zses USE_INTERCEPT=TRUE
/RANDOM USE_INTERCEPT=TRUE SUBJECTS=schcode COVARIANCE_TYPE= VARIANCE_
  COMPONENTS
/BUILD_OPTIONS TARGET_CATEGORY_ORDER=ASCENDING INPUTS_CATEGORY_
  ORDER=ASCENDING MAX_ITERATIONS=100
CONFIDENCE_LEVEL=95 DF_METHOD=RESIDUAL COVB=ROBUST
/EMMEANS_OPTIONS SCALE=ORIGINAL PADJUST=LSD.
```

Model 4.5: Table 6.38 (ch6data2.sav)

```
GENLINMIXED
/DATA_STRUCTURE SUBJECTS=schcode
/FIELDS TARGET=persist3 TRIALS=NONE OFFSET=NONE
/TARGET_OPTIONS REFERENCE=2 DISTRIBUTION=MULTINOMIAL LINK=LOGIT
/FIXEDEFFECTS=zses zgpa USE_INTERCEPT=TRUE
/RANDOM USE_INTERCEPT=TRUE SUBJECTS=schcode COVARIANCE_TYPE=IDENTITY
/BUILD_OPTIONS TARGET_CATEGORY_ORDER=ASCENDING INPUTS_CATEGORY_
  ORDER=ASCENDING MAX_ITERATIONS=100
CONFIDENCE_LEVEL=95 DF_METHOD=RESIDUAL COVB=ROBUST
/EMMEANS_OPTIONS SCALE=ORIGINAL PADJUST=LSD.
```

Chapter 7: Two-Level Models With Count Data (ch7data1.sav, ch7data2.sav)

Preliminary Single-Level Models
Model 1.1: Tables 7.3, 7.4, 7.5 (ch7data1.sav)

```
GENLIN fail WITH lowses male math age
/MODEL lowses male math age INTERCEPT=YES
DISTRIBUTION=POISSON LINK=LOG
/CRITERIA METHOD=FISHER(1) SCALE=1 COVB=ROBUST MAXITERATIONS=100
  MAXSTEPHALVING=5
PCONVERGE=1E-006(ABSOLUTE) SINGULAR=1E-012 ANALYSISTYPE=3(WALD)
  CILEVEL=95 CITYPE=WALD
LIKELIHOOD=FULL
/MISSING CLASSMISSING=EXCLUDE
/PRINT CPS DESCRIPTIVES MODELINFO FIT SUMMARY SOLUTION
  (EXPONENTIATED).
```

Preliminary Single-Level Models
Model 1.2: Tables 7.6, 7.7 (ch7data1.sav)

```
GENLIN fail WITH lowses male gmmath gmage
/MODEL lowses male gmmath gmage INTERCEPT=YES
DISTRIBUTION=POISSON LINK=LOG
/CRITERIA METHOD=FISHER(1) SCALE=1 COVB=ROBUST MAXITERATIONS=100
  MAXSTEPHALVING=5
PCONVERGE=1E-006(ABSOLUTE) SINGULAR=1E-012 ANALYSISTYPE=3(WALD)
  CILEVEL=95 CITYPE=WALD
LIKELIHOOD=FULL
/MISSING CLASSMISSING=EXCLUDE
/PRINT CPS DESCRIPTIVES MODELINFO FIT SUMMARY SOLUTION
  (EXPONENTIATED).
```

Tables 7.10, 7.11: Parameter Estimates With zmath and Goodness of Fit (ch7data1-verB.sav)

```
GENLIN fail WITH lowses male gmage zmath
/MODEL lowses male gmage zmath INTERCEPT=YES
DISTRIBUTION=POISSON LINK=LOG
/CRITERIA METHOD=FISHER(1) SCALE=1 COVB=ROBUST MAXITERATIONS=100
   MAXSTEPHALVING=5
PCONVERGE=1E-006(ABSOLUTE) SINGULAR=1E-012 ANALYSISTYPE=3(WALD)
   CILEVEL=95 CITYPE=WALD
LIKELIHOOD=FULL
/MISSING CLASSMISSING=EXCLUDE
/PRINT CPS DESCRIPTIVES MODELINFO FIT SUMMARY SOLUTION
   (EXPONENTIATED).
```

Considering Possible Overdispersion

Model 1.3: Tables 7.12, 7.13 (Negative Binomial Distribution, Log Link, Specify Value "0"; ch7data1.sav)

```
GENLIN fail WITH lowses male gmmath gmage
/MODEL lowses male gmmath gmage INTERCEPT=YES
DISTRIBUTION=NEGBIN(0) LINK=LOG
/CRITERIA METHOD=FISHER(1) SCALE=1 COVB=ROBUST MAXITERATIONS=100
   MAXSTEPHALVING=5
PCONVERGE=1E-006(ABSOLUTE) SINGULAR=1E-012 ANALYSISTYPE=3(WALD)
   CILEVEL=95 CITYPE=WALD
LIKELIHOOD=FULL
/MISSING CLASSMISSING=EXCLUDE
/PRINT CPS DESCRIPTIVES MODELINFO FIT SUMMARY SOLUTION
   (EXPONENTIATED).
```

Model 1.4: Tables 7.14, 7.15 (Negative Binomial Distribution, Log Link, Specify Value "1"; ch7data1.sav)

```
GENLIN fail WITH lowses male gmmath gmage
/MODEL lowses male gmmath gmage INTERCEPT=YES
DISTRIBUTION=NEGBIN(1) LINK=LOG
/CRITERIA METHOD=FISHER(1) SCALE=1 COVB=ROBUST MAXITERATIONS=100
   MAXSTEPHALVING=5
PCONVERGE=1E-006(ABSOLUTE) SINGULAR=1E-012 ANALYSISTYPE=3(WALD)
   CILEVEL=95 CITYPE=WALD
LIKELIHOOD=FULL
/MISSING CLASSMISSING=EXCLUDE
/PRINT CPS DESCRIPTIVES MODELINFO FIT SUMMARY SOLUTION
   (EXPONENTIATED).
```

Model 1.5: Table 7.16 (Negative Binomial Distribution, Log Link, Estimate Value; ch7data1.sav)

```
GENLIN fail WITH lowses male gmmath gmage
/MODEL lowses male gmmath gmage INTERCEPT=YES
DISTRIBUTION=NEGBIN(MLE) LINK=LOG
/CRITERIA METHOD=FISHER(1) SCALE=1 COVB=ROBUST MAXITERATIONS=100
   MAXSTEPHALVING=5
PCONVERGE=1E-006(ABSOLUTE) SINGULAR=1E-012 ANALYSISTYPE=3(WALD)
   CILEVEL=95 CITYPE=WALD
LIKELIHOOD=FULL
/MISSING CLASSMISSING=EXCLUDE
/PRINT CPS DESCRIPTIVES MODELINFO FIT SUMMARY SOLUTION
   (EXPONENTIATED).
```

Estimating Two-Level Count Data With GENLIN MIXED

Model 2.1: Table 7.18 (Poisson Distribution, Log Link; ch7data1.sav)

```
GENLINMIXED
/DATA_STRUCTURE SUBJECTS=nschcode
/FIELDS TARGET=fail TRIALS=NONE OFFSET=NONE
/TARGET_OPTIONS DISTRIBUTION=POISSON LINK=LOG
/FIXEDEFFECTS=lowses male gmmath gmage USE_INTERCEPT=TRUE
/BUILD_OPTIONS TARGET_CATEGORY_ORDER=ASCENDING INPUTS_CATEGORY_
   ORDER=ASCENDING MAX_ITERATIONS=100
CONFIDENCE_LEVEL=95 DF_METHOD=RESIDUAL COVB=ROBUST
/EMMEANS_OPTIONS SCALE=ORIGINAL PADJUST=LSD.
```

Unconditional Model

Model 2.2: Tables 7.19, 7.20, 7.21 (ch7data1.sav)

```
GENLINMIXED
/DATA_STRUCTURE SUBJECTS=nschcode
/FIELDS TARGET=fail TRIALS=NONE OFFSET=NONE
/TARGET_OPTIONS DISTRIBUTION=POISSON LINK=LOG
/FIXED USE_INTERCEPT=TRUE
/RANDOM USE_INTERCEPT=TRUE SUBJECTS=nschcode COVARIANCE_TYPE=
   VARIANCE_COMPONENTS
/BUILD_OPTIONS TARGET_CATEGORY_ORDER=ASCENDING INPUTS_CATEGORY_
   ORDER=ASCENDING MAX_ITERATIONS=100
CONFIDENCE_LEVEL=95 DF_METHOD=RESIDUAL COVB=ROBUST
/EMMEANS_OPTIONS SCALE=ORIGINAL PADJUST=LSD.
```

Within-Schools Model
Model 2.3: Tables 7.22, 7.23, 7.24 (ch7data1.sav)

```
GENLINMIXED
/DATA_STRUCTURE SUBJECTS=nschcode
/FIELDS TARGET=fail TRIALS=NONE OFFSET=NONE
/TARGET_OPTIONS DISTRIBUTION=POISSON LINK=LOG
/FIXEDEFFECTS=lowses male gmmath gmage USE_INTERCEPT=TRUE
/RANDOM USE_INTERCEPT=TRUE SUBJECTS=nschcode COVARIANCE_TYPE=
  VARIANCE_COMPONENTS
/BUILD_OPTIONS TARGET_CATEGORY_ORDER=ASCENDING INPUTS_CATEGORY_
  ORDER=ASCENDING MAX_ITERATIONS=100
CONFIDENCE_LEVEL=95 DF_METHOD=RESIDUAL COVB=ROBUST
/EMMEANS_OPTIONS SCALE=ORIGINAL PADJUST=LSD.
```

Examining Whether the Negative Binomial Distribution Is a Better Choice
Model 2.4: Tables 7.25, 7.26 (ch7data1.sav)

```
GENLINMIXED
/DATA_STRUCTURE SUBJECTS=nschcode
/FIELDS TARGET=fail TRIALS=NONE OFFSET=NONE
/TARGET_OPTIONS DISTRIBUTION=NEGATIVE_BINOMIAL LINK=LOG
/FIXEDEFFECTS=lowses male gmmath gmage USE_INTERCEPT=TRUE
/RANDOM USE_INTERCEPT=TRUE SUBJECTS=nschcode COVARIANCE_TYPE=
  VARIANCE_COMPONENTS
/BUILD_OPTIONS TARGET_CATEGORY_ORDER=ASCENDING INPUTS_CATEGORY_
  ORDER=ASCENDING MAX_ITERATIONS=100
CONFIDENCE_LEVEL=95 DF_METHOD=RESIDUAL COVB=ROBUST
/EMMEANS_OPTIONS SCALE=ORIGINAL PADJUST=LSD.
```

Does the SES-Failure Slope Vary Across Schools?
Model 2.5: Tables 7.28, 7.29 (ch7data1.sav)

```
GENLINMIXED
/DATA_STRUCTURE SUBJECTS=nschcode
/FIELDS TARGET=fail TRIALS=NONE OFFSET=NONE
/TARGET_OPTIONS DISTRIBUTION=NEGATIVE_BINOMIAL LINK=LOG
/FIXEDEFFECTS=lowses male gmmath gmage USE_INTERCEPT=TRUE
/RANDOM EFFECTS=lowses USE_INTERCEPT=TRUE SUBJECTS=nschcode
  COVARIANCE_TYPE=VARIANCE_COMPONENTS
/BUILD_OPTIONS TARGET_CATEGORY_ORDER=ASCENDING INPUTS_CATEGORY_
  ORDER=ASCENDING MAX_ITERATIONS=100
CONFIDENCE_LEVEL=95 DF_METHOD=RESIDUAL COVB=ROBUST
/EMMEANS_OPTIONS SCALE=ORIGINAL PADJUST=LSD.
```

Modeling Variability at Level 2
Model 2.6: Tables 7.30, 7.31, 7.32 (ch7data1.sav)

```
GENLINMIXED
/DATA_STRUCTURE SUBJECTS=nschcode
/FIELDS TARGET=fail TRIALS=NONE OFFSET=NONE
/TARGET_OPTIONS DISTRIBUTION=NEGATIVE_BINOMIAL LINK=LOG
/FIXEDEFFECTS=lowses male gmmath gmage gmAvgYearsExp gmlicensedper
  USE_INTERCEPT=TRUE
/RANDOM EFFECTS=lowses USE_INTERCEPT=TRUE SUBJECTS=nschcode
  COVARIANCE_TYPE=VARIANCE_COMPONENTS
/BUILD_OPTIONS TARGET_CATEGORY_ORDER=ASCENDING INPUTS_CATEGORY_
  ORDER=ASCENDING MAX_ITERATIONS=100
CONFIDENCE_LEVEL=95 DF_METHOD=RESIDUAL COVB=ROBUST
/EMMEANS_OPTIONS SCALE=ORIGINAL PADJUST=LSD.
```

Adding the Cross-Level Interactions
Model 2.7: Tables 7.33, 7.34, 7.35 (ch7data1.sav)

```
GENLINMIXED
/DATA_STRUCTURE SUBJECTS=nschcode
/FIELDS TARGET=fail TRIALS=NONE OFFSET=NONE
/TARGET_OPTIONS DISTRIBUTION=NEGATIVE_BINOMIAL LINK=LOG
/FIXEDEFFECTS=lowses male gmmath gmage gmAvgYearsExp gmlicensedper
  lowses*gmAvgYearsExp lowses*gmlicensedper USE_INTERCEPT=TRUE
/RANDOM EFFECTS=lowses USE_INTERCEPT=TRUE SUBJECTS=nschcode
  COVARIANCE_TYPE=UNSTRUCTURED
/BUILD_OPTIONS TARGET_CATEGORY_ORDER=ASCENDING INPUTS_CATEGORY_
  ORDER=ASCENDING MAX_ITERATIONS=100
CONFIDENCE_LEVEL=95 DF_METHOD=RESIDUAL COVB=ROBUST
/EMMEANS_OPTIONS SCALE=ORIGINAL PADJUST=LSD.
```

Specifying a Single-Level Model
Model 3.1: Table 7.38 (ch7data2.sav)

```
GENLIN fail WITH ses male gmmath gmgpa twoyear
/MODEL ses male gmmath gmgpa twoyear INTERCEPT=YES
DISTRIBUTION=POISSON LINK=LOG
/CRITERIA METHOD=FISHER(1) SCALE=1 COVB=ROBUST MAXITERATIONS=100
  MAXSTEPHALVING=5
PCONVERGE=1E-006(ABSOLUTE) SINGULAR=1E-012 ANALYSISTYPE=3(WALD)
  CILEVEL=95 CITYPE=WALD
LIKELIHOOD=FULL
/MISSING CLASSMISSING=EXCLUDE
/PRINT CPS DESCRIPTIVES MODELINFO FIT SUMMARY SOLUTION
  (EXPONENTIATED).
```

Adding the Offset
Model 3.2: Table 7.39 (ch7data2.sav)

```
GENLIN fail WITH ses male gmmath gmgpa twoyear
/MODEL ses male gmmath gmgpa twoyear INTERCEPT=YES OFFSET=Lnsemester
DISTRIBUTION=POISSON LINK=LOG
/CRITERIA METHOD=FISHER(1) SCALE=1 COVB=ROBUST MAXITERATIONS=100
  MAXSTEPHALVING=5
PCONVERGE=1E-006(ABSOLUTE) SINGULAR=1E-012 ANALYSISTYPE=3(WALD)
  CILEVEL=95 CITYPE=WALD
LIKELIHOOD=FULL
/MISSING CLASSMISSING=EXCLUDE
/PRINT CPS DESCRIPTIVES MODELINFO FIT SUMMARY SOLUTION
  (EXPONENTIATED).
```

Adding the Offset (Changing the Scale Parameter Method to Pearson Chi-Square)
Model 3.3: Tables 7.40, 7.41 (ch7data2.sav)

```
GENLIN fail WITH ses male gmmath gmgpa twoyear
/MODEL ses male gmmath gmgpa twoyear INTERCEPT=YES OFFSET=Lnsemester
DISTRIBUTION=POISSON LINK=LOG
/CRITERIA METHOD=FISHER(1) SCALE=PEARSON COVB=ROBUST MAXITERATIONS=100
  MAXSTEPHALVING=5
PCONVERGE=1E-006(ABSOLUTE) SINGULAR=1E-012 ANALYSISTYPE=3(WALD)
  CILEVEL=95 CITYPE=WALD
LIKELIHOOD=FULL
/MISSING CLASSMISSING=EXCLUDE
/PRINT CPS DESCRIPTIVES MODELINFO FIT SUMMARY SOLUTION
  (EXPONENTIATED).
```

Adding the Offset (Adding *Lnsemester* Into Model as Covariate)
Model 3.4: Table 7.42 (ch7data2.sav)

```
GENLIN fail WITH ses male gmmath gmgpa twoyear Lnsemester
/MODEL Lnsemester ses male gmmath gmgpa twoyear INTERCEPT=YES
DISTRIBUTION=POISSON LINK=LOG
/CRITERIA METHOD=FISHER(1) SCALE=1 COVB=ROBUST MAXITERATIONS=100
  MAXSTEPHALVING=5
PCONVERGE=1E-006(ABSOLUTE) SINGULAR=1E-012 ANALYSISTYPE=3(WALD)
  CILEVEL=95 CITYPE=WALD
LIKELIHOOD=FULL
/MISSING CLASSMISSING=EXCLUDE
/PRINT CPS DESCRIPTIVES MODELINFO FIT SUMMARY SOLUTION
  (EXPONENTIATED).
```

Estimating the Model With GENLIN MIXED
Model 4.1: Tables 7.43, 7.44 (ch7data2.sav)

```
GENLINMIXED
/DATA_STRUCTURE SUBJECTS=nschcode
/FIELDS TARGET=fail TRIALS=NONE OFFSET=FIELD(Lnsemester)
/TARGET_OPTIONS DISTRIBUTION=POISSON LINK=LOG
/FIXEDEFFECTS=ses male gmmath gmgpa twoyear gmacadprocess gminstqual
   USE_INTERCEPT=TRUE
/RANDOM USE_INTERCEPT=TRUE SUBJECTS=nschcode COVARIANCE_TYPE=
   VARIANCE_COMPONENTS
/BUILD_OPTIONS TARGET_CATEGORY_ORDER=ASCENDING INPUTS_CATEGORY_
   ORDER=ASCENDING MAX_ITERATIONS=100
CONFIDENCE_LEVEL=95 DF_METHOD=RESIDUAL COVB=ROBUST
/EMMEANS_OPTIONS SCALE=ORIGINAL PADJUST=LSD.
```

Model 4.2: Table 7.46 (Probability of Failing One or More Courses; ch7data2.sav)

```
GENLINMIXED
/DATA_STRUCTURE SUBJECTS=nschcode
/FIELDS TARGET=fail TRIALS=NONE OFFSET=FIELD(Lnsemester)
/TARGET_OPTIONS DISTRIBUTION=NEGATIVE_BINOMIAL LINK=LOG
/FIXEDEFFECTS=ses male gmmath gmgpa twoyear gmacadprocess gminstqual
   USE_INTERCEPT=TRUE
/RANDOM USE_INTERCEPT=TRUE SUBJECTS=nschcode COVARIANCE_TYPE=
   VARIANCE_COMPONENTS
/BUILD_OPTIONS TARGET_CATEGORY_ORDER=ASCENDING INPUTS_CATEGORY_
   ORDER=ASCENDING MAX_ITERATIONS=100
CONFIDENCE_LEVEL=95 DF_METHOD=RESIDUAL COVB=ROBUST
/EMMEANS_OPTIONS SCALE=ORIGINAL PADJUST=LSD.
```

Appendix B: Model Comparisons Across Software Applications

For comparative purposes, as mentioned in Chapter 1, we provide the results generated by IBM SPSS and for HLM software. First, the estimation procedures provide substantial agreement of fixed effects, robust standard errors, and variance components in the model.

TABLE B.1 IBM SPSS and HLM Estimates for Student Proficiency (Coded 1 = Yes, 0 = No)

	IBM SPSS estimates		HLM estimates	
	Coefficient	SE	Coefficient	SE
Between levels				
Model for proficiency, β_0				
Intercept, γ_{00}	1.294^a	0.090	1.293^a	0.090
Schcomp, γ_{01}	-0.463^a	0.064	-0.463^a	0.064
Smallsch, γ_{02}	0.135	0.085	0.135	0.085
Within level				
SES slope, β_1				
Intercept, γ_{10}	-0.769^a	0.065	-0.769^a	0.065
Female slope, β_2				
Intercept, γ	0.474^a	0.059	0.474^a	0.059
Minority slope, β_3				
Intercept, γ_{30}	-0.623^a	0.078	-0.623^a	0.078
Schcomp, γ_{31}	0.145^b	0.075	0.145^b	0.075
Residual variances				
Readprof, u_o	0.224^a	0.068	0.223^a	NA
Slope, u_3	0.199^a	0.084	0.199^a	NA
Covariance	-0.163^a	0.066	-0.163	

Note: NA = not applicable; HLM produces the standard deviation.
[a] $p < .05$.
[b] $p < .10$.

TABLE B.2 IBM SPSS and Mplus Probit Estimates for Student Proficiency (Coded 1 = Yes, 0 = No)

	IBM SPSS estimates		Mplus estimates	
	Coefficient	SE	Coefficient	SE
Between levels				
Model for proficiency, β_0				
Intercept, γ_{00}	0.779[a]	0.053	−0.781[a]	0.053
Schcomp, γ_{01}	−0.266[a]	0.037	−0.269[a]	0.038
Smallsch, γ_{02}	0.083	0.051	0.090[b]	0.051
Within level				
SES slope, β_1				
Intercept, γ_{10}	−0.466[a]	0.038	−0.464[a]	0.039
Female slope, β_2				
Intercept, γ_{20}	0.286[a]	0.035	0.274[a]	0.036
Minority slope, β_3				
Intercept, γ_{30}	−0.368[a]	0.046	−0.371[a]	0.046
Schcomp, γ_{31}	0.070	0.044	0.073	0.045
Residual variances				
Readprof, u_0	0.074[a]	0.022	0.074[a]	0.023
Slope, u_3	0.068[a]	0.029	0.063[a]	0.026
Covariance	−0.053[a]	0.022	−0.052[a]	0.021

[a] $p < .05$.
[b] $p < .10$.

Author Index

Subject Index